Springer Theses

Recognizing Outstanding Ph.D. Research

Aims and Scope

The series "Springer Theses" brings together a selection of the very best Ph.D. theses from around the world and across the physical sciences. Nominated and endorsed by two recognized specialists, each published volume has been selected for its scientific excellence and the high impact of its contents for the pertinent field of research. For greater accessibility to non-specialists, the published versions include an extended introduction, as well as a foreword by the student's supervisor explaining the special relevance of the work for the field. As a whole, the series will provide a valuable resource both for newcomers to the research fields described, and for other scientists seeking detailed background information on special questions. Finally, it provides an accredited documentation of the valuable contributions made by today's younger generation of scientists.

Theses are accepted into the series by invited nomination only and must fulfill all of the following criteria

- They must be written in good English.
- The topic should fall within the confines of Chemistry, Physics, Earth Sciences, Engineering and related interdisciplinary fields such as Materials, Nanoscience, Chemical Engineering, Complex Systems and Biophysics.
- The work reported in the thesis must represent a significant scientific advance.
- If the thesis includes previously published material, permission to reproduce this must be gained from the respective copyright holder.
- They must have been examined and passed during the 12 months prior to nomination.
- Each thesis should include a foreword by the supervisor outlining the significance of its content.
- The theses should have a clearly defined structure including an introduction accessible to scientists not expert in that particular field.

More information about this series at http://www.springer.com/series/8790

Ryuji Ukai

Multi-Step Multi-Input One-Way Quantum Information Processing with Spatial and Temporal Modes of Light

Doctoral Thesis accepted by
The University of Tokyo, Tokyo, Japan

Springer

Author
Dr. Ryuji Ukai
Department of Applied Physics, School of Engineering
The University of Tokyo
Tokyo
Japan

Supervisor
Prof. Akira Furusawa
Department of Applied Physics, School of Engineering
The University of Tokyo
Tokyo
Japan

ISSN 2190-5053 ISSN 2190-5061 (electronic)
ISBN 978-4-431-55018-1 ISBN 978-4-431-55019-8 (eBook)
DOI 10.1007/978-4-431-55019-8

Library of Congress Control Number: 2014946963

Springer Tokyo Heidelberg New York Dordrecht London

Printed on acid-free paper

Springer is part of Springer Science+Business Media (www.springer.com)

Parts of this thesis have been published in some of the following journal articles:

1. M. Yukawa, R. Ukai, P. van Loock, and A. Furusawa, *Experimental generation of four-mode continuous-variable cluster states*, Phys. Rev. A **78**, 012301 (2008)
2. R. Ukai, J. Yoshikawa, N. Iwata, P. van Loock, and A. Furusawa, *Universal linear Bogoliubov transformations through one-way quantum computation*, Phys. Rev. A **81**, 032315 (2010)
3. Y. Miwa, R. Ukai, J. Yoshikawa, R. Filip, P. van Loock, and A. Furusawa, *Demonstration of cluster-state shaping and quantum erasure for continuous variables*, Phys. Rev. A **82**, 032305 (2010)
4. Y. Miwa, J. Yoshikawa, R. Ukai, R. Filip, P. van Loock, and A. Furusawa, *Demonstration of Universal Quantum Erasing for Continuous Variables*, arXiv:1007.0314 [quant-ph]
5. R. Ukai, N. Iwata, Y. Shimokawa, S.C. Armstrong, A. Politi, J. Yoshikawa, P. van Loock, and A. Fu-rusawa, *Demonstration of Unconditional One-Way Quantum Computations for Continuous Variables*, Phys. Rev. Lett. **106**, 240504 (2011)
6. R. Ukai, S. Yokoyama, J. Yoshikawa, P. van Loock, and A. Furusawa, *Demonstration of a Controlled-Phase Gate for Continuous-Variable One-Way Quantum Computation*, Phys. Rev. Lett. **107**, 250501 (2011)
7. S. Yokoyama, R. Ukai, S.C. Armstrong, C. Sornphiphatphong, T. Kaji, S. Suzuki, J. Yoshikawa, H. Yonezawa, N.C. Menicucci, and A. Furusawa, *Ultra-Large-Scale continuous-variable cluster states multiplexed in the time domain*, Nature Photonics, **7**, 982 (2013)
8. R.N. Alexander, S.C. Armstrong, R. Ukai, and N.C. Menicucci, *Noise analysis of single-qumode Gaussian operations using continuous-variable cluster states*, arXiv:1311.3538 [quant-ph]
9. S. Yokoyama, R. Ukai, J. Yoshikawa, P. Marek, R. Filip, and A. Furusawa, *Nonlocal quantum gate on quantum continuous variables with minimum resources*, Phys. Rev. A. **90**, 012311 (2014)

Supervisor's Foreword

Large-scale quantum information processing (QIP) is one of the biggest challenges in the field of quantum information science. Toward that goal, researchers in the world have been trying to test many methodologies. To date, one of the most promising ways in that direction is one-way QIP or cluster-state QIP, where QIP is performed based on measurements of subsystems in a large-scale entangled state called a cluster state. The important property of the cluster states is that the overall entanglement does not disappear when a subsystem is measured, in contrast to most multipartite entangled states like Greenberger–Horn–Zeilinger (GHZ) states, which collapse when a subsystem is measured. In the cluster-state QIP, the back-action of measurements changes the cluster-state, facilitating QIP.

In the present thesis work by Ryuji Ukai, optical realizations of cluster-state QIP are systematically investigated. There are two directions for the realizations. One is in the spatial domain, and the other is in the time domain. This work shows solutions for both directions. As for the spatial domain, the author shows the experimental results on multistep and multi-input cluster-state QIP and proves the powerfulness of this scheme. Moreover, he shows the complete theoretical analysis on this scheme. As for the time domain, the author presents a new scheme of cluster-state QIP to handle time-domain multiplexed ultra-large cluster states.

This work offers a comprehensive guideline for a deep understanding of large-scale cluster-state QIP.

Tokyo, Japan, March 2014 Prof. Akira Furusawa Ph.D.

Acknowledgments

This thesis reports my research in collaboration with Mr. Seiji C. Armstrong and Mr. Shota Yokoyama in the Furusawa–Yonezawa laboratory of The University of Tokyo.

First and foremost, I would like to thank my supervisor, Prof. Akira Furusawa, for his continuous research guidance and encouragement in the course of this work. The Furusawa–Yonezawa laboratory is the best place in the world to study quantum physics.

I would like to thank Dr. Hidehiro Yonezawa for his great advice, founded on optics and electronic circuits. His excellent SHG cavity led us to the success of all our experiments in this work.

I would like to thank Dr. Jun-ichi Yoshikawa for the good discussions on both theories and experiments.

I am very grateful to research collaborators Mr. Seiji C. Armstrong (visiting researcher from The Australian National University) and Mr. Shota Yokoyama. This work is supported by their outstanding intelligence. Seiji's unlimited imagination led to breakthroughs in our experiments. As the successor of the cluster group, Shota will make a remarkable achievement in the near future!

I thank Mr. Chanond Sornphiphatphong and Mr. Toshiyuki Kaji for their considerable efforts toward the experiment on generation of temporal-mode cluster states.

I also thank all members of the Furusawa laboratory: Assoc. Prof. Genta Masada, Dr. Shige-nari Suzuki, Dr. Yoshichika Miwa, Dr. Mitsuyoshi Yukawa, Mr. Noriaki Iwata, Mr. Yuji Shimokawa, Mr. Shuntaro Takeda, Mr. Kenzo Makino, Mr. Kazunori Miyata, Mr. Takahiro Mizuta, Mr. Priyasheel Prasad, Ms. Maria Fuwa, Mr. Hisashi Ogawa, Mr. Yosuke Hashimoto, and Ms. Yumiko Yoshikawa.

I would like to thank Prof. Makoto Gonokami, Prof. Hidetoshi Katori, Prof. Masato Koashi, and Assoc. Prof. Naoki Yamamoto for valuable comments and recommendations on this thesis.

Finally, I express my gratitude to my family for their support and cheerfulness over the years.

Tokyo, Japan, March 2014 Ryuji Ukai

Contents

Chapter 1
Introduction

1.1 One-Way Quantum Computation

1.1.1 Quantum Computation

The historical origin of quantum mechanics was the *quantum hypothesis* proposed by Planck in 1900 [1]. From that time, considerable efforts have been devoted to the research of quantum mechanics. Since it predicted curious phenomena contradictory to classical mechanics, several researchers were skeptical about the new-coming theories. In order to show the incompleteness of quantum mechanics, several thought experiments were proposed. They included the paradox of nonlocal correlations proposed by Einstein et al. [2]. Afterward, it was experimentally demonstrated that such curious phenomena are observable in the real world. After long discussions on quantum mechanics, it is now accepted and hailed as one of the great scientific theories.

Recently, it was found that quantum mechanics can be applied to information processing. The birth of quantum information processing was the proposal of the quantum Turing machine by Deutsch in 1985 [3]. Although the superiority of quantum computation over classical computation had not initially been demonstrated, it was later proved that a quantum computer can solve some problems more efficiently than a classical computer [4–8]. A well-known example is the factoring of integers. Its discovery expedited research on quantum computers.

In the same way as the classical computation, the input of a quantum computer is classical information (Fig. 1.1). It is firstly encoded into a quantum state $|\psi_{in}\rangle$. A quantum computer has the ability to apply a unitary operator $\hat{U}$ onto the quantum state, leading to the output quantum state $|\psi_{out}\rangle = \hat{U}|\psi_{in}\rangle$. This quantum state manipulation is the key part of the quantum computation. The final answer of the quantum computation is also classical information, which will be read out by performing a measurement on the output quantum state.

The *superposition* of quantum states and quantum *entangled* states are characteristic traits of in quantum theory. Consider a two-level system labeled by $|0\rangle$ and $|1\rangle$, such as photon number states, or ground and excited states of an atom. An arbitrary one qubit state is represented by

R. Ukai, *Multi-Step Multi-Input One-Way Quantum Information Processing with Spatial and Temporal Modes of Light*, Springer Theses,
DOI 10.1007/978-4-431-55019-8_1

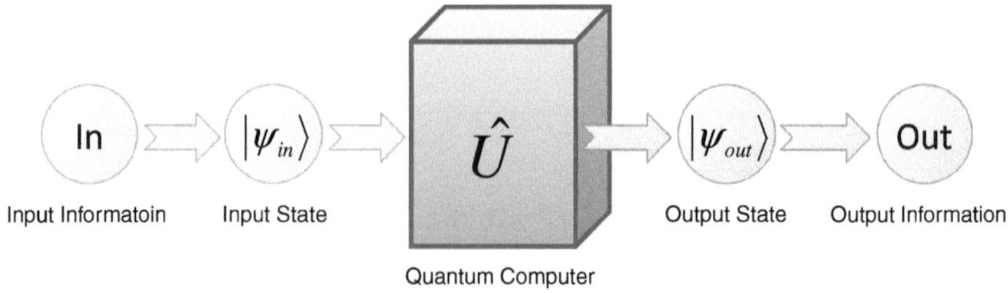

Fig. 1.1 Quantum computation

$$a|0\rangle + b|1\rangle. \tag{1.1}$$

This is called a superposition of $|0\rangle$ and $|1\rangle$. Different from a statistical mixture of two quantum states, it leads to quantum interference.

An entangled state (Sect. 3.7.1) refers to a quantum state which cannot be decomposed into two independent subsystems. The following is an example of entangled states in a two-qubit system:

$$\frac{|0\rangle_A|0\rangle_B + |1\rangle_A|1\rangle_B}{\sqrt{2}}. \tag{1.2}$$

This state cannot be represented by a tensor product of a state in the subsystem A and another state in the subsystem B. As mentioned above, entanglement was originally proposed by Einstein, Podolsky, and Rosen as a paradox which pointed out the apparent incompleteness of quantum mechanics. However, it was later demonstrated experimentally [9, 10], and it is now understood as a characteristic property of quantum mechanics. The quantum state proposed by Einstein, Podolsky, and Rosen is called the Einstein-Podolsky-Rosen state (*EPR state*). It is thought that the superiority of quantum computations originate from the parallelism of computation, based on superposition and entanglement [11].

1.1.2 Quantum Teleportation

Another application of quantum entanglement is *quantum teleportation* [12–21] (Sect. 4.1.1). It is a protocol with which one can transmit an unknown quantum state to a receiver at a distance (Fig. 1.2). The sender and the receiver are usually named "Alice" and "Bob", respectively.

For this purpose, Alice and Bob share an EPR state in advance. The procedure of quantum teleportation is as follows. Firstly, Alice entangles the quantum state to be transmitted and half of the EPR state which belongs to Alice. Alice measures the two outcomes in an appropriate measurement basis. The measurement results are sent to Bob through classical channels. By performing correction operations on the other half of the EPR state, Bob can reconstruct the quantum state which was initially prepared by Alice.

Quantum teleportation has features in common with quantum computation, including that both handle entanglement, and require (some) quantum state transformations. Quantum teleportation can be considered as an identity operation on an input state, since the quantum state which Bob reconstructs is equivalent to that which was initially owned by Alice.

1.1.3 Application of Quantum Teleportation to Quantum Computation (Gate Teleportation, Offline Scheme)

Although quantum teleportation was initially proposed as a protocol to transmit a quantum state, it was later found that it can be applied to implement quantum computation. In quantum teleportation, an EPR state $|EPR\rangle$ is utilized as a resource for its protocol (Fig. 1.2). The first scheme of its application is to replace the resource state $|EPR\rangle$ with another state $\hat{D}|EPR\rangle$ [22, 23] (Fig. 1.3, Sect. 4.1.2).

By changing the resource state, Bob reconstructs $\hat{D}|\psi\rangle$, which is a unitary transformed version of the initial state $|\psi\rangle$. The unitary operator $\hat{D}$ is determined by the resource state $\hat{D}|EPR\rangle$. This scheme is called a *gate teleportation* since the unitary operator $\hat{D}$, which was initially applied to the resource state, becomes applied to the input state through the quantum teleportation. In addition, it is also called an *offline scheme* of quantum computation. This is because the unitary operator $\hat{D}$ can be considered to be applied to the offline resource state $|EPR\rangle$, leading to the revised resource state $\hat{D}|EPR\rangle$. During the online computation, it is applied to the input state through the quantum teleportation.

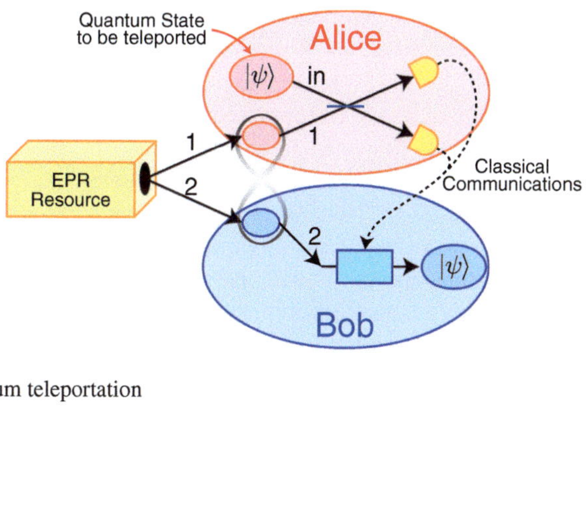

Fig. 1.2 Quantum teleportation

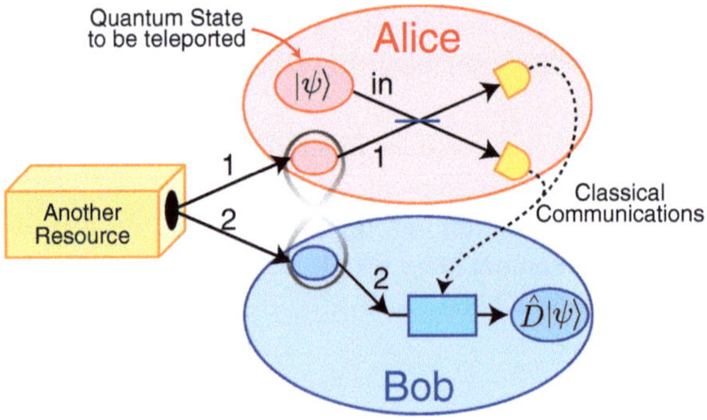

Fig. 1.3 Gate teleportation, offline scheme

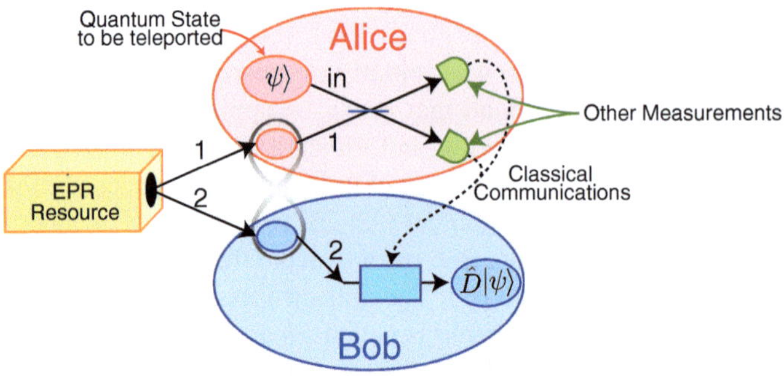

Fig. 1.4 One-way quantum computation

1.1.4 Application of Quantum Teleportation to Quantum Computation (One-Way Quantum Computation)

Another application of quantum teleportation to quantum computation was later proposed. It is called *one-way quantum computation* [24–29] (Sect. 4.2 and Chap. 5), which is the main topic of this thesis. In one-way quantum computation, we change the measurements from the quantum teleportation (Fig. 1.4).

Since the quantum state is projected to another basis, the state after the measurements is dependent on the measurement basis. Therefore, the state which Bob reconstructs becomes $\hat{D}|\psi\rangle$, where $\hat{D}$ is determined by the basis of the measurements Alice has performed. One-way quantum computation is a model of quantum computation where operations are controlled by measurement bases. It is also called the *cluster model* of quantum computation, or *cluster computation*, since a multi-partite entangled state, called the *cluster state*, is used as a resource for quantum computation.

1.1.5 Circuit Model and Cluster Model

Quantum computation is usually studied by using the *circuit model* [11, 30], where computation is described by the order of unitary gates onto the initial states. The circuit model is a universal model of quantum computation: an arbitrary quantum computation can be described in this model.

The cluster model is an alternative to the standard circuit model. In this model, the actual process of computation is focused on: operations are achieved by a succession of measurements on an entangled state. In the example of Fig. 1.4, a bipartite entangled state (EPR state) is utilized as a resource for the computation. Since the number of measurements is equivalent to the number of resource modes, two measurements are involved. In addition, it is also equivalent to the degrees of freedom (DOF) of unitary transformations since each measurement provides a DOF. Therefore, Fig. 1.4 has the ability to perform unitary transformations with two DOF.[1] In this manner, the DOF of unitary operations achieved by one-way quantum computation is determined by the number of resource modes. By using a larger-scale resource state, we can perform unitary operations with more DOF.

The procedure of cluster-model quantum computation is summarized by the following:

- Prepare a multi-partite entangled state (cluster state, Sect. 5.1), which will be used as a resource for quantum computation.
- Couple an input state with the cluster state (Sect. 5.3).
- Perform reshaping of the cluster state based on the operation to be achieved (Sect. 5.4).
- Perform unitary operations through measurements (Sect. 5.5).
- Read out the unmeasured modes, which give us the solution of the computation.

A sufficiently large cluster state can be used as a universal resource for one-way quantum computation, that is, an arbitrary computation is achieved by using the same cluster state. Once a requested unitary operation is determined, the cluster state is transformed so that it can be efficiently implemented through the one-way quantum computation. The operation determines the set of measurement bases. By choosing an appropriate set of measurement bases, we can implement an arbitrary unitary transformation. Since operations can be switched by adjusting measurement bases using the same resource cluster state, one-way quantum computation can be considered as a *software-based* quantum computation. Note that the resource cluster state is consumed irreversibly during the computation. This is the reason why it is

[1] To be precise, Fig. 1.4 shows the teleportation-based input coupling scheme (Sect. 5.3.3), where both measurements are homodyne measurements. A homodyne measurement has one DOF: the relative phase θ between the signal beam and the local oscillator beam. Thus, Fig. 1.4 has two DOF. In general, we can implement a transformation with multiple DOF using an elementary one-mode one-way gate (Sect. 4.2.3) by choosing a complicated measurement. However, we usually assume that we construct a large-scale quantum computation by concatenating several kinds of elementary gates with limited DOF. Therefore, the total DOF of operations still increase as well by concatenating elementary gates.

called the *one-way* scheme. However, the transformation of the input state to the output state is unitary, thus reversible.

1.1.6 Continuous-Variables and Universality

1.1.6.1 Continuous-Variable Quantum Computation

We have so far used two-level systems with $|0\rangle$ and $|1\rangle$ in order to show several examples. However, quantum computation is also discussed by using more high-dimensional systems, including continuous-variable (CV) systems where computational bases are continuously varying quantum states, such as eigenstates of momentum operators $|p = s\rangle$ (Sect. 3.1).

In recent experimental demonstrations of quantum computations using quantum states of light (Chap. 2), not only discrete-variable (DV) systems but also CV systems are utilized. We use optical CV systems in this thesis. The main merit of optical CV systems is that entangled states can be generated deterministically (success probability is equal to 100 %), and thus operations can also be implemented deterministically. It is stark contrast to optical DV systems where entangled states are generated probabilistically and experiments are verified via postselections.

We mention here that there also exists a drawback in optical CV systems: ideal computational basis states such as $|p = s\rangle$ require an infinite amount of energy. It is different from the DV systems where computational basis states are physical states. Since we cannot employ such unphysical states in actual experiments, they are approximated by other physical states. For example, the zero eigenstate of the momentum operator $|p = 0\rangle$ is approximated by a p-squeezed vacuum state (Sect. 2.2.3). Approximations of these states lead to unavoidable errors in CV quantum computations.[2]

1.1.6.2 Universality

It is known that an arbitrary digital classical computation can be achieved by cascading NAND gates in an appropriate order. In a similar manner, an arbitrary CV quantum computation can be achieved by cascading the following elementary gates in an appropriate order (Sect. 3.6):

- Gate which can perform an arbitrary one-mode Gaussian operation.
- Gate which can perform a two-mode Gaussian operation.
- Gate which can perform a one-mode non-Gaussian operation.

---

[2] Since both optical DV and CV systems have merits and demerits of their own, a hybrid scheme is also being studied [31], where DV gates are implemented using optical CV setups. It is thought that the imperfections of quantum states may be compensated by limiting the dimension of computational bases.

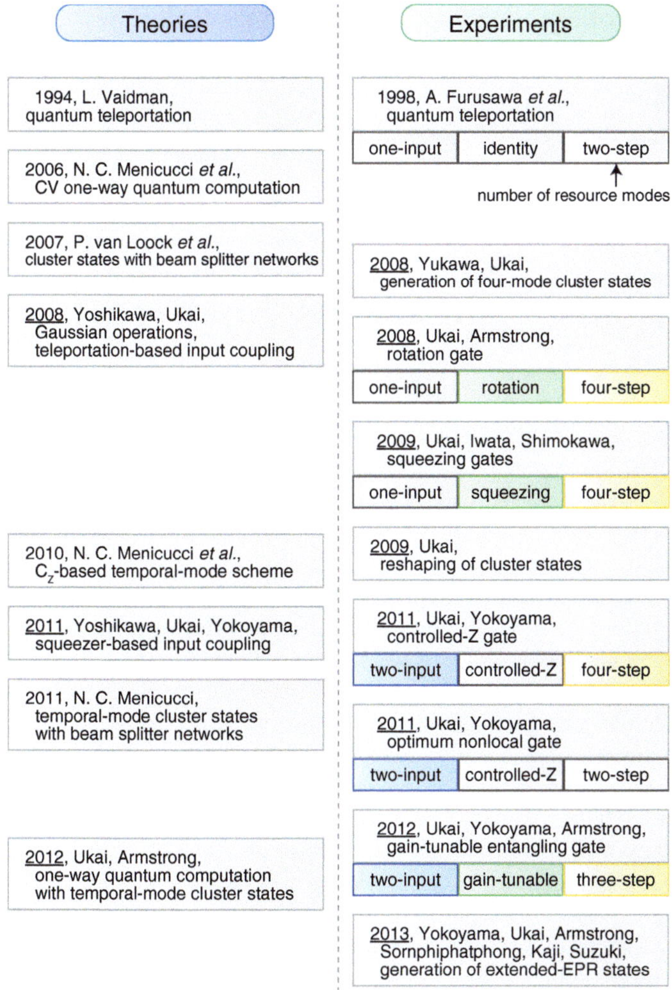

Fig. 1.5 History of CV one-way quantum computation. Works in our laboratory are shown with *underlines*

Here, Gaussian operations correspond to unitary transformations with inhomogeneous quadratic polynomial Hamiltonians in the canonical operators (Sect. 3.4.3). Similar to the classical computation, multi-step operations should also become possible because a total quantum computation is achieved by cascading these elementary gates.

1.2 History of One-Way Quantum Computation and This Thesis

1.2.1 History of One-Way Quantum Computation

Figure 1.5 shows a brief history of CV one-way quantum computation and the status of this thesis. In this figure, we include CV quantum teleportation as well since it has a connection with CV one-way quantum computation (Sect. 1.1).

Although quantum teleportation was originally proposed in the DV systems [12], it was later extended to the CV systems [13, 15]. An experimental realization of CV quantum teleportation was first reported in 1998 [17].

Since quantum teleportation is a protocol to transmit a one-mode quantum state without applying any unitary transformations, we can consider it as an identity operation on a one-mode state. Note that a two-mode entangled state called EPR state is utilized as a resource for quantum teleportation. Since we equate the number of resource modes with the number of operations in one-way quantum computation, we can consider that a quantum teleportation is a two-step operation.

Similar to quantum teleportation, one-way quantum computation was first defined in the DV systems [24, 32]. Several years after the original proposal, DV one-way quantum computation experiments were also reported [33, 34].

The CV cluster state was introduced in 2006 [26] by extending the DV counterpart. Shortly after, one-way quantum computation was proposed by Menicucci [27]. They were basic theories of CV cluster states and one-way quantum computations where controlled-Z gates are utilized as entangling gates for the generation of a cluster state and the coupling of an input state with the cluster state[3] (Sects. 5.2.1 and 5.3.2). However, they were not desirable theories especially for experimentalists because a controlled-Z gate required a large-scale experimental setup [36, 37], and thus it was not realistic to prepare multiple controlled-Z gates for generations of cluster states and demonstrations of one-way quantum computations.

One of these problems was solved by van Loock et al. [35] (Sect. 5.2.2). They showed that, compared to the original proposal that an ideal cluster state is generated by entangling eigenstates of momentum operators by using controlled-Z gates, an approximation of the ideal cluster state can be generated by combining squeezed vacuum states by using beam splitters. In the limit of infinite squeezing, it becomes identical to the ideal cluster state. It made it easier to generate cluster states, leading to several experimental reports including four-mode cluster state generations [38, 39] and eight-mode cluster state generations [40]. In addition, an experimental demonstration of cluster state reshaping (Sect. 5.4) was also reported [41].

As for the coupling of an input state with a cluster state, we found that a quantum teleportation can be applied for this purpose, where a beam splitter, not a controlled-Z gate, plays the role of input coupling [42] (Sect. 5.3.3). Together with the cluster

[3] In Ref. [26], it was already mentioned that special shapes of cluster states can be generated by using beam splitter networks. However, its generalization was given in Ref. [35].

state generation with beam splitter networks, it became possible to implement one-way quantum computations without using controlled-Z gates. Afterward, we found that not only quantum teleportation but also a squeezing operation can be applied to input coupling in one-way quantum computations as well (Sect. 5.3.4).

We then move onto implementations of one-way quantum computations. In the original paper of CV one-way quantum computations [27], it was already given how the operation is determined by a specific measurement basis. However, conversely, it was an open question how the set of measurement bases should be chosen for a specific operation using a multi-partite cluster state. An example of multi-step one-way quantum computation was first given in Ref. [29]: approximate *squeezing operations*, which are members of one-mode Gaussian operations, are achieved by using four-mode cluster states with homodyne measurements. Note that the operation was inevitably an approximate version of the ideal squeezing operation even though the resource cluster state was an ideal state. This is because the sets of measurement bases were not optimum for the squeezing operations.

In 2009, we reported several general answers for choices of measurement bases [42] (Sect. 5.5). One main result was that arbitrary one-mode Gaussian operations, including squeezing operations, can be achieved by using a four-mode linear cluster state with homodyne measurements. Different from the earlier proposal [29], we have shown the set of measurement bases with which we can achieve an ideal one-mode Gaussian operation when we can utilize an ideal cluster state as a resource. We have also proved that several one-mode Gaussian operations cannot be implemented by using a three-mode linear cluster state as a resource, thus the four-mode linear cluster state is the minimum resource which satisfies the universality of one-mode Gaussian operations. In addition, we have also proposed the set of measurement bases for arbitrary multi-mode Gaussian operations though it was not the optimal choice of measurement bases.

By combining the methodology of one-mode Gaussian operations with the teleportation-based input coupling scheme [42], we have experimentally demonstrated several one-mode Gaussian operations (rotation and squeezing operations) using four-mode linear cluster states as resources. Although we have demonstrated only several members of one-mode Gaussian operations, the experimental setup had the ability to implement an arbitrary one-mode Gaussian operation.

As a next step for universal one-way quantum computation, demonstrations of multi-mode Gaussian operations were highly anticipated. Its first trial was reported in Ref. [43]. Although they tried to demonstrate a two-mode gate which had an ability to entangle two separable states, they could not observe entanglement at the output. The main reason was that the level of entanglement present in the cluster state was not sufficient. Entanglement at the output is the key criterion of the entangling gate, thus the operation was not successful.

1.2.2 This Thesis

1.2.2.1 Spatial-Mode Experiments

In this thesis, we show the first successful demonstration of a two-mode operation (controlled-Z gate) in one-way quantum computation (Chap. 7), which has already been reported in Ref. [44] as well. Though it was a similar experiment to the previous report [43], we have successfully observed entanglement at the output, which was sufficient to show the nonclassical nature of the gate.

By combining with the demonstrations of one-mode Gaussian operations, we now have all the tools to implement an arbitrary multi-mode Gaussian operation in a framework of one-way quantum computation.

Including the controlled-Z gate experiment, three experimental demonstrations of one-way quantum computations on two-mode input states are reported in this thesis [44]. They have a common property that they are demonstrations of two-input quantum gates which have the ability to entangle two independent quantum states. In addition, each mode of the quantum states is distinguished from the others by its spatial location (spatial modes).

The inherent features of each experiment are summarized as follows.

Controlled-Z Gate Experiment

This is an experiment of a two-mode Gaussian operation using a four-mode linear cluster state as a resource. The operation which is implemented is the *unity-gain* controlled-Z gate [27]: it does not impart an additional squeezing operation on each quantum mode. It can be considered as an experimental demonstration of gate teleportations [22, 23] on a two-mode system, because it is nothing but the circuit which is acquired by exchanging the order of the controlled-Z gate in the circuit where a controlled-Z gate is applied to the outcomes of two quantum teleportations.

Optimum Nonlocal Gate Experiment

This is an experiment of a two-mode Gaussian operation using a two-mode cluster state as a resource. The *nonlocal* controlled-Z gate (with additional squeezing operations) is implemented using the minimum resource: a bipartite entangled state shared in advance, and a classical channel in each direction (two channels in total) [45]. Here, a *nonlocal* gate refers to a quantum gate whose target two modes are located at a distance from each other [36, 45–49].

Gain-Tunable Entangling Gate Experiment

This is an experiment of two-mode Gaussian operations using a three-mode cluster state as a resource. In contrast to the other two experiments where operations are fixed, we can control the on-off switching as well as the gain of the entangling interaction by changing the measurement basis [42]. It shows the property of one-way quantum computations where operations are controlled by measurement bases onto cluster states.

Temporal-Mode Theories

In this thesis, one-way quantum computation using *temporal-mode* cluster states is also studied [50, 51]. It will lead to the experimental generation of ultra-large entangled states and experimental demonstrations of many-step one-way quantum computations.

Similar to classical computations where an arbitrary computation is achieved by concatenating NAND gates, an arbitrary CV quantum computation can be achieved by concatenating several basic quantum gates. For that purpose, many-step operations should be implemented. Nonetheless, the current schemes for one-way quantum computation experiments lack extensibility. This is because each mode of cluster states is encoded *spatially*, thus the experimental setup becomes larger in proportion to the number of operations. Although it is proposed that one can encode each mode in a different frequency using a frequency comb [52, 53], it would not be easy to implement one-way quantum computations using the frequency-mode cluster states.

Recently, one-way quantum computation using *temporal modes* was proposed [50, 51]. In this scheme, each mode is encoded in a different time, instead of its spatial location. It enables us to implement many-step one-way quantum computations without enlarging our experimental setup.

In Ref. [51], the schematic for the generation of temporal-mode cluster states was proposed, with which one can generate cluster states of an arbitrary size by using a limited and fixed optical setup (Sect. 10.1.3). It was a giant step toward the realization of many-step one-way quantum computation.

However, since the cluster states to be generated had complex structures, it was not known how all resource modes can be utilized for computation. It was proposed that quantum computation can be achieved by using a half or a quarter of the generated cluster states after eliminating the other modes. Although it was sufficient to show superiority of the temporal-mode scheme theoretically, it was expected that a strategy for implementing one-way quantum computations without disposing cluster modes would be discovered (Sect. 10.1.4).

In this thesis, we propose a solution to this problem. We show that computation using a temporal-mode cluster state is equivalent to a repetition of quantum teleportations (Sect. 10.3.3). Since we can implement one-mode Gaussian operations with two DOF by controlling the measurement basis in a quantum teleportation (Sect. 5.3.3), we can utilize all modes of the temporal-mode cluster state in one-way quantum computations without eliminating any of them. In addition, we show that non-Gaussian operations and multi-mode Gaussian operations are also achieved without eliminating resource modes (Sects. 10.5 and 10.6) by considering repetitions of a one-way quantum computation circuit with the three-mode linear cluster state, and the quantum computation circuit of the controlled-Z gate experiment (Chap. 7). It is expected that these findings will lead to the development of both theories and experiments over temporal-mode one-way quantum computation.

1.3 Structure of This Thesis

In Chap. 2, we briefly review quantum optics. In Chap. 3, we describe CV quantum states and quantum state manipulations.

Theories of one-way quantum computation is described in Chaps. 4 and 5. In the former section, we will introduce one-way quantum computation by comparing it with the offline scheme. The details of one-way quantum computation are discussed in the latter section.

Through Chaps. 6–9, we show experiments of one-way quantum computations. Chapter 6 describes the generation of cluster states which are used in the following three demonstrations of quantum gates. In Chap. 7, we show the theory, experimental setup, and results of the controlled-Z gate experiment using a four-mode linear cluster state as a resource. In Chap. 8, we show the optimum nonlocal gate experiment using a two-mode cluster state. In Chap. 9, we show the gain-tunable entangling gate using a three-mode cluster state as a resource.

Before concluding this thesis, we show theories of one-way quantum computation using temporal-mode cluster states in Chap. 10.

References

1. Planck, M.: Ueber das Gesetz der Energieverteilung im Normalspectrum. Annalen der Physik **309**, 553 (1901)
2. Einstein, A., Podolsky, B., Rosen, N.: Can quantum-mechanical description of physical reality be considered complete? Phys. Rev. **47**, 777 (1935)
3. Deutsch, D.: Proc. Roy. Soc. London, Ser. A **400**, 96 (1985)
4. Shor, P.W.: In: Proceedings, 35th Annual Symposium on Foundations of Computer Science. IEEE Press, Los Alamitos, CA (1994)
5. Shor, P.W.: Polynomial-time algorithms for prime factorization and discrete logorithms on a quantum computer. SIAM J. Computing **26**, 1484 (1997)
6. Grover, L.K.: Quantum mechanics helps in searching for a needle in a Haystack. Phys. Rev. Lett. **79**, 325 (1997)
7. S. J. Lomonaco, Jr., and L. H. Kauffman, Quantum Hidden Subgroup Problems: A Mathematical Perspective, e-print arXiv:0201095 [quant-ph]
8. Lomonaco, S.J. Jr., Kauffman, L.H.: A Continuous Variable Shor Algorithm, e-print arXiv:0210141 [quant-ph]

9. Aspect, A., Grangier, P., Roger, G.: Experimental tests of realistic local theories via Bell's theorem. Phys. Rev. Lett. **47**, 460 (1981)
10. Aspect, A., Dalibard, J., Roger, G.: Experimental test of Bell's inequalities using time-varying analyzers. Phys. Rev. Lett. **49**, 1804 (1982)
11. Nielsen, M.A., Chuang, I.L.: Quantum Computation and Quantum Information. Cambridge University Press, Cambridge (2000)
12. Bennett, C.H., Brassard, G., Crépeau, C., Jozsa, R., Peres, A., Wootters, W.K.: Teleporting an unknown quantum state via dual classical and Einstein-Podolsky-Rosen channels. Phys. Rev. Lett **70**, 1895 (1993)
13. Vaidman, L.: Teleportation of quantum states. Phys. Rev. A **49**, 1473 (1994)
14. Bouwmeester, D., Pan, J.-W., Mattle, K., Eibl, M., Weinfurter, H., Zeilinger, A.: Experimental quantum teleportation. Nature **390**, 575 (1997)
15. Braunstein, S.L., Kimble, H.J.: Teleportation of continuous quantum variables. Phys. Rev. Lett. **80**, 869 (1998)
16. Boschi, D., Branca, S., De Martini, F., Hardy, L., Popescu, S.: Experimental realization of teleporting an unknown pure quantum state via dual classical and Einstein-Podolsky-Rosen channels. Phys. Rev. Lett. **80**, 1121 (1998)
17. Furusawa, A., Sørensen, J.L., Braunshtein, S.L., Fuchs, C.A., Kimble, H.J., Polzik, E.S.: Unconditional quantum teleportation. Science **282**, 706 (1998)
18. Takei, N., Yonezawa, H., Aoki, T., Furusawa, A.: High-fidelity teleportation beyond the no-cloning limit and entanglement swapping for continuous variables. Phys. Rev. Lett. **94**, 220502 (2005)
19. Yonezawa, H., Furusawa, A., van Loock, P.: Sequential quantum teleportation of optical coherent states. Phys. Rev. A **76**, 032305 (2007)
20. Yonezawa, H., Braunstein, S.L., Furusawa, A.: Experimental demonstration of quantum teleportation of broadband squeezing. Phys. Rev. Lett. **99**, 110503 (2007)
21. Yukawa, M., Benichi, H., Furusawa, A.: High-fidelity continuous-variable quantum teleportation toward multistep quantum operations. Phys. Rev. A **77**, 022314 (2008)
22. Gottesman, D., Chuang, I.L.: Demonstrating the viability of universal quantum computation using teleportation and single-qubit operations. Nature **402**, 390 (1999)
23. Bartlett, S.D., Munro, W.J.: Quantum teleportation of optical quantum gates. Phys. Rev. Lett. **90**, 117901 (2003)
24. Raussendorf, R., Briegel, H.J.: A one-way quantum computer. Phys. Rev. Lett. **86**, 5188 (2001)
25. Gottesman, D., Kitaev, A., Preskill3, J.: Encoding a qubit in an oscillator. Phys. Rev. A **64**, 012310 (2001)
26. Zhang, J., Braunstein, S.L.: Continuous-variable Gaussian analog of cluster states. Phys. Rev. A **73**, 032318 (2006)
27. Menicucci, N.C., van Loock, P., Gu, M., Weedbrook, C., Ralph, T.C., Nielsen, M.A.: Universal quantum computation with continuous-variable cluster states. Phys. Rev. Lett. **97**, 110501 (2006)
28. Nielsen, M.A.: Cluster-state quantum computation. Rep. Math. Phys. **57**, 147 (2006)
29. van Loock, P.: Examples of Gaussian cluster computation. J. Opt. Soc. Am. B **24**, 340 (2007)
30. Braunstein, S.L., Pati, A.K.: Quantum information theory with continuous variables. Kluwer, Dordrecht (2003)
31. Furusawa, A., van Loock, P.: Quantum teleportation and entanglement. Wiley -VCH (2011)
32. Briegel, H.J., Raussendorf, R.: Persistent entanglement in arrays of interacting particles. Phys. Rev. Lett. **86**, 910 (2001)
33. Walther, P., Resch, K.J., Rudolph, T., Schenck, E., Weinfurter, H., Vedral, V., Aspelmeyer, M., Zeilinger, A.: Experimental one-way quantum computing. Nature **434**, 169 (2005)
34. Prevedel, R., Walther, P., Tiefenbacher, F., Böhi, P., Kaltenbaek, R., Jennewein, T., Zeilinger, A.: High-speed linear optics quantum computing using active feed-forward. Nature **445**, 65 (2007)
35. van Loock, Peter, Weedbrook, Christian, Mile, Gu: Building Gaussian cluster states by linear optics. Phys. Rev. A **76**, 032321 (2007)

36. Filip, R., Marek, P., Andersen, U.L.: Measurement-induced continuous-variable quantum interactions. Phys. Rev. A **71**, 042308 (2005)
37. Yoshikawa, J., Miwa, Y., Huck, A., Andersen, U.L., van Loock, P., Furusawa, A.: Demonstration of a quantum nondemolition sum gate. Phys. Rev. Lett. **101**, 250501 (2008)
38. Su, X., Tan, A., Jia, X., Zhang, J., Xie, C., Peng, K.: Experimental preparation of quadripartite cluster and Greenberger-Horne-Zeilinger entangled states for continuous variables. Phys. Rev. Lett. **98**, 070502 (2007)
39. Yukawa, M., Ukai, R., van Loock, P., Furusawa, A.: Experimental generation of four-mode continuous-variable cluster states. Phys. Rev. A **78**, 012301 (2008)
40. Su, X., Zhao, Y., Hao, S., Jia, X., Xie, C., Peng, K.: Experimental preparation of eight-partite linear and two-diamond shape cluster states for photonic qumodes, arXiv:1205.0590 [quant-ph]
41. Miwa, Y., Ukai, R., Yoshikawa, J., Filip, R., van Loock, P., Furusawa, A.: Demonstration of cluster-state shaping and quantum erasure for continuous variables. Phys. Rev. A **82**, 032305 (2010)
42. Ukai, R., Yoshikawa, J., Iwata, N., van Loock, P., Furusawa, A.: Universal linear Bogoliubov transformations through one-way quantum computation. Phys. Rev. A **81**, 032315 (2010)
43. Wang, Y., Su, X., Shen, H., Tan, A., Xie, C., Peng, K.: Toward demonstrating controlled-X operation based on continuous-variable four-partite cluster states and quantum teleporters. Phys. Rev. A **81**, 022311 (2010)
44. Ukai, R., Yokoyama, S., Yoshikawa, J., van Loock, P., Furusawa, A.: Demonstration of a controlled-phase gate for continuous-variable one-way quantum computation. Phys. Rev. Lett. **107**, 250501 (2011)
45. Eisert, J., Jacobs, K., Papadopoulos, P., Plenio, B.: Optimal local implementation of nonlocal quantum gates. Phys. Rev. A **62**, 052317 (2000)
46. Huang, Y.-F., Ren, X.-F., Zhang, Y.-S., Duan, L.-M., Guo, G.-C.: Experimental teleportation of a quantum controlled-NOT gate. Phys. Rev. Lett. **93**, 240501 (2004)
47. Gao, W.-B., Goebel, A.M., Lu, C.-Y., Dai, H.-N., Wagenknecht, C., Zhang, Q., Zhao, B., Peng, C.-Z., Chen, Z.-B., Chen, Y.-A., Pan, J.-W.: Teleportation-based realization of an optical quantum two-qubit entangling gate. In: Proceedings of the National Academy of Sciences of the United States of America vol. 107, p. 20869 (2010)
48. Gottesman, D.: The Heisenberg representation of quantum computers. In: Corney, S.P., Delbourgo, R., Jarvis, P.D. (eds.) Group22: Proceedings of the XXII International Colloquium on Group Theoretical Methods in Physics, pp. 32–43. International Press, Cambridge, MA (1999), e-print arXiv:9807006 [quant-ph]
49. Zhou, X., Leung, D.W., Chuang, I.L.: Methodology for quantum logic gate construction. Phys. Rev. A **62**, 052316 (2000)
50. Menicucci, N.C., Ma, X., Ralph, T.C.: Arbitrarily large continuous-variable cluster states from a single quantum nondemolition gate. Phys. Rev. Lett. **104**, 250503 (2010)
51. Nicolas, C.: Menicucci, temporal-mode continuous-variable cluster states using linear optics. Phys. Rev. A **83**, 062314 (2011)
52. Menicucci, N.C., Flammia, S.T., Zaidi, H., Pfister, O.: Ultracompact generation of continuous-variable cluster states. Phys. Rev. A **76**, 010302(R) (2007)
53. Zaidi, H., Menicucci, N.C., Flammia, S.T., Bloomer, R., Pysher, M., Pfister, O.: Entangling the optical frequency comb: simultaneous generation of multiple 2×2 and 2×3 continuous-variable cluster states in a single optical parametric oscillator. Laser Phys. **18**, 659 (2008)

Chapter 2
Quantum Optics

2.1 Quantum Mechanics and Quantum Optics

2.1.1 Uncertainty Principle

2.1.1.1 Variance and Standard Deviation

Let $\hat{A}$ and $|\psi\rangle$ be a Hermitian operator and a state vector, respectively. The variance and standard deviation of the operator $\hat{A}$ for the state $|\psi\rangle$ are defined as

$$\langle\Delta^2\hat{A}\rangle = \langle(\hat{A} - \langle\hat{A}\rangle)^2\rangle = \langle\hat{A}^2\rangle - \langle\hat{A}\rangle^2, \tag{2.1}$$

$$\langle\Delta\hat{A}\rangle_{sd} = \sqrt{\langle\Delta^2\hat{A}\rangle} = \sqrt{\langle\hat{A}^2\rangle - \langle\hat{A}\rangle^2}, \tag{2.2}$$

where

$$\langle\hat{A}\rangle = \langle\psi|\hat{A}|\psi\rangle, \quad \langle\hat{A}^2\rangle = \langle\psi|\hat{A}^2|\psi\rangle \tag{2.3}$$

are expectation values of $\hat{A}$ and $\hat{A}^2$, respectively.

2.1.1.2 Uncertainty Principle

Consider two Hermitian operators $\hat{A}$ and $\hat{B}$ with a state $|\psi\rangle$. They satisfy

$$\langle\Delta\hat{A}\rangle_{sd}\langle\Delta\hat{B}\rangle_{sd} \geqq \frac{1}{2}\left|\langle[\hat{A}, \hat{B}]\rangle\right|, \tag{2.4}$$

which is called the uncertainty principle. In the special case of $[\hat{A}, \hat{B}] = i\hbar$, Eq. (2.4) becomes

$$\langle\Delta\hat{A}\rangle_{sd}\langle\Delta\hat{B}\rangle_{sd} \geqq \frac{1}{2}\hbar. \tag{2.5}$$

A state is referred to as a *minimum uncertainty state* when it has the minimum product of fluctuations.

R. Ukai, *Multi-Step Multi-Input One-Way Quantum Information Processing with Spatial and Temporal Modes of Light*, Springer Theses,
DOI 10.1007/978-4-431-55019-8_2

2.1.2 Quantized Electromagnetic Field

2.1.2.1 Electric Field, Magnetic Field, and Hamiltonian

The electric field operator and the magnetic field operator [1, 2] are given by

$$\hat{E}(\boldsymbol{r},t) = i\left(\frac{1}{2\pi}\right)^{\frac{3}{2}} \int \mathrm{d}^3 V_k \sum_{\sigma=1}^{2} \sqrt{\frac{\hbar\omega_k}{2\varepsilon_0}} \boldsymbol{e}_{k\sigma} \left(\hat{a}_{k\sigma} e^{i(\boldsymbol{k}\cdot\boldsymbol{r}-\omega_k t)} - \hat{a}^{\dagger}_{k\sigma} e^{-i(\boldsymbol{k}\cdot\boldsymbol{r}-\omega_k t)}\right), \tag{2.6}$$

$$\hat{B}(\boldsymbol{r},t) = i\left(\frac{1}{2\pi}\right)^{\frac{3}{2}} \int \mathrm{d}^3 V_k \sum_{\sigma=1}^{2} \sqrt{\frac{\hbar}{2\varepsilon_0\omega_k}} \boldsymbol{k}\times\boldsymbol{e}_{k\sigma} \left(\hat{a}_{k\sigma} e^{i(\boldsymbol{k}\cdot\boldsymbol{r}-\omega_k t)} - \hat{a}^{\dagger}_{k\sigma} e^{-i(\boldsymbol{k}\cdot\boldsymbol{r}-\omega_k t)}\right). \tag{2.7}$$

By using these two operators, the Hamiltonian of the electromagnetic field becomes

$$\hat{H}_{sys} = \int \mathrm{d}^3 V_k \sum_{\sigma=1}^{2} \frac{\hbar\omega_k}{2} \left(\hat{a}^{\dagger}_{k\sigma}\hat{a}_{k\sigma} + \hat{a}_{k\sigma}\hat{a}^{\dagger}_{k\sigma}\right). \tag{2.8}$$

These three operators satisfy

$$\hat{\boldsymbol{E}}^{\dagger}(\boldsymbol{r},t) = \hat{\boldsymbol{E}}(\boldsymbol{r},t), \quad \hat{\boldsymbol{B}}^{\dagger}(\boldsymbol{r},t) = \hat{\boldsymbol{B}}(\boldsymbol{r},t), \quad \hat{H}^{\dagger}_{sys} = \hat{H}_{sys}. \tag{2.9}$$

Therefore, they are Hermitian operators and thus observables. In these equations, $\hat{a}_{k\sigma}$ and $\hat{a}^{\dagger}_{k\sigma}$ are annihilation and creation operators of photons with the wave vector $\boldsymbol{k}$ and the polarization σ, respectively. They satisfy

$$[\hat{a}_{k\sigma}, \hat{a}^{\dagger}_{k'\sigma'}] = \delta_{\sigma\sigma'}\delta^{(3)}(\boldsymbol{k}-\boldsymbol{k}'), \quad [\hat{a}_{k\sigma}, \hat{a}_{k'\sigma'}] = [\hat{a}^{\dagger}_{k\sigma}, \hat{a}^{\dagger}_{k'\sigma'}] = 0. \tag{2.10}$$

$\boldsymbol{e}_{k\sigma}$ and ω_k represent the unit vector and the angular frequency, respectively.

2.1.2.2 Quadratures

The annihilation and creation operators are not Hermitian operators, and thus not observables. By using these two operators, the following quadrature operators are defined:

$$\hat{x}_{k\sigma} = \sqrt{\frac{\hbar}{2\omega_k}} \left(\hat{a}_{k\sigma} + \hat{a}^{\dagger}_{k\sigma}\right), \tag{2.11}$$

$$\hat{p}_{k\sigma} = -i\sqrt{\frac{\hbar\omega_k}{2}} \left(\hat{a}_{k\sigma} - \hat{a}^{\dagger}_{k\sigma}\right). \tag{2.12}$$

They correspond to the position and momentum operators of quantized harmonic oscillators. They satisfy the commutation relation:

$$[\hat{x}_{\boldsymbol{k}\sigma}, \hat{p}_{\boldsymbol{k}'\sigma'}] = i\hbar\delta_{\sigma\sigma'}\delta^{(3)}(\boldsymbol{k}-\boldsymbol{k}'). \tag{2.13}$$

They are Hermitian operators $\hat{x}^{\dagger}_{\boldsymbol{k}\sigma} = \hat{x}_{\boldsymbol{k}\sigma}$ and $\hat{p}^{\dagger}_{\boldsymbol{k}\sigma} = \hat{p}_{\boldsymbol{k}\sigma}$, and thus observables. $\hat{a}_{\boldsymbol{k}\sigma}$ and $\hat{a}^{\dagger}_{\boldsymbol{k}\sigma}$ are conversely acquired as

$$\hat{a}_{\boldsymbol{k}\sigma} = \sqrt{\frac{\omega_{\boldsymbol{k}}}{2\hbar}}\hat{x}_{\boldsymbol{k}\sigma} + i\sqrt{\frac{1}{2\hbar\omega_{\boldsymbol{k}}}}\hat{p}_{\boldsymbol{k}\sigma}, \tag{2.14}$$

$$\hat{a}^{\dagger}_{\boldsymbol{k}\sigma} = \sqrt{\frac{\omega_{\boldsymbol{k}}}{2\hbar}}\hat{x}_{\boldsymbol{k}\sigma} - i\sqrt{\frac{1}{2\hbar\omega_{\boldsymbol{k}}}}\hat{p}_{\boldsymbol{k}\sigma}. \tag{2.15}$$

2.1.2.3 Single Mode and Simplification

Although all wave vectors $\boldsymbol{k}$ and polarizations σ are so far integrated and added, only a single mode with a specific wave vector and polarization is considered in the following discussion.[1] In addition, all constants except for the reduced Planck constant $\hbar$ are omitted.[2] As a result, the electric and magnetic field operators become

$$\hat{E}(\boldsymbol{r},t) = i\sqrt{\frac{\hbar}{2}}\boldsymbol{e}\left(\hat{a}e^{i(\boldsymbol{k}\cdot\boldsymbol{r}-\omega t)} - \hat{a}^{\dagger}e^{-i(\boldsymbol{k}\cdot\boldsymbol{r}-\omega t)}\right), \tag{2.16}$$

$$\hat{B}(\boldsymbol{r},t) = i\sqrt{\frac{\hbar}{2}}\boldsymbol{k}\times\boldsymbol{e}\left(\hat{a}e^{i(\boldsymbol{k}\cdot\boldsymbol{r}-\omega t)} - \hat{a}^{\dagger}e^{-i(\boldsymbol{k}\cdot\boldsymbol{r}-\omega t)}\right), \tag{2.17}$$

respectively. The quadrature operators are

$$\hat{x} = \sqrt{\frac{\hbar}{2}}\left(\hat{a}+\hat{a}^{\dagger}\right), \tag{2.18}$$

$$\hat{p} = -i\sqrt{\frac{\hbar}{2}}\left(\hat{a}-\hat{a}^{\dagger}\right), \tag{2.19}$$

or conversely

$$\hat{a} = \frac{1}{\sqrt{2\hbar}}\left(\hat{x}+i\hat{p}\right), \tag{2.20}$$

$$\hat{a}^{\dagger} = \frac{1}{\sqrt{2\hbar}}\left(\hat{x}-i\hat{p}\right). \tag{2.21}$$

The commutation relations become

---

[1] To be precise, a mode is defined by integrating wave vectors with a mode function which has distinctive nonzero values around a specific wave vector. Therefore, the commutation relations (Eqs. (2.22) and (2.23)) do not become delta functions.

[2] $\hbar$ is abbreviated to $\hbar = \frac{1}{2}$, 1 and 2 depending on papers, which sometimes leads to confusion.

$$[\hat{x}, \hat{p}] = i\hbar, \tag{2.22}$$

and

$$[\hat{a}, \hat{a}^\dagger] = 1. \tag{2.23}$$

By using Eqs. (2.16), (2.17), (2.20) and (2.21), the electric and magnetic field operators are described as

$$\hat{E}(\boldsymbol{r}, t) = -\boldsymbol{e}\left[\hat{x}\sin(\boldsymbol{k}\cdot\boldsymbol{r}-\omega t) + \hat{p}\cos(\boldsymbol{k}\cdot\boldsymbol{r}-\omega t)\right], \tag{2.24}$$

$$\hat{B}(\boldsymbol{r}, t) = -\boldsymbol{k}\times\boldsymbol{e}\left[\hat{x}\sin(\boldsymbol{k}\cdot\boldsymbol{r}-\omega t) + \hat{p}\cos(\boldsymbol{k}\cdot\boldsymbol{r}-\omega t)\right]. \tag{2.25}$$

2.1.2.4 Uncertainty Principle for Quadratures

By applying the uncertainty principle to the quadrature operators $\hat{x}$ and $\hat{p}$, we get

$$\langle\Delta\hat{x}\rangle_{sd}\langle\Delta\hat{p}\rangle_{sd} \geqq \frac{1}{2}\hbar. \tag{2.26}$$

Therefore, $\hat{x}$ and $\hat{p}$ cannot be determined simultaneously.

2.2 Several Quantum States

2.2.1 Coherent State

2.2.1.1 Definition

A coherent state $|\alpha\rangle$ is defined to be an eigenstate of the annihilation operator $\hat{a}$ with the eigenvalue $\alpha = |\alpha|e^{i\theta}$:

$$\hat{a}|\alpha\rangle = \alpha|\alpha\rangle, \quad \alpha \in \mathbb{C}. \tag{2.27}$$

2.2.1.2 Averages and Fluctuations

For the coherent state $|\alpha\rangle$,

$$\langle\hat{x}\rangle = \sqrt{2\hbar}\Re[\alpha], \quad \langle\hat{x}^2\rangle = 2\hbar(\Re[\alpha])^2 + \frac{\hbar}{2}, \quad \langle\Delta^2\hat{x}\rangle = \frac{\hbar}{2}, \quad \langle\Delta\hat{x}\rangle_{sd} = \sqrt{\frac{\hbar}{2}}. \tag{2.28}$$

Similarly,

$$\langle \hat{p} \rangle = \sqrt{2\hbar}\Im[\alpha], \quad \langle \hat{p}^2 \rangle = 2\hbar(\Im[\alpha])^2 + \frac{\hbar}{2}, \quad \langle \Delta^2 \hat{p} \rangle = \frac{\hbar}{2}, \quad \langle \Delta \hat{p} \rangle_{sd} = \sqrt{\frac{\hbar}{2}}. \tag{2.29}$$

Thus, $\hat{x}$ and $\hat{p}$ have the same fluctuations. In addition, the coherent state is a minimum uncertainty state.

2.2.1.3 Representation in Phase Space

A single mode quantum state can be represented in a two-dimensional phase space. Let $\hat{x}$ be the horizontal axis, with $\hat{p}$ the vertical axis. Averages and fluctuations of position and momentum operators are described by the location and shade of a circle (Fig. 2.1).

2.2.1.4 Representation in Time

By using Eq. (2.24), the mean value of the electric field operator $\langle \hat{E} \rangle$ is given by

$$\langle \hat{E} \rangle = \langle \alpha | \hat{E} | \alpha \rangle = -\boldsymbol{e}[\langle \alpha | \hat{x} | \alpha \rangle \sin(\boldsymbol{k} \cdot \boldsymbol{r} - \omega t) + \langle \alpha | \hat{p} | \alpha \rangle \cos(\boldsymbol{k} \cdot \boldsymbol{r} - \omega t)] \tag{2.30}$$

$$= -\boldsymbol{e}\sqrt{2\hbar}|\alpha| \sin(\boldsymbol{k} \cdot \boldsymbol{r} - \omega t + \theta). \tag{2.31}$$

In a similar manner, $\langle \hat{E}^2 \rangle$ becomes

$$\langle \hat{E}^2 \rangle = \langle \hat{E} \rangle^2 + \frac{\hbar}{2}, \tag{2.32}$$

where

$$\langle \hat{x}\hat{p} + \hat{p}\hat{x} \rangle = -i\hbar(\alpha^2 - \alpha^{*2}) = 4\hbar\Re[\alpha]\Im[\alpha] \tag{2.33}$$

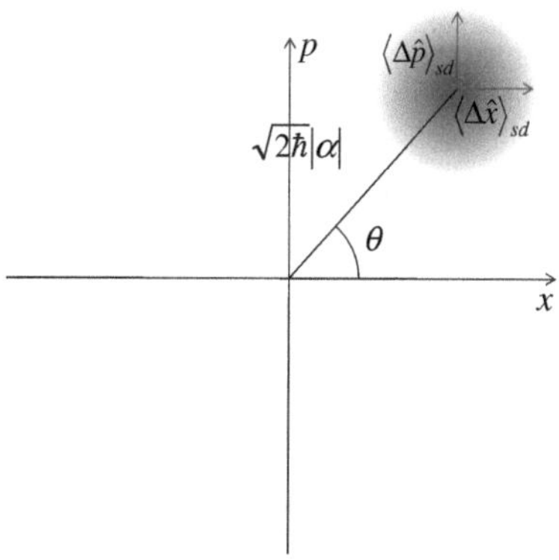

Fig. 2.1 Representation in phase space

is utilized. Therefore, the variance is given by

$$\langle \Delta^2 \hat{E} \rangle = \frac{\hbar}{2}. \tag{2.34}$$

Figure 2.2 shows a graph of the electric field in the time domain. The mean value is a sine wave, while the fluctuations do not depend on time.

2.2.1.5 Displacement Operator

The coherent state $|\alpha\rangle$ is represented as

$$|\alpha\rangle = \hat{D}(\alpha)|0\rangle, \tag{2.35}$$

where

$$\hat{D}(\alpha) = e^{\alpha \hat{a}^\dagger - \alpha^* \hat{a}}, \quad \alpha \in \mathbb{C} \tag{2.36}$$

is a displacement operator, while $|0\rangle$ is the vacuum state (see the next subsection).

2.2.2 *Vacuum State*

2.2.2.1 Definition

The vacuum state is a coherent state with $\alpha = 0$.

2.2.2.2 Averages and Fluctuations

Since the vacuum state is an example of coherent states, it shares all the properties described in 2.2.1.

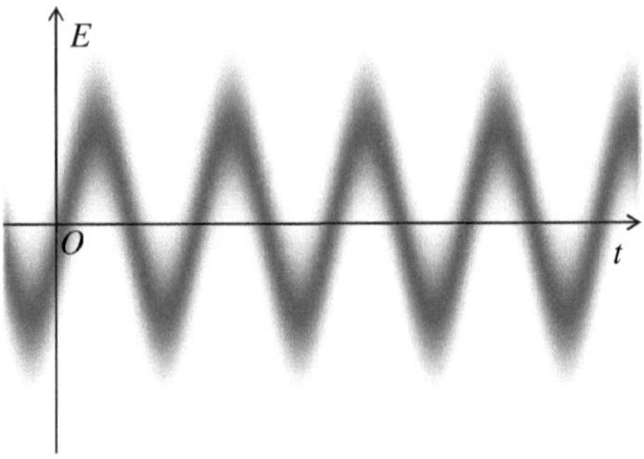

Fig. 2.2 Representation in time

$$\langle \hat{x} \rangle = \langle \hat{p} \rangle = 0, \quad \langle \hat{x}^2 \rangle = \langle \hat{p}^2 \rangle = \frac{\hbar}{2}, \quad \langle \Delta \hat{x} \rangle_{sd} = \langle \Delta \hat{p} \rangle_{sd} = \sqrt{\frac{\hbar}{2}}. \tag{2.37}$$

These equations show that the mean values of $\hat{x}$ and $\hat{p}$ are equal to zero. The vacuum state is a minimum uncertainty state with finite fluctuations derived from the uncertainty principle.

2.2.2.3 Representation in Phase Space

Phase space representation of the vacuum state is shown in Fig. 2.3.

2.2.2.4 Representation in Time

By using

$$\langle \hat{E} \rangle = 0, \quad \langle \Delta^2 \hat{E} \rangle = \frac{\hbar}{2}, \tag{2.38}$$

for the vacuum state, we get the graph of the electric field in the time domain (Fig. 2.4). The mean value is constantly zero, while the fluctuations do not depend on time.

2.2.3 *Squeezed Vacuum State*

2.2.3.1 Definition

We define a squeezing operator:

$$\hat{S}(r) = e^{\frac{r}{2}(\hat{a}^2 - \hat{a}^{\dagger 2})}, \quad r \in \mathbb{R}, \tag{2.39}$$

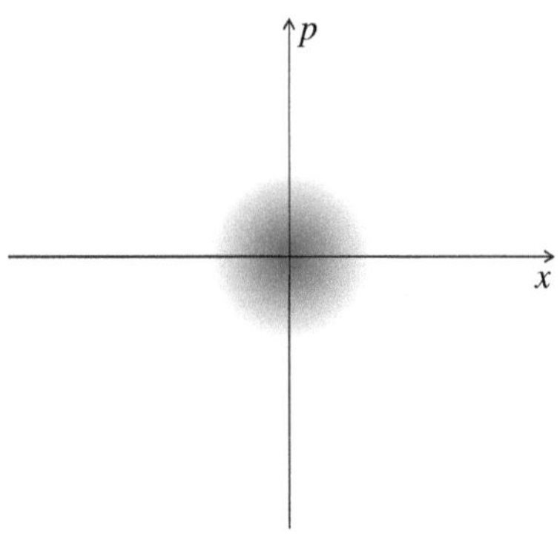

Fig. 2.3 Representation in phase space

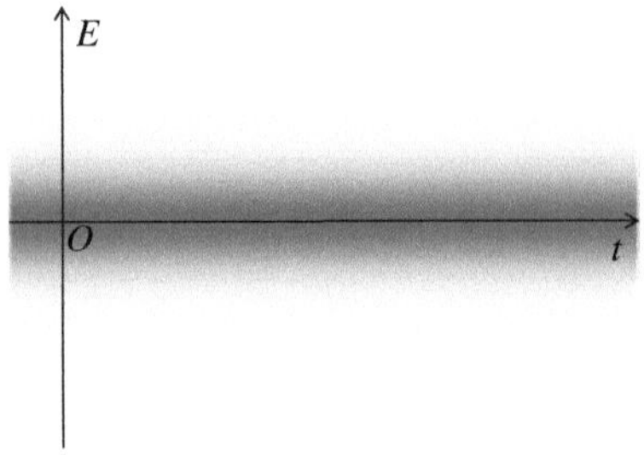

Fig. 2.4 Representation in time

where r is called a squeezing parameter. A squeezed vacuum state,[3] or a squeezed state, is defined as

$$\hat{S}(r)|0\rangle, \tag{2.40}$$

where $|0\rangle$ is the vacuum state.

2.2.3.2 Averages and Fluctuations

By using

$$\hat{S}^\dagger(r)\hat{a}\hat{S}(r) = \hat{a}\cosh r - \hat{a}^\dagger \sinh r, \tag{2.41}$$

$$\hat{S}^\dagger(r)\hat{a}^\dagger\hat{S}(r) = \hat{a}^\dagger\cosh r - \hat{a} \sinh r, \tag{2.42}$$

$$\hat{S}^\dagger(r)\hat{x}\hat{S}(r) = e^{-r}\hat{x}, \quad \hat{S}^\dagger(r)\hat{p}\hat{S}(r) = e^{r}\hat{p}, \tag{2.43}$$

we get

$$\langle\hat{x}\rangle = \langle\hat{p}\rangle = 0, \quad \langle\hat{x}^2\rangle = \frac{\hbar}{2}e^{-2r}, \quad \langle\hat{p}^2\rangle = \frac{\hbar}{2}e^{2r}, \tag{2.44}$$

and thus

$$\langle\Delta^2\hat{x}\rangle = \frac{\hbar}{2}e^{-2r}, \quad \langle\Delta^2\hat{p}\rangle = \frac{\hbar}{2}e^{2r}, \quad \langle\Delta\hat{x}\rangle_{sd} = \sqrt{\frac{\hbar}{2}}e^{-r}, \quad \langle\Delta\hat{p}\rangle_{sd} = \sqrt{\frac{\hbar}{2}}e^{r}. \tag{2.45}$$

Note that the variance of $\hat{x}$ is not equal to that of $\hat{p}$ for $r \neq 0$. However, they satisfy

$$\langle\Delta\hat{x}\rangle_{sd}\langle\Delta\hat{p}\rangle_{sd} = \frac{\hbar}{2}. \tag{2.46}$$

Therefore, the squeezed vacuum state is also a minimum uncertainty state.

---

[3] When $r > 0$, it becomes an x-squeezed state. When $r = 0$, it is the vacuum state. When $r < 0$, it becomes a p-squeezed state.

2.2.3.3 Representation in Phase Space

Figures 2.5 and 2.6 show phase space representations of squeezed states with $r > 0$ and $r < 0$.

In these figures, dotted circles show phase space representations of the vacuum states. Fluctuations of $\hat{x}$ and $\hat{p}$ are smaller than those of the vacuum states in Figs. 2.5 and 2.6, respectively.

2.2.3.4 Representation in Time

By using Eq. (2.24), the mean value of the electric field operator $\langle\hat{E}\rangle$ for the squeezed state $\hat{S}(r)|0\rangle$ is given by

$$\langle\hat{E}\rangle = \langle 0|\hat{S}(r)^{\dagger}\hat{E}\hat{S}(r)|0\rangle = 0. \tag{2.47}$$

Therefore, $\langle\Delta^2\hat{E}\rangle = \langle\hat{E}^2\rangle - \langle\hat{E}\rangle^2 = \langle\hat{E}^2\rangle$ becomes

$$\langle\Delta^2\hat{E}\rangle = \frac{\hbar}{2}e^{-2r}\sin^2(\boldsymbol{k}\cdot\boldsymbol{r} - \omega t) + \frac{\hbar}{2}e^{2r}\cos^2(\boldsymbol{k}\cdot\boldsymbol{r} - \omega t). \tag{2.48}$$

Figures 2.7 and 2.8 show graphs of the squeezed states at $\boldsymbol{r} = 0$ in the time domain.

In both cases, the mean value of the electric field operator is constantly zero. When $r > 0$, the graph is considered as a superposition (in a classical sense) of cosine waves with different amplitudes, because the fluctuation of $\hat{p}$ is larger than that of $\hat{x}$. To the contrary, when $r < 0$, it is considered as a superposition of sine waves, because the fluctuation of $\hat{x}$ is larger than that of $\hat{p}$.

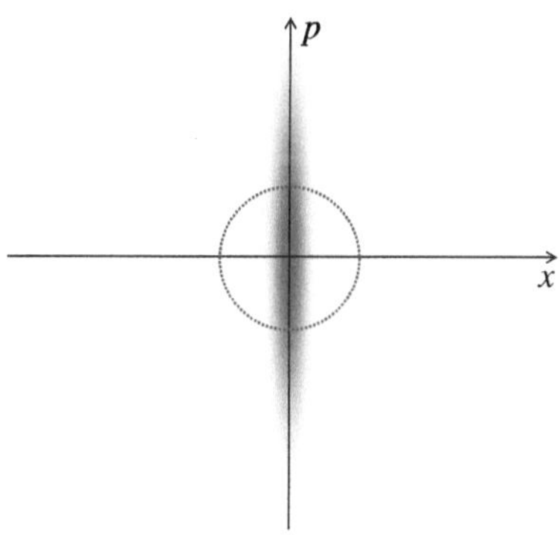

Fig. 2.5 Representation in phase space ($r > 0$)

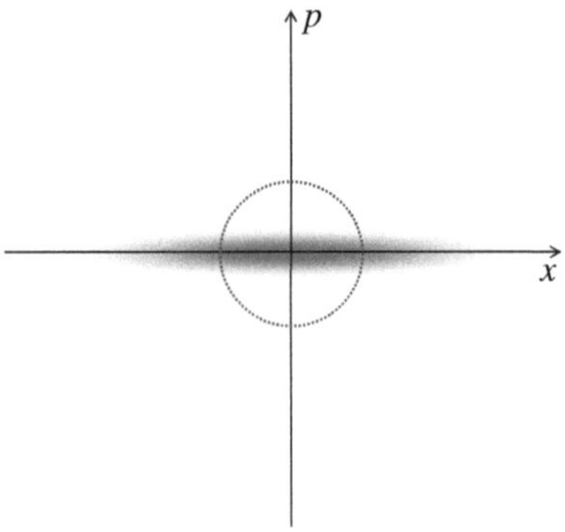

Fig. 2.6 Representation in phase space ($r < 0$)

2.3 Optical Parametric Oscillator

In all experiments in this thesis, each squeezed state is generated by using an optical parametric oscillator (OPO). An OPO is an optical cavity with a nonlinear medium (Fig. 2.9). The generation process of the squeezed state is fully formulated by the quantum Langevin equation [3, 4]. In this section, we briefly review the quantum Langevin equation.

2.3.1 *Quantum Langevin Equation*

We define $\hat{a}_{in}$ and $\hat{a}_{out}$ as the annihilation operators of the main input and output fields, while γ_1 ($\in \mathbb{R}$) is the damping constant. In a similar manner, we define $\hat{b}_{in}$ and $\hat{b}_{out}$ as the annihilation operators of the loss-mode input and output fields, while γ_2 ($\in \mathbb{R}$) is the damping constant. The output field of the OPO $\hat{a}_{out}$ is given by

Fig. 2.7 Representation in time ($r > 0$)

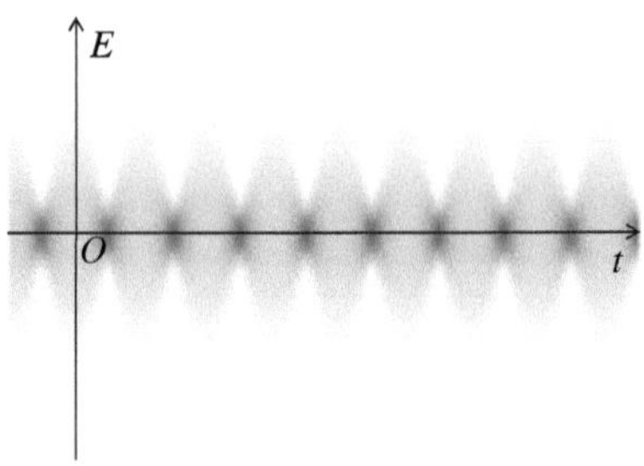

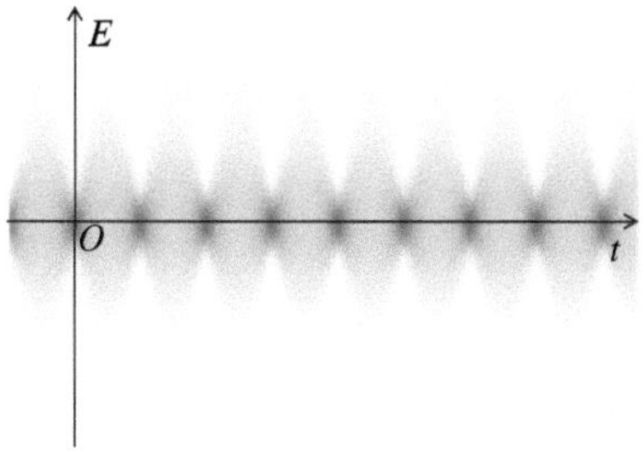

Fig. 2.8 Representation in time ($r < 0$)

$$\hat{a}_{out}(\Omega+\omega) = \frac{[(\frac{\gamma_1}{2})^2 - (\frac{\gamma_2}{2} - i\omega)^2 + |\varepsilon|^2]\hat{a}_{in}(\Omega+\omega) - \varepsilon\gamma_1\hat{a}^\dagger_{in}(\Omega-\omega)}{(\frac{\gamma_1+\gamma_2}{2} - i\omega)^2 - |\varepsilon|^2} + \frac{\sqrt{\gamma_1\gamma_2}(\frac{\gamma_1+\gamma_2}{2} - i\omega)\hat{b}_{in}(\Omega+\omega) - \varepsilon\sqrt{\gamma_1\gamma_2}\hat{b}^\dagger_{in}(\Omega-\omega)}{(\frac{\gamma_1+\gamma_2}{2} - i\omega)^2 - |\varepsilon|^2}, \tag{2.49}$$

where Ω is the center angular frequency of the laser, while ω is a relative angular frequency. ε ($\in \mathbb{C}$) shows a constant which is determined by the amplitude of the pump laser and the second order susceptibility of the nonlinear medium $\chi^{(2)}$. In the following, we omit Ω for simplicity. By defining

$$\hat{a}_p(\omega) = \frac{\hat{a}_{out}(\omega) + \hat{a}_{out}(-\omega)}{\sqrt{2}}, \quad \hat{a}_m(\omega) = \frac{\hat{a}_{out}(\omega) - \hat{a}_{out}(-\omega)}{\sqrt{2}}, \tag{2.50}$$

we get

$$\langle \hat{x}_p(\omega)\hat{x}_p(\omega')\rangle = \langle \hat{p}_m(\omega)\hat{p}_m(\omega')\rangle = \frac{\hbar}{2}\delta(\omega-\omega')\left[Sq(\omega)\cos^2\frac{\theta}{2} + Asq(\omega)\sin^2\frac{\theta}{2}\right], \tag{2.51}$$

$$\langle \hat{x}_m(\omega)\hat{x}_m(\omega')\rangle = \langle \hat{p}_p(\omega)\hat{p}_p(\omega')\rangle = \frac{\hbar}{2}\delta(\omega-\omega')\left[Asq(\omega)\cos^2\frac{\theta}{2} + Sq(\omega)\sin^2\frac{\theta}{2}\right], \tag{2.52}$$

where

$$Sq(\omega) = 1 - \frac{2|\varepsilon|\gamma_1}{(\frac{\gamma_1+\gamma_2}{2} + |\varepsilon|)^2 + \omega^2}, \tag{2.53}$$

$$Asq(\omega) = 1 + \frac{2|\varepsilon|\gamma_1}{(\frac{\gamma_1+\gamma_2}{2} - |\varepsilon|)^2 + \omega^2} \tag{2.54}$$

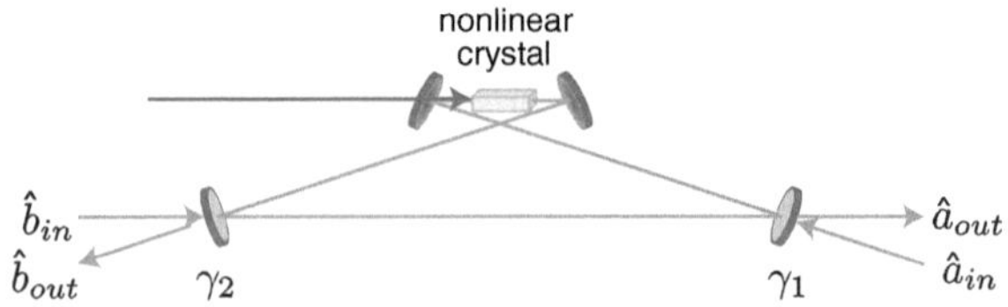

Fig. 2.9 Optical parametric oscillator

show levels of squeezing and antisqueezing. Here, we have defined $|\varepsilon|$ and θ as $\varepsilon = e^{i\theta}|\varepsilon|$. These equations show that the operator with angular frequency $\Omega + \omega$ has correlations only with the operator with angular frequency $\Omega - \omega$. We can consider that there exist two independent modes $\hat{a}_p(\omega)$ and $\hat{a}_m(\omega)$, which are in $\hat{x}_p$-squeezed and $\hat{p}_m$-squeezed states when $\theta = 0$.

2.3.2 Relationship to Experimental Setups

The damping constants γ_1 and γ_2 are approximated by

$$\gamma_1 = \frac{Tc}{l}, \quad \gamma_2 = \frac{Lc}{l}, \tag{2.55}$$

where c is the speed of light, while l, T, and L are the cavity length, the energy transmissivity, and the intracavity energy loss of the OPO, respectively. We also define

$$\varepsilon_{th} = \gamma_{opo} = \frac{\gamma_1 + \gamma_2}{2}, \quad x = \frac{|\varepsilon|}{\varepsilon_{th}}, \tag{2.56}$$

where ε_{th} and γ_{opo} represent the oscillation threshold and the half width at half maximum (HWHM) of the OPO, respectively. x shows the ratio between the pump amplitude and the oscillation threshold. It is related to the maximum (G_+) and minimum (G_-) amplification factors (parametric gains) of the classical parametric process:

$$G_\pm = \frac{1}{(1 \mp x)^2}. \tag{2.57}$$

By using these parameters, we get

$$Sq(\omega) = 1 - \frac{T}{T+L}\frac{4x}{(1+x)^2 + (\frac{\omega}{\gamma_{opo}})^2}, \tag{2.58}$$

$$Asq(\omega) = 1 + \frac{T}{T+L}\frac{4x}{(1-x)^2 + (\frac{\omega}{\gamma_{opo}})^2}. \tag{2.59}$$

2.4 Homodyne Measurement

Homodyne measurement [1] is the basic measurement in quantum optics. In all experimental demonstrations reported in this thesis, all measurements on quantum states are homodyne measurements.

The basic structure of the homodyne measurement is described in Fig. 2.10.

It detects interferences between the signal beam and a local oscillator (LO) beam, which is an intense coherent beam. Since the LO beam works as a reference, we can measure a quantum state in specific spatial and frequency modes. In addition, we can measure quantum fluctuations of the signal beam without being obstructed by thermal noise.

We define $\hat{a}$ as the annihilation operator of the signal mode, while $\alpha = |\alpha|e^{i\phi}$ is the complex amplitude of the LO beam. Here, we have assumed that the intensity of the LO beam is so high that its quantum nature is negligible. These two modes are combined using a 50:50 beam splitter (half beam splitter, HBS), leading to the annihilation operators

$$\hat{a}_1 = \frac{1}{\sqrt{2}}(\hat{a} + \alpha), \quad \hat{a}_2 = \frac{1}{\sqrt{2}}(-\hat{a} + \alpha). \tag{2.60}$$

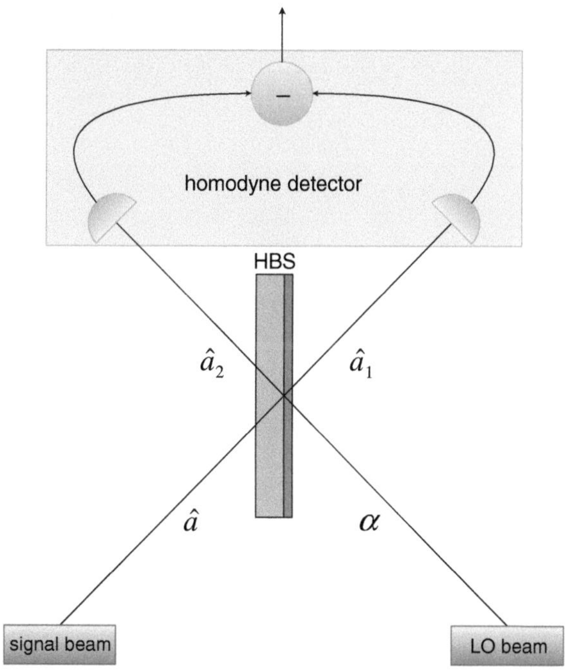

Fig. 2.10 Homodyne measurement

These outcomes are converted to photocurrents. The output of the homodyne detection is

$$\hat{I}(t) = \hat{a}_1^\dagger \hat{a}_1 - \hat{a}_2^\dagger \hat{a}_2 = \sqrt{\frac{2}{\hbar}}|\alpha|[\hat{x}(t)\cos\phi + \hat{p}(t)\sin\phi]. \tag{2.61}$$

It shows that the measurement variable of the homodyne detection is a linear combination of position and momentum operators. It is controlled by the phase ϕ of the LO beam. We can also get

$$\hat{I}(t) = \int_{-\infty}^{\infty} d\omega[\alpha^* \hat{a}(\omega)e^{-i\omega t} + \alpha \hat{a}^\dagger(\omega)e^{i\omega t}] \tag{2.62}$$

$$= \int_{0}^{\infty} d\omega[\alpha^* \hat{a}(-\omega)e^{i\omega t} + \alpha \hat{a}^\dagger(-\omega)e^{-i\omega t} + \alpha^* \hat{a}(\omega)e^{-i\omega t} + \alpha \hat{a}^\dagger(\omega)e^{i\omega t}]. \tag{2.63}$$

By using $\alpha = |\alpha|e^{i\phi}$, it becomes

$$\hat{I}(t) = |\alpha|\sqrt{\frac{2}{\hbar}}\int_{0}^{\infty} d\omega\Big[\cos\omega t[(\hat{x}(\omega) + \hat{x}(-\omega))\cos\phi + (\hat{p}(\omega) + \hat{p}(-\omega))\sin\phi]$$
$$+ \sin\omega t[(\hat{p}(\omega) - \hat{p}(-\omega))\cos\phi - (\hat{x}(\omega) - \hat{x}(-\omega))\sin\phi]\Big]. \tag{2.64}$$

By defining

$$\hat{a}_p(\omega) = \frac{\hat{a}(\omega) + \hat{a}(-\omega)}{\sqrt{2}}, \quad \hat{a}_m(\omega) = \frac{\hat{a}(\omega) - \hat{a}(-\omega)}{\sqrt{2}}, \tag{2.65}$$

we get

$$\hat{I}(t) = |\alpha|\frac{2}{\sqrt{\hbar}}\int_{0}^{\infty} d\omega\Big[\cos\omega t[\hat{x}_p(\omega)\cos\phi + \hat{p}_p(\omega)\sin\phi] + \sin\omega t[\hat{p}_m(\omega)\cos\phi - \hat{x}_m(\omega)\sin\phi]\Big] \tag{2.66}$$

$$= |\alpha|\frac{2}{\sqrt{\hbar}}\int_{0}^{\infty} d\omega\Big[\hat{x}_p(\omega,\phi)\cos\omega t + \hat{x}_m(\omega,\phi + \tfrac{\pi}{2})\sin\omega t\Big]. \tag{2.67}$$

Here, we have used

$$\hat{x}(\omega,\phi) = \hat{x}(\omega)\cos\phi + \hat{p}(\omega)\sin\phi. \tag{2.68}$$

In all experiments reported in this thesis, the outcomes of homodyne detections are measured by using a spectrum analyzer. In general, a spectrum analyzer shows the power of electrical signals with a specific angular frequency ω, which corresponds to a measurement of the power, or the variance, of $\hat{x}_p(\omega, \phi)$ and $\hat{x}_m(\omega, \phi + \frac{\pi}{2})$. Note that it shows the total power of these two operators because it cannot distinguish between $\hat{x}_p(\omega, \phi)$ and $\hat{x}_m(\omega, \phi + \frac{\pi}{2})$.

References

1. Leonhardt, U.: Measuring the Quantum State of Light. Cambridge University Press, Cambridge (1997)
2. Furusawa, A., van Loock, P.: Quantum Teleportation and Entanglement. WILEY-VCH, New York (2011)
3. Collett, M.J., Gardiner, C.W.: Squeezing of intracavity and traveling-wave light fields produced in parametric amplification. Phys. Rev. A **30**, 1386 (1984)
4. Gardiner, C.W., Collett, M.J.: Input and output in damped quantum systems: Quantum stochastic differential equations and the master equation. Phys. Rev. A **31**, 3761 (1985)

Chapter 3
Quantum States and Quantum State Manipulations

3.1 Computational Basis

Quantum computation with discrete variables (DV) utilizes a discrete and finite number of quantum states as its computational basis [1]. On the contrary, quantum computation with continuous variables (CV) involves a continuous and infinite number of quantum states as its computational basis. In this thesis, we focus on CV quantum computation.

In CV quantum computation with bosonic fields, two types of computational bases are traditionally used—eigenstates of position and momentum operators $|x = a\rangle$ and $|p = b\rangle$ with $a, b \in \mathbb{R}$,

$$\hat{x}|x = a\rangle = a|x = a\rangle, \quad a \in \mathbb{R}, \tag{3.1}$$

$$\hat{p}|p = b\rangle = b|p = b\rangle, \quad b \in \mathbb{R}, \tag{3.2}$$

where a and b represent their eigenvalues. In the following, we sometimes describe the eigenstate of the position operator as $|x\rangle$ or $|a\rangle_x$, while that of the momentum operator as $|p\rangle$ or $|b\rangle_p$, for simplicity. They are complete:

$$\int_{-\infty}^{\infty} \mathrm{d}x\, |x\rangle\langle x| = 1, \quad \int_{-\infty}^{\infty} \mathrm{d}p\, |p\rangle\langle p| = 1, \tag{3.3}$$

and orthogonal:

$$\langle x|x'\rangle = \delta(x - x'), \quad \langle p|p'\rangle = \delta(p - p'). \tag{3.4}$$

Two eigenstates $|x\rangle$ and $|p\rangle$ of position and momentum operators are related to each other by the Fourier transformation:

R. Ukai, *Multi-Step Multi-Input One-Way Quantum Information Processing with Spatial and Temporal Modes of Light*, Springer Theses,
DOI 10.1007/978-4-431-55019-8_3

$$|x\rangle = \frac{1}{\sqrt{2\pi\hbar}} \int_{-\infty}^{\infty} \mathrm{d}p\, e^{-\frac{ixp}{\hbar}} |p\rangle, \tag{3.5}$$

$$|p\rangle = \frac{1}{\sqrt{2\pi\hbar}} \int_{-\infty}^{\infty} \mathrm{d}x\, e^{\frac{ixp}{\hbar}} |x\rangle. \tag{3.6}$$

The inner product of $|x\rangle$ and $|p\rangle$ is given by

$$\langle x|p\rangle = \frac{1}{\sqrt{2\pi\hbar}} e^{\frac{ixp}{\hbar}}. \tag{3.7}$$

3.2 Representations of Quantum States

In this section, we summarize several methods for representing quantum states.

3.2.1 State Vector

3.2.1.1 One-Mode State

We set the eigenstate $|x\rangle$ of the position operator $\hat{x}$ as the computational basis. An arbitrary one-mode pure quantum state $|\psi\rangle$ can be expanded as

$$|\psi\rangle = \int_{-\infty}^{\infty} \mathrm{d}x\, |x\rangle\langle x|\psi\rangle = \int_{-\infty}^{\infty} \mathrm{d}x\, f(x)|x\rangle, \tag{3.8}$$

where $f(x) = \langle x|\psi\rangle$ is normalized:

$$\int_{-\infty}^{\infty} \mathrm{d}x\, f^*(x)f(x) = \int_{-\infty}^{\infty} \mathrm{d}x\, \langle\psi|x\rangle\langle x|\psi\rangle = 1. \tag{3.9}$$

When we set the eigenstate $|p\rangle$ of the momentum operator $\hat{p}$ as the computational basis, $|\psi\rangle$ can be expanded as

$$|\psi\rangle = \int_{-\infty}^{\infty} \mathrm{d}p\, g(p)|p\rangle, \tag{3.10}$$

$$\int_{-\infty}^{\infty} \mathrm{d}p\, g^*(p)g(p) = 1. \tag{3.11}$$

3.2.1.2 Multi-Mode State

As a natural extension of the one-mode case above, the N-mode infinite-dimensional Hilbert space is spanned by the direct product of eigenstates of position or momentum operators. By taking the former case for example, the computational basis for the N-mode quantum state is given by

$$|x_1\rangle \otimes |x_2\rangle \otimes \cdots \otimes |x_N\rangle = |x_1\rangle|x_2\rangle \cdots |x_N\rangle. \tag{3.12}$$

An arbitrary N-mode pure quantum state $|\psi_N\rangle$ can be expanded in the position basis as

$$|\psi_N\rangle = \int_{-\infty}^{\infty} \mathrm{d}x_1 \int_{-\infty}^{\infty} \mathrm{d}x_2 \cdots \int_{-\infty}^{\infty} \mathrm{d}x_N f(x_1, x_2, \ldots, x_N)|x_1\rangle|x_2\rangle \cdots |x_N\rangle, \tag{3.13}$$

where $f(x_1, x_2, \ldots, x_N)$ is normalized:

$$\int_{-\infty}^{\infty} \mathrm{d}x_1 \int_{-\infty}^{\infty} \mathrm{d}x_2 \ldots \int_{-\infty}^{\infty} \mathrm{d}x_N f^*(x_1, x_2, \ldots, x_N) f(x_1, x_2, \ldots, x_N) = 1. \tag{3.14}$$

3.2.2 *Density Operator*

Although all pure quantum states can be described by state vectors, mixed states cannot be described in this manner. In order to formulate mixed states, we introduce density operators.

3.2.2.1 Definition of Density Operator

Consider an ensemble of pure states $|\psi_n\rangle$ with statistical probabilities p_n. The density operator $\hat{\rho}$ of the mixed quantum state is defined as

$$\hat{\rho} = \sum_n p_n|\psi_n\rangle\langle\psi_n|. \tag{3.15}$$

Since p_n represents the probability, we have

$$p_n > 0, \quad \sum_n p_n = 1. \tag{3.16}$$

In the case of a pure quantum state $|\psi\rangle$, the density operator $\hat{\rho}$ is given by

$$\hat{\rho} = |\psi\rangle\langle\psi|. \tag{3.17}$$

Any density operator is Hermitian with trace of one, and positive definite:

$$\hat{\rho}^\dagger = \hat{\rho}, \quad \mathrm{tr}\hat{\rho} = 1, \quad \hat{\rho} \geq 0. \tag{3.18}$$

3.2.2.2 Observable and Its Expectation Value

In general, any observable can be formulated by a Hermitian operator $\hat{A}$. It can be decomposed into

$$\hat{A} = \sum_a a|a\rangle\langle a|, \tag{3.19}$$

where $|a\rangle$ shows the eigenstate of $\hat{A}$ with eigenvalue a (spectral decomposition). The expectation value of the observable $\hat{A}$ in the state with the density operator $\hat{\rho}$ is given by

$$\langle\hat{A}\rangle = \sum_n p_n\langle\psi_n|\hat{A}|\psi_n\rangle = \sum_a \langle a|\sum_n p_n|\psi_n\rangle\langle\psi_n|\hat{A}|a\rangle = \mathrm{tr}\hat{\rho}\hat{A}. \tag{3.20}$$

Here, the trace operation "tr" satisfies

$$\mathrm{tr}\hat{A}\hat{B} = \mathrm{tr}\hat{B}\hat{A}, \tag{3.21}$$

for any operators $\hat{A}$ and $\hat{B}$.

3.2.3 *Stabilizer*

Consider a pure one-mode quantum state $|\psi\rangle$. Consider the set $\{\hat{S}\}$ of unitary operators $\hat{S}$ which satisfy

$$\hat{S}|\psi\rangle = |\psi\rangle. \tag{3.22}$$

The set forms a subgroup (stabilizer group) of the unitary group which consists of unitary operators. We refer to $\{\hat{S}\}$ as the stabilizer of the quantum state $|\psi\rangle$ [2, 3].

For example, the set $\{\hat{F}(\theta)\}$ of the rotation operators $\hat{F}(\theta)$ (Sect. 3.5.3):

$$\hat{F}(\theta) = e^{i\theta\hat{a}^\dagger\hat{a}} \tag{3.23}$$

is the stabilizer of the vacuum state $|0\rangle$ since

$$\hat{F}(\theta)|0\rangle = |0\rangle. \tag{3.24}$$

Accordingly, the quantum state which is not changed by the rotation operator $\hat{F}(\theta)$ with an arbitrary θ including the global phase is nothing but the vacuum state.

An arbitrary one-mode pure quantum state $|\psi\rangle$ is given by the vacuum state $|0\rangle$ and a unitary operator $\hat{U}$ as

$$|\psi\rangle = \hat{U}|0\rangle. \tag{3.25}$$

Since we have

$$\hat{U}\hat{F}(\theta)\hat{U}^{\dagger}|\psi\rangle = \hat{U}\hat{F}(\theta)\hat{U}^{\dagger}[\hat{U}|0\rangle] = |\psi\rangle, \tag{3.26}$$

$\{\hat{U}\hat{F}(\theta)\hat{U}^{\dagger}\}$ is the stabilizer of the quantum state $|\psi\rangle$. Accordingly, the quantum state which is stabilized by $\{\hat{U}\hat{F}(\theta)\hat{U}^{\dagger}\}$ is nothing but $|\psi\rangle$. Therefore, the quantum state can be uniquely determined by its stabilizer.

The discussion above can be easily extended to N-mode pure quantum states since the stabilizer of the N-mode vacuum state is given by $\{\hat{F}_k(\theta_k)\}$, where $\hat{F}_k(\theta_k)$ is the rotation operator on mode k.

We lastly consider the eigenstates $|x = a\rangle$ and $|p = b\rangle$ of position operator $\hat{x}$ and momentum operator $\hat{p}$. By using

$$e^{is\frac{\hat{x}-a}{\hbar}}|x = a\rangle = |x = a\rangle, \quad e^{-is\frac{\hat{p}-b}{\hbar}}|p = b\rangle = |p = b\rangle, \tag{3.27}$$

we find that the stabilizers of these quantum states are $\{e^{is\frac{\hat{x}-a}{\hbar}}\}$ and $\{e^{-is\frac{\hat{p}-b}{\hbar}}\}$. Note that they are the momentum displacement operator and the position displacement operator up to global phases. When the eigenvalues are $a = 0$ and $b = 0$, they become the momentum displacement operator $\hat{Z}(s)$ and the position displacement operator $\hat{X}(s)$ (Sects. 3.5.1 and 3.5.2).

3.2.4 Nullifier

Consider a pure one-mode quantum state $|\psi\rangle$. Consider an operator $\hat{\delta}$ ($\neq 0$) which satisfies

$$\hat{\delta}|\psi\rangle = 0. \tag{3.28}$$

We refer to $\hat{\delta}$ as a nullifier of the quantum state $|\psi\rangle$ [2, 3].

For example, the annihilation operator $\hat{a}$ is the nullifier of the vacuum state $|0\rangle$ since $\hat{a}|0\rangle = 0$. The nullifier of $|\psi\rangle = \hat{U}|0\rangle$ is given by $\hat{U}\hat{a}\hat{U}^{\dagger}$.

Similarly, since the nullifiers of the N-mode vacuum state $|0\rangle^{\otimes N}$ are $\{\hat{a}_k\}$, $k = 1, \ldots, N$, the nullifiers of $|\psi\rangle = \hat{U}|0\rangle^{\otimes N}$ are given by $\{\hat{U}\hat{a}_k\hat{U}^{\dagger}\}$, $k = 1, \ldots, N$.

Any N-mode pure quantum state can be uniquely determined by its N independent nullifiers.

We sometimes write

$$\hat{\delta} = 0, \tag{3.29}$$

in order to show that $\hat{\delta}$ is a nullifier of the quantum state, for simplicity.

Consider that an operator $\hat{\delta}$ is a nullifier of an N-mode quantum state $|\psi\rangle$ ($\hat{\delta}|\psi\rangle = 0$). Since

$$(a\hat{\delta})|\psi\rangle = 0, \tag{3.30}$$

$a\hat{\delta}$ is also a nullifier of the quantum state $|\psi\rangle$.

Consider that two operators $\hat{\delta}_1$ and $\hat{\delta}_2$ are nullifiers of a quantum state $|\psi\rangle$ ($\hat{\delta}_1|\psi\rangle = 0$, $\hat{\delta}_2|\psi\rangle = 0$). Since

$$(a\hat{\delta}_1 + b\hat{\delta}_2)|\psi\rangle = 0, \quad \hat{\delta}_1\hat{\delta}_2|\psi\rangle = 0, \tag{3.31}$$

$(a\hat{\delta}_1 + b\hat{\delta}_2)$ and $\hat{\delta}_1\hat{\delta}_2$ are also nullifiers of the quantum state $|\psi\rangle$.

3.2.5 *Moment*

3.2.5.1 *xpxp* Notation and *xxpp* Notation

We utilize two types of notations in order to formulate vectors of quadrature operators based on the order of position operators and momentum operators.

In *xpxp* notation, we describe N-mode quadrature operators as

$$\hat{\boldsymbol{\xi}} = (\hat{x}_1, \hat{p}_1, \ldots, \hat{x}_N, \hat{p}_N)^T. \tag{3.32}$$

On the other hand, in *xxpp* notation, we describe N-mode quadrature operators as

$$\hat{\boldsymbol{q}} = (\hat{x}_1, \ldots, \hat{x}_N, \hat{p}_1, \ldots, \hat{p}_N)^T. \tag{3.33}$$

These two notations are related as

$$\begin{pmatrix} \hat{x}_1 \\ \vdots \\ \hat{x}_N \\ \hat{p}_1 \\ \vdots \\ \hat{p}_N \end{pmatrix} = \begin{pmatrix} 1\,0 & & & & \\ & 1\,0 & & & \\ & & \ddots & & \\ & & & 1\,0 & \\ 0\,1 & & & & \\ & 0\,1 & & & \\ & & \ddots & & \\ & & & 0\,1 & \end{pmatrix} \begin{pmatrix} \hat{x}_1 \\ \hat{p}_1 \\ \vdots \\ \hat{x}_N \\ \hat{p}_N \end{pmatrix} \equiv T \begin{pmatrix} \hat{x}_1 \\ \hat{p}_1 \\ \vdots \\ \hat{x}_N \\ \hat{p}_N \end{pmatrix}. \tag{3.34}$$

Here, T satisfies

$$T^{-1} = T^T. \tag{3.35}$$

3.2.5.2 Definition of Moment

We assume that $\hat{\rho}$ is the density operator of an N-mode mixed quantum state. We define the k-th order moment in *xpxp* notation as

$$M_{j_1,\ldots,j_k} = \mathrm{tr}\left[\hat{\rho}\hat{\xi}_{j_1}\cdots\hat{\xi}_{j_k}\right]. \tag{3.36}$$

Similarly, we can define k-th order moments in *xxpp* notation.

3.2.6 Expectation Value and Covariance Matrix of Quadrature Operators

3.2.6.1 Expectation Value of Quadrature Operators

We consider the first-order moments in *xpxp* notation. By definition $m_j = \mathrm{tr}\left[\hat{\rho}\hat{\xi}_j\right]$, they show expectation values of quadrature operators $m_j = \langle\hat{\xi}_j\rangle$. Similarly, the first-order moments in *xxpp* notation are given by $n_j = \langle\hat{q}_j\rangle$.

3.2.6.2 Covariance Matrix in *xpxp* Notation

We consider the second-order moments in *xpxp* notation. By definition, they are $M_{j_1,j_2} = \mathrm{tr}\left[\hat{\rho}\hat{\xi}_{j_1}\hat{\xi}_{j_2}\right]$. Note that $M_{2j-1,2j} = \mathrm{tr}\left[\hat{\rho}\hat{x}_j\hat{p}_j\right]$, which shows correlation between position and momentum operators of single mode, does not become a real number. Therefore, by subtracting the expectation values, and by symmetrizing with respect to position and momentum operators, we define the covariance matrix V in *xpxp* notation as

$$V_{j,k} = \mathrm{tr}\left[\hat{\rho}\left\{\Delta\hat{\xi}_j, \Delta\hat{\xi}_k\right\}\right], \quad \Delta\hat{\xi}_j = \hat{\xi}_j - \langle\hat{\xi}_j\rangle, \quad \left\{\Delta\hat{\xi}_j, \Delta\hat{\xi}_k\right\} = \frac{\Delta\hat{\xi}_j\Delta\hat{\xi}_k + \Delta\hat{\xi}_k\Delta\hat{\xi}_j}{2}. \tag{3.37}$$

The matrix V can also be described as

$$V = \left\langle(\hat{\boldsymbol{\xi}} - \langle\hat{\boldsymbol{\xi}}\rangle)(\hat{\boldsymbol{\xi}} - \langle\hat{\boldsymbol{\xi}}\rangle)^T\right\rangle - \frac{i\hbar}{2}\Omega. \tag{3.38}$$

Here, $\Omega = \bigoplus_{k=1}^{N} \omega$, where $\omega = \begin{pmatrix} 0 & 1 \\ -1 & 0 \end{pmatrix}$.

By definition, V is a real symmetric matrix. The uncertainty principle is described as

$$V + \frac{i\hbar}{2}\Omega \geq 0. \tag{3.39}$$

It is a necessary and sufficient condition for a real symmetric matrix V that it shows a covariance matrix of a physical quantum state:

A real symmetric matrix V represents a covariance matrix

$$\text{of a physical quantum state} \Longleftrightarrow V + \frac{i\hbar}{2}\Omega \geq 0. \tag{3.40}$$

It shows that V is a positive-definite matrix:

$$V + \frac{i\hbar}{2}\Omega \geq 0 \Longrightarrow V > 0. \tag{3.41}$$

3.2.6.3 Covariance Matrix in *xxpp* Notation

We can define the covariance matrix Σ in *xxpp* notation as well:

$$\Sigma = \left\langle(\hat{\boldsymbol{q}} - \langle\hat{\boldsymbol{q}}\rangle)(\hat{\boldsymbol{q}} - \langle\hat{\boldsymbol{q}}\rangle)^T\right\rangle - \frac{i\hbar}{2}\Omega', \quad \Omega' = \begin{pmatrix} O & I \\ -I & O \end{pmatrix}, \tag{3.42}$$

where I and O represent the $N \times N$ unit matrix and zero matrix, respectively.

3.2.6.4 Williamson's Theorem

Here, we use *xpxp* notation. An arbitrary $2N \times 2N$ positive-definite real symmetric matrix V can be diagonalized by a symplectic matrix $S \in Sp(2N, \mathbb{R})$[1]:

$$SVS^T = \begin{pmatrix} \nu_1 & & & & \\ & \nu_1 & & & \\ & & \ddots & & \\ & & & \nu_N & \\ & & & & \nu_N \end{pmatrix}, \quad \nu_i > 0. \tag{3.43}$$

This is called the *Williamson's theorem*. Here, ν_k is called a symplectic eigenvalue. It can be computed by acquiring the eigenvalues of $i\Omega V$, which are equivalent to $\pm\nu_k$. Note that the diagonal matrix $\mathrm{diag}[\nu_1, \nu_1, \ldots, \nu_N, \nu_N]$ represents the covariance matrix of an N-mode thermal state. Thus, we get the following statement:

A real symmetric matrix V represents a covariance matrix

$$\text{of a physical quantum state} \Longleftrightarrow \min \nu_k \geq \frac{\hbar}{2} \text{ and } V > 0. \tag{3.44}$$

In addition, we have

$$V > 0 \Longleftrightarrow \det V_k > 0 \text{ for all } k, \tag{3.45}$$

for a real symmetric matrix V, where V_k is the k-th order principal submatrix of V. As a result,

A real symmetric matrix V represents a covariance matrix

$$\text{of a physical quantum state} \Longleftrightarrow \min \nu_k \geq \frac{\hbar}{2} \text{ and } \det V_i > 0 \text{ for all } i. \tag{3.46}$$

In the special case of a two-mode state, we describe two symplectic eigenvalues as $\nu_\pm$ $(\nu_+ \geq \nu_-)$.

3.2.7 Wigner Function and Gaussian State

3.2.7.1 Wigner Function

We consider an N-mode mixed quantum state with density operator $\hat{\rho}$. By using the Wigner characteristic function:

---

[1] Note that V does not necessarily show a covariance matrix of a physical quantum state. For more details of symplectic matrices and symplectic groups, see Refs. [4, 5].

$$\chi(\boldsymbol{\eta}) = \mathrm{tr}\left[\hat{\rho}e^{\frac{i}{\hbar}\eta^T\hat{\xi}}\right], \tag{3.47}$$

we define the Wigner function [6] of the N-mode state as

$$W(\boldsymbol{\xi}) = \frac{1}{(2\pi\hbar)^{2N}} \int_{-\infty}^{\infty} \mathrm{d}^{2N}\eta\, \chi(\eta) e^{-\frac{i}{\hbar}\eta^T\xi}, \tag{3.48}$$

where

$$\boldsymbol{\xi} = (x_1, p_1, \ldots, x_N, p_N)^T, \quad \boldsymbol{\eta} = (u_1, v_1, \ldots, u_N, v_N)^T. \tag{3.49}$$

It is a real-valued function whose range might contain negative values.

3.2.7.2 Gaussian State

Consider a quantum state whose Wigner function is a Gaussian function. It is called a *Gaussian state*. Any Gaussian state is uniquely determined by its first-order and second-order moments, or equivalently, the expectation values $\boldsymbol{m}$ and the covariance matrix V. Note that here we use *xpxp* notation. By using $\boldsymbol{m}$ and V, the Wigner function is given by

$$W(\boldsymbol{\xi}) = \frac{1}{(2\pi)^N\sqrt{\det V}} e^{-\frac{1}{2}(\boldsymbol{\xi}-\boldsymbol{m})^T V^{-1}(\boldsymbol{\xi}-\boldsymbol{m})}. \tag{3.50}$$

The vacuum state, squeezed state, and coherent state are one-mode Gaussian states. In addition, all the quantum states in the experiments in this thesis are Gaussian states.

3.2.8 Quadrature Operator Which Specifies a Particular Quantum State

We sometimes show that a quantum mode is in a particular quantum state by adding a superscript to its quadrature operators. For example, consider that mode 1 is in a vacuum state. By adding superscripts (0) to quadrature operators, we describe them as

$$\hat{x}_1^{(0)} + i\hat{p}_1^{(0)}. \tag{3.51}$$

Similarly, a p-squeezed state is formulated as

$$e^{r_1}\hat{x}_1^{(0)} + ie^{-r_1}\hat{p}_1^{(0)}. \tag{3.52}$$

3.3 Representations of Quantum State Manipulations

In this section, we summarize several methods for the representation of quantum state manipulations.

3.3.1 Unitary Operator, Transformation of State Vector, Schrödinger Picture

A quantum state manipulation corresponds to a transformation of a quantum state $|\psi\rangle$ to $|\psi'\rangle$. By using a unitary operator $\hat{U}$ with $\hat{U}^{\dagger}\hat{U} = \hat{I}$, it can be formulated as

$$|\psi'\rangle = \hat{U}|\psi\rangle. \tag{3.53}$$

3.3.2 Transformation of Density Operator

We assume that the density operator of an initial state is given by $\hat{\rho} = \sum_i p_i|\psi_i\rangle\langle\psi_i|$. By applying a unitary operator $\hat{U}$, it becomes

$$\hat{\rho}' = \sum_i p_i \hat{U}|\psi_i\rangle\langle\psi_i|\hat{U}^{\dagger} = \hat{U}\hat{\rho}\hat{U}^{\dagger}. \tag{3.54}$$

3.3.3 Transformation of Stabilizer

We assume that the set of unitary operators $\{\hat{S}\}$ is the stabilizer of a quantum state $|\psi\rangle$ ($\hat{S}|\psi\rangle = |\psi\rangle$). We apply a unitary operator $\hat{U}$ to the quantum state. By using

$$\hat{U}\hat{S}\hat{U}^{\dagger}[\hat{U}|\psi\rangle] = \hat{U}|\psi\rangle, \tag{3.55}$$

we find that the stabilizer of the resulting quantum state $\hat{U}|\psi\rangle$ is given by $\{\hat{U}\hat{S}\hat{U}^{\dagger}\}$.

3.3.4 Transformation of Nullifier

We assume that the operator $\hat{\delta}$ is a nullifier of a quantum state $|\psi\rangle$ ($\hat{\delta}|\psi\rangle = 0$). We apply a unitary operator $\hat{U}$ to the quantum state. By using

$$\hat{U}\hat{\delta}\hat{U}^{\dagger}[\hat{U}|\psi\rangle] = 0, \tag{3.56}$$

we find that $\hat{U}\hat{\delta}\hat{U}^{\dagger}$ is a nullifier of the resulting quantum state $\hat{U}|\psi\rangle$.

3.3.5 Transformation of Annihilation Operator, Heisenberg Picture

3.3.5.1 General Operation

An arbitrary N-mode Hermitian operator $\hat{F}$, which shows an observable, can be formulated as a function of annihilation and creation operators as $\hat{F}(\hat{a}_1, \hat{a}_1^{\dagger}, \ldots, \hat{a}_N, \hat{a}_N^{\dagger})$. In the Heisenberg picture, it is transformed by a unitary transformation with a unitary operator $\hat{U}$ to

$$\hat{U}^{\dagger}\hat{F}(\hat{a}_1, \hat{a}_1^{\dagger}, \ldots, \hat{a}_N, \hat{a}_N^{\dagger})\hat{U} = \hat{F}(\hat{U}^{\dagger}\hat{a}_1\hat{U}, \hat{U}^{\dagger}\hat{a}_1^{\dagger}\hat{U}, \ldots, \hat{U}^{\dagger}\hat{a}_N\hat{U}, \hat{U}^{\dagger}\hat{a}_N^{\dagger}\hat{U}). \tag{3.57}$$

Note that the creation operators are the Hermitian conjugates of the annihilation operators. Therefore, the transformation of an arbitrary Hermitian operator can be traced only by tracing the transformation of annihilation operators:

$$\hat{a}_k \to \hat{a}_k' = \hat{U}^{\dagger}\hat{a}_k\hat{U}, \tag{3.58}$$

where $\hat{a}_k$ and $\hat{a}_k'$ show annihilation operators of mode k before and after the quantum state manipulation, respectively.

3.3.5.2 Gaussian Operation

In the case of Gaussian operations, transformations of annihilation and creation operators are formulated by matrices (for more details, see Sect. 3.4):

$$\hat{a}_j \to \hat{a}_j' = \hat{U}^{\dagger}\hat{a}_j\hat{U} = \sum_{k=1}^{N}(E_{jk}\hat{a}_k + F_{jk}\hat{a}_k^{\dagger}) + \beta_j, \quad E_{jk} \in \mathbb{C}, F_{jk} \in \mathbb{C}, \beta_j \in \mathbb{C}. \tag{3.59}$$

In *xpxp* notation with

$$\hat{\boldsymbol{\xi}} = (\hat{x}_1, \hat{p}_1, \ldots, \hat{x}_N, \hat{p}_N)^T, \tag{3.60}$$

it is formulated as

$$\hat{\boldsymbol{\xi}}' = M\hat{\boldsymbol{\xi}} + \boldsymbol{d}, \quad M_{ij} \in \mathbb{R}, \ d_i \in \mathbb{R}, \tag{3.61}$$

or equivalently,

$$\begin{pmatrix} \hat{x}' \\ \hat{p}' \end{pmatrix} = \begin{pmatrix} A & B \\ C & D \end{pmatrix} \begin{pmatrix} \hat{x} \\ \hat{p} \end{pmatrix} + \begin{pmatrix} e \\ f \end{pmatrix}, \quad A_{ij}, B_{ij}, C_{ij}, D_{ij}, e_i, f_i \in \mathbb{R}, \tag{3.62}$$

in *xxpp* notation. A, B, C, D, E, F, M have to satisfy some condition in order to preserve commutation relations (see Sect. 3.4).

3.3.6 Transformation of Expectation Values and Covariance Matrix for Gaussian Operations

3.3.6.1 *xpxp* Notation

We assume that the expectation values of quadrature operators and the covariance matrix of an N-mode quantum state are given by

$$\boldsymbol{m} = \langle \hat{\boldsymbol{\xi}} \rangle, \quad V = \left\langle (\hat{\boldsymbol{\xi}} - \boldsymbol{m})(\hat{\boldsymbol{\xi}} - \boldsymbol{m})^T \right\rangle - \frac{i\hbar}{2}\Omega, \tag{3.63}$$

in *xpxp* notation with

$$\hat{\boldsymbol{\xi}} = (\hat{x}_1, \hat{p}_1, \ldots, \hat{x}_N, \hat{p}_N)^T. \tag{3.64}$$

By performing a Gaussian operation with

$$\hat{\boldsymbol{\xi}}' = M\hat{\boldsymbol{\xi}} + \boldsymbol{d}, \quad M_{ij} \in \mathbb{R}, \; d_i \in \mathbb{R}, \tag{3.65}$$

in *xpxp* notation, the expectation values of quadrature operators and the covariance matrix become

$$\boldsymbol{m}' = M\boldsymbol{m} + \boldsymbol{d}, \quad V' = MVM^T. \tag{3.66}$$

3.3.6.2 *xxpp* Notation

We assume that the expectation values of quadrature operators and the covariance matrix of an N-mode quantum state are given by

$$\boldsymbol{n} = \langle \hat{\boldsymbol{q}} \rangle, \quad \Sigma = \left\langle (\hat{\boldsymbol{q}} - \langle \hat{\boldsymbol{q}} \rangle)(\hat{\boldsymbol{q}} - \langle \hat{\boldsymbol{q}} \rangle)^T \right\rangle - \frac{i\hbar}{2}\Omega', \tag{3.67}$$

in *xxpp* notation with

$$\hat{\boldsymbol{q}} = (\hat{x}_1, \ldots, \hat{x}_N, \hat{p}_1, \ldots, \hat{p}_N)^T. \tag{3.68}$$

By performing a Gaussian operation with

$$\hat{\boldsymbol{q}}' = M\hat{\boldsymbol{q}} + \boldsymbol{d}, \quad M_{ij} \in \mathbb{R},\ d_i \in \mathbb{R}, \tag{3.69}$$

in *xxpp* notation, the expectation values of quadrature operators and the covariance matrix become

$$\boldsymbol{n}' = M\boldsymbol{n} + \boldsymbol{d}, \quad \Sigma' = M\Sigma M^T. \tag{3.70}$$

3.3.7 Comparison Between Transformation of Nullifier and Transformation of Annihilation Operator

Although the transformation of nullifiers seems to be the inverse of the annihilation-operator transformation, there exists a difference between them when we consider multi-step quantum state manipulations.

Consider a one-mode quantum state $|\psi_0\rangle$. We apply a unitary operator $\hat{U}_1(\hat{x},\hat{p})$, followed by $\hat{U}_2(\hat{x},\hat{p})$. to the state $|\psi_0\rangle$.

First, we consider the transformation of annihilation operators. We define $\hat{x}_0, \hat{p}_0$ as the quadrature operators of the initial mode (before unitary transformations). By the first unitary operator, they are transformed to

$$\hat{x}_1 = \hat{U}_1^\dagger(\hat{x}_0,\hat{p}_0)\hat{x}_0\hat{U}_1(\hat{x}_0,\hat{p}_0), \quad \hat{p}_1 = \hat{U}_1^\dagger(\hat{x}_0,\hat{p}_0)\hat{p}_0\hat{U}_1(\hat{x}_0,\hat{p}_0). \tag{3.71}$$

By the second unitary operator, the quadrature operators $\hat{x}_1, \hat{p}_1$ are transformed to

$$\hat{x}_2 = \hat{U}_2^\dagger(\hat{x}_1,\hat{p}_1)\hat{x}_1\hat{U}_2(\hat{x}_1,\hat{p}_1), \quad \hat{p}_2 = \hat{U}_2^\dagger(\hat{x}_1,\hat{p}_1)\hat{p}_1\hat{U}_2(\hat{x}_1,\hat{p}_1). \tag{3.72}$$

Next, we consider the transformation of nullifiers. We define $\hat{\delta}_0(\hat{x}_0,\hat{p}_0)$ to be the nullifier of the initial state (before unitary transformations). By the first unitary operator, it is transformed to

$$\hat{\delta}_1(\hat{x}_0,\hat{p}_0) = \hat{U}_1(\hat{x}_0,\hat{p}_0)\hat{\delta}_0(\hat{x}_0,\hat{p}_0)\hat{U}_1^\dagger(\hat{x}_0,\hat{p}_0). \tag{3.73}$$

By the second unitary operator, it is transformed to

$$\hat{\delta}_2(\hat{x}_0,\hat{p}_0) = \hat{U}_2(\hat{x}_0,\hat{p}_0)\hat{\delta}_1(\hat{x}_0,\hat{p}_0)\hat{U}_2^\dagger(\hat{x}_0,\hat{p}_0). \tag{3.74}$$

3.3.8 Transformation of Quadrature Operator Which Specifies a Particular Quantum State

Since Sect. 3.3.5 gives us the general rule for transformations from the annihilation operators *before* operations to those *after* operations, we can obtain the quadrature operators which specify the output quantum state by substituting quadrature operators which specify the input state for the input-mode quadratures in the transformation rule of the annihilation operators.

We take an example of a p-squeezing operation on a vacuum state. By using quadrature operators which specify a particular quantum state, the initial vacuum state is described as

$$\hat{x}^{(0)} + i\hat{p}^{(0)}. \tag{3.75}$$

The p-squeezing operation is formulated as:

$$\hat{x}' + i\hat{p}' = e^r\hat{x} + ie^{-r}\hat{p}. \tag{3.76}$$

By substituting $\hat{x}^{(0)}$ for $\hat{x}$, and $\hat{p}^{(0)}$ for $\hat{p}$, we get the quadrature operators which specify the output state:

$$\hat{x}' + i\hat{p}' = e^r\hat{x}^{(0)} + ie^{-r}\hat{p}^{(0)}. \tag{3.77}$$

3.4 Group of Operator

In general, an arbitrary unitary operator $\hat{U}$ can be described as

$$\hat{U} = e^{-i\frac{\hat{H}}{\hbar}}, \tag{3.78}$$

where $\hat{H}$ is the corresponding Hamiltonian which is a polynomial of position operators $\hat{x}_k$ and momentum operators $\hat{p}_k$. In this section, we classify unitary operators based on the order of polynomials.

3.4.1 *Pauli Group (Heisenberg-Weyl Group)*

3.4.1.1 Operator

Consider a Hamiltonian which is a linear combination of n-mode position operators $\hat{x}_j$ and momentum operators $\hat{p}_j$:

$$\hat{H} = \sum_{j=1}^{n}(\alpha_j\hat{x}_j + \beta_j\hat{p}_j) + \gamma. \tag{3.79}$$

The set of unitary operators:

$$P(n) = \left\{ e^{-\frac{i}{\hbar}[\sum_{j=1}^{n}(\alpha_j \hat{x}_j + \beta_j \hat{p}_j) + \gamma]} \middle| \alpha_j, \beta_j, \gamma \in \mathbb{R} \right\} \tag{3.80}$$

forms a group called the Pauli group or the Heisenberg-Weyl group. Note that γ represents a global phase.

3.4.1.2 Matrix

We define $\hat{x}_k$ and $\hat{p}_k$ as the position and momentum operators of mode k. The Pauli group transformation with $\{\alpha_k, \beta_k\}$ is formulated in the Heisenberg picture as

$$\hat{x}'_k + i\hat{p}'_k = \hat{x}_k + i\hat{p}_k - \alpha_k i + \beta_k. \tag{3.81}$$

By defining

$$\hat{\boldsymbol{x}} = \begin{pmatrix} \hat{x}_1 \\ \vdots \\ \hat{x}_n \end{pmatrix}, \quad \hat{\boldsymbol{p}} = \begin{pmatrix} \hat{p}_1 \\ \vdots \\ \hat{p}_n \end{pmatrix}, \quad \hat{\boldsymbol{\alpha}} = \begin{pmatrix} \hat{\alpha}_1 \\ \vdots \\ \hat{\alpha}_n \end{pmatrix}, \quad \hat{\boldsymbol{\beta}} = \begin{pmatrix} \hat{\beta}_1 \\ \vdots \\ \hat{\beta}_n \end{pmatrix}, \tag{3.82}$$

we can represent it by using vectors in *xxpp* notation:

$$\begin{pmatrix} \hat{\boldsymbol{x}}' \\ \hat{\boldsymbol{p}}' \end{pmatrix} = \begin{pmatrix} \hat{\boldsymbol{x}} \\ \hat{\boldsymbol{p}} \end{pmatrix} + \begin{pmatrix} \boldsymbol{\beta} \\ -\boldsymbol{\alpha} \end{pmatrix}. \tag{3.83}$$

Thus, the Pauli group operator is nothing but a displacement operator (Sects. 3.5.1 and 3.5.2):

$$\hat{X}_1(\beta_1) \cdots \hat{X}_n(\beta_n) \hat{Z}_1(-\alpha_1) \cdots \hat{Z}_n(-\alpha_n). \tag{3.84}$$

3.4.2 Symplectic Group

3.4.2.1 Operator

Consider an n-mode unitary operator $\hat{U} = e^{-\frac{i}{\hbar}\hat{H}}$ whose Hamiltonian $\hat{H}$ is a homogeneous quadratic polynomial in the quadrature operators. The set of the unitary operators $\hat{U}$ forms a group called the Symplectic group.

3.4.2.2 Matrix (*xxpp* Notation)

We define $\hat{x}_k$ and $\hat{p}_k$ as the position and momentum operators of mode k. The Symplectic group transformation is formulated in the Heisenberg picture as

$$\begin{pmatrix} \hat{x}' \\ \hat{p}' \end{pmatrix} = \begin{pmatrix} A & B \\ C & D \end{pmatrix} \begin{pmatrix} \hat{x} \\ \hat{p} \end{pmatrix}, \tag{3.85}$$

in *xxpp* notation, where A, B, C, and D are $n \times n$ real matrices. In order to preserve commutation relations, they satisfy

$$AB^T - BA^T = O, \quad CD^T - DC^T = O, \quad AD^T - BC^T = I, \tag{3.86}$$

where O and I are $n \times n$ zero and unit matrices, respectively. Equation (3.86) shows that the matrix $\begin{pmatrix} A & B \\ C & D \end{pmatrix}$ is a member of the real symplectic group $Sp(2n, \mathbb{R})$.[2]

3.4.2.3 Matrix (*xpxp* Notation)

Symplectic groups can be formulated in *xpxp* notation as well.

$$\hat{\xi}' = M\hat{\xi}, \quad M = T^{-1} \begin{pmatrix} A & B \\ C & D \end{pmatrix} T, \tag{3.87}$$

where T is defined in Eq. (3.34).

3.4.2.4 Matrix (Annihilation and Creation Operators)

An arbitrary Symplectic group operation can be formulated as

$$\hat{a}_j \to \hat{a}'_j = \hat{U}^\dagger \hat{a}_j \hat{U} = \sum_{k=1}^{n} (E_{jk}\hat{a}_k + F_{jk}\hat{a}_k^\dagger), \quad E_{jk} \in \mathbb{C}, F_{jk} \in \mathbb{C}, \tag{3.88}$$

or equivalently,

$$\begin{pmatrix} \hat{a}_1 \\ \vdots \\ \hat{a}_n \end{pmatrix} \to \begin{pmatrix} \hat{a}'_1 \\ \vdots \\ \hat{a}'_n \end{pmatrix} = \begin{pmatrix} E_{11} & \cdots & E_{1n} \\ \vdots & \ddots & \vdots \\ E_{n1} & \cdots & E_{nn} \end{pmatrix} \begin{pmatrix} \hat{a}_1 \\ \vdots \\ \hat{a}_n \end{pmatrix} + \begin{pmatrix} F_{11} & \cdots & F_{1n} \\ \vdots & \ddots & \vdots \\ F_{n1} & \cdots & F_{nn} \end{pmatrix} \begin{pmatrix} \hat{a}_1^\dagger \\ \vdots \\ \hat{a}_n^\dagger \end{pmatrix}. \tag{3.89}$$

In order to preserve commutation relations, $n \times n$ real matrices E and F satisfy

$$EF^T = FE^T, \quad EE^\dagger = FF^\dagger + I. \tag{3.90}$$

[2] For more details of symplectic matrices and symplectic groups, see Refs. [4, 5].

3.4.2.5 Inverse Transformation

The inverse transformation of

$$\hat{b}_j = \sum_k (E_{jk}\hat{a}_k + F_{jk}\hat{a}_k^\dagger) \tag{3.91}$$

is given by

$$\hat{a}_j = \sum_k (E_{kj}^*\hat{b}_k - F_{kj}\hat{b}_k^\dagger). \tag{3.92}$$

By considering preservation of the commutation relations, E and F satisfy

$$E^\dagger F = (E^\dagger F)^T, \quad E^\dagger E = (F^\dagger F)^T + I. \tag{3.93}$$

3.4.3 *Clifford Group (Gaussian Operation)*

3.4.3.1 Operator

Consider an n-mode unitary operator $\hat{U} = e^{-\frac{i}{\hbar}\hat{H}}$ whose Hamiltonian $\hat{H}$ is an inhomogeneous quadratic polynomial in the quadrature operators. The set of unitary operators $\hat{U}$ forms a group called the Clifford group. Operations by the Clifford group operators are called Clifford operations or Gaussian operations. The Clifford group $C(n)$ is a semidirect product of the Pauli group $P(n)$ and the Symplectic group $Sp(2n, \mathbb{R})$:

$$C(n) = Sp(2n, \mathbb{R}) \ltimes P(n). \tag{3.94}$$

3.4.3.2 Matrix (*xxpp* Notation)

We define $\hat{x}_k$ and $\hat{p}_k$ as the position and momentum operators of mode k. The Symplectic group transformation is formulated in the Heisenberg picture as

$$\begin{pmatrix} \hat{\boldsymbol{x}}' \\ \hat{\boldsymbol{p}}' \end{pmatrix} = \begin{pmatrix} A & B \\ C & D \end{pmatrix} \begin{pmatrix} \hat{\boldsymbol{x}} \\ \hat{\boldsymbol{p}} \end{pmatrix} + \begin{pmatrix} \boldsymbol{e} \\ \boldsymbol{f} \end{pmatrix} \tag{3.95}$$

in *xxpp* notation, where A, B, C, and D are $n \times n$ real matrices, while $\boldsymbol{e}$ and $\boldsymbol{f}$ are n-dimensional real vectors. In order to preserve commutation relations, they satisfy

$$AB^T - BA^T = O, \quad CD^T - DC^T = O, \quad AD^T - BC^T = I. \tag{3.96}$$

Thus, the matrix $\begin{pmatrix} A & B \\ C & D \end{pmatrix}$ is a member of the real symplectic group $Sp(2n, \mathbb{R})$.

3.4.3.3 Matrix (Annihilation and Creation Operators)

An arbitrary Clifford group operation can be formulated as

$$\hat{a}_j \rightarrow \hat{a}'_j = \hat{U}^\dagger \hat{a}_j \hat{U} = \sum_{k=1}^{n} (E_{jk}\hat{a}_k + F_{jk}\hat{a}_k^\dagger) + \beta_j, \quad E_{jk} \in \mathbb{C}, F_{jk} \in \mathbb{C}, \beta_j \in \mathbb{C}, \tag{3.97}$$

or equivalently,

$$\begin{pmatrix} \hat{a}_1 \\ \vdots \\ \hat{a}_n \end{pmatrix} \rightarrow \begin{pmatrix} \hat{a}'_1 \\ \vdots \\ \hat{a}'_n \end{pmatrix} = \begin{pmatrix} E_{11} & \cdots & E_{1n} \\ \vdots & \ddots & \vdots \\ E_{n1} & \cdots & E_{nn} \end{pmatrix} \begin{pmatrix} \hat{a}_1 \\ \vdots \\ \hat{a}_n \end{pmatrix} + \begin{pmatrix} F_{11} & \cdots & F_{1n} \\ \vdots & \ddots & \vdots \\ F_{n1} & \cdots & F_{nn} \end{pmatrix} \begin{pmatrix} \hat{a}_1^\dagger \\ \vdots \\ \hat{a}_n^\dagger \end{pmatrix} + \begin{pmatrix} \beta_1 \\ \vdots \\ \beta_n \end{pmatrix}. \tag{3.98}$$

This shows that the Gaussian operation is equivalent to the linear unitary Bogoliubov (LUBO) transformation.

In order to preserve commutation relations, $n \times n$ real matrices E and F satisfy

$$EF^T = FE^T, \quad EE^\dagger = FF^\dagger + I. \tag{3.99}$$

3.4.4 Gaussian Operation as Clifford Group

3.4.4.1 Clifford Group and Pauli Group

The Clifford group C is the normalizer of the Pauli group:

$$C = N(P) = \{\hat{b} \in U | \hat{b}P\hat{b}^\dagger = P\}, \tag{3.100}$$

where U is the unitary group. For an arbitrary set of a Clifford group operator $\hat{C}$ and a Pauli group operator $\hat{P}_1$, there exists a Pauli group operator $\hat{P}_2$ which satisfies

$$\hat{C}\hat{P}_1\hat{C}^\dagger = \hat{P}_2, \tag{3.101}$$

and thus

$$\hat{C}\hat{P}_1 = \hat{P}_2\hat{C}. \tag{3.102}$$

Therefore, the order of Clifford group operator and Pauli group operator can be swapped by changing the Pauli group operator to another Pauli group operator. Importantly, the Clifford group operator is not changed.

In the following, we write down the operation directly in the Heisenberg picture.

We define a Clifford group transformation with a Clifford operator $\hat{C}$ as

$$\begin{pmatrix} \hat{x}' \\ \hat{p}' \end{pmatrix} = \begin{pmatrix} A & B \\ C & D \end{pmatrix} \begin{pmatrix} \hat{x} \\ \hat{p} \end{pmatrix} + \begin{pmatrix} e \\ f \end{pmatrix}. \tag{3.103}$$

Its inverse transformation $\hat{C}^\dagger$ is given by

$$\begin{pmatrix} \hat{x}' \\ \hat{p}' \end{pmatrix} = \begin{pmatrix} A & B \\ C & D \end{pmatrix}^{-1} \begin{pmatrix} \hat{x} \\ \hat{p} \end{pmatrix} - \begin{pmatrix} A & B \\ C & D \end{pmatrix}^{-1} \begin{pmatrix} e \\ f \end{pmatrix}. \tag{3.104}$$

We define a Pauli group transformation with a Pauli operator $\hat{P}_1$ as

$$\begin{pmatrix} \hat{x}' \\ \hat{p}' \end{pmatrix} = \begin{pmatrix} \hat{x} \\ \hat{p} \end{pmatrix} + \begin{pmatrix} g \\ h \end{pmatrix}. \tag{3.105}$$

By using these equations, $\hat{P}_1 \hat{C}^\dagger$ is

$$\begin{pmatrix} \hat{x}' \\ \hat{p}' \end{pmatrix} = \begin{pmatrix} A & B \\ C & D \end{pmatrix}^{-1} \begin{pmatrix} \hat{x} \\ \hat{p} \end{pmatrix} - \begin{pmatrix} A & B \\ C & D \end{pmatrix}^{-1} \begin{pmatrix} e \\ f \end{pmatrix} + \begin{pmatrix} g \\ h \end{pmatrix}. \tag{3.106}$$

Therefore, $\hat{C}\hat{P}_1\hat{C}^\dagger$ becomes

$$\begin{pmatrix} \hat{x}' \\ \hat{p}' \end{pmatrix} = \begin{pmatrix} A & B \\ C & D \end{pmatrix} \left(\begin{pmatrix} A & B \\ C & D \end{pmatrix}^{-1} \begin{pmatrix} \hat{x} \\ \hat{p} \end{pmatrix} - \begin{pmatrix} A & B \\ C & D \end{pmatrix}^{-1} \begin{pmatrix} e \\ f \end{pmatrix} + \begin{pmatrix} g \\ h \end{pmatrix} \right) + \begin{pmatrix} e \\ f \end{pmatrix} \tag{3.107}$$

$$= \begin{pmatrix} \hat{x} \\ \hat{p} \end{pmatrix} + \begin{pmatrix} A & B \\ C & D \end{pmatrix} \begin{pmatrix} g \\ h \end{pmatrix}, \tag{3.108}$$

which corresponds to a member of Pauli group operator $\hat{P}_2$.

3.4.5 Summary of Names

Table 3.1 shows summary of names for groups.

Table 3.1 Summary of names for groups

Name of group	Pauli group	Symplectic group	Clifford group
Symbol	$P(n)$	$Sp(2n, \mathbb{R})$	$C(n)$ $C(n) = Sp(2n, \mathbb{R}) \ltimes P(n)$
Other names	Displacement operation Heisenberg-Weyl group		Gaussian operation LUBO transformation
Hamiltonian	Linear	Homogeneous quadratic	Inhomogeneous quadratic
Heisenberg picture	Constant	$Sp(2n, \mathbb{R})$	$Sp(2n, \mathbb{R})$ and Constant

3.5 Operators

In this section, we review several major operators.

3.5.1 *Position Displacement Operator*

3.5.1.1 Definition

The position displacement operator $\hat{X}(s)$ is defined as

$$\hat{X}(s) = e^{-\frac{i}{\hbar}s\hat{p}}, \quad s \in \mathbb{R}. \tag{3.109}$$

It displaces a state in phase space by s in position.

3.5.1.2 Schrödinger Picture

By applying the position displacement operator $\hat{X}(s)$ on eigenstates of position operator $|x = a\rangle$ and momentum operator $|p = b\rangle$, we get

$$\hat{X}(s)|x = a\rangle = |x = a + s\rangle, \quad \hat{X}(s)|p = b\rangle = e^{-\frac{isb}{\hbar}}|p = b\rangle. \tag{3.110}$$

3.5.1.3 Heisenberg Picture

We define $\hat{x}$ and $\hat{p}$ as the position and momentum operators of the mode on which the operator works. The transformation by the position displacement operator $\hat{X}(s)$ in the Heisenberg picture is given by

$$\hat{X}^{\dagger}(s)\hat{x}\hat{X}(s) = \hat{x} + s, \quad \hat{X}^{\dagger}(s)\hat{p}\hat{X}(s) = \hat{p}, \tag{3.111}$$

or equivalently,

$$\begin{pmatrix} \hat{x}' \\ \hat{p}' \end{pmatrix} = \begin{pmatrix} \hat{x} \\ \hat{p} \end{pmatrix} + \begin{pmatrix} s \\ 0 \end{pmatrix}. \tag{3.112}$$

3.5.1.4 Transformation of Nullifier

We show the transformation of quadrature operators by the Hermitian conjugate of the operator.[3]

$$\hat{X}(s)\hat{x}\hat{X}^\dagger(s) = \hat{x} - s, \quad \hat{X}(s)\hat{p}\hat{X}^\dagger(s) = \hat{p}. \tag{3.113}$$

We will utilize these equations when we trace quantum state manipulations by using transformations of nullifiers.

3.5.2 Momentum Displacement Operator

3.5.2.1 Definition

The momentum displacement operator $\hat{Z}(s)$ is defined as

$$\hat{Z}(s) = e^{\frac{i}{\hbar}s\hat{x}}, \quad s \in \mathbb{R}. \tag{3.114}$$

It displaces a state in phase space by s in momentum.

3.5.2.2 Schrödinger Picture

By applying the momentum displacement operator $\hat{Z}(s)$ on eigenstates of position operator $|x = a\rangle$ and momentum operator $|p = b\rangle$, we get

$$\hat{Z}(s)|x = a\rangle = e^{\frac{isa}{\hbar}}|x = a\rangle, \quad \hat{Z}(s)|p = b\rangle = |p = b + s\rangle. \tag{3.115}$$

3.5.2.3 Heisenberg Picture

We define $\hat{x}$ and $\hat{p}$ as the position and momentum operators of the mode on which the operator works. The transformation by the momentum displacement operator $\hat{Z}(s)$ in the Heisenberg picture is given by

[3] In the following subsections, we show it as transformation rules for nullifiers.

$$\hat{Z}^\dagger(s)\hat{x}\hat{Z}(s) = \hat{x}, \quad \hat{Z}^\dagger(s)\hat{p}\hat{Z}(s) = \hat{p} + s, \tag{3.116}$$

or equivalently,

$$\begin{pmatrix} \hat{x}' \\ \hat{p}' \end{pmatrix} = \begin{pmatrix} \hat{x} \\ \hat{p} \end{pmatrix} + \begin{pmatrix} 0 \\ s \end{pmatrix}. \tag{3.117}$$

3.5.2.4 Transformation of Nullifier

The transformation rule for nullifiers is given by

$$\hat{Z}(s)\hat{x}\hat{Z}^\dagger(s) = \hat{x}, \quad \hat{Z}(s)\hat{p}\hat{Z}^\dagger(s) = \hat{p} - s. \tag{3.118}$$

3.5.3 Rotation Operator

3.5.3.1 Definition

The rotation operator $\hat{F}(\theta)$ is defined as

$$\hat{F}(\theta) = e^{\frac{i\theta}{2\hbar}(\hat{x}^2+\hat{p}^2)} = e^{i\theta(\hat{a}^\dagger\hat{a}+\frac{1}{2})}, \quad \theta \in \mathbb{R}. \tag{3.119}$$

It rotates a state in phase space by θ in a counter-clockwise direction.

3.5.3.2 Heisenberg Picture

We define $\hat{x}$ and $\hat{p}$ as the position and momentum operators of the mode on which the operator works. The transformation by the rotation operator $\hat{F}(\theta)$ in the Heisenberg picture is given by

$$\hat{F}^\dagger(\theta)\hat{x}\hat{F}(\theta) = \hat{x}\cos\theta - \hat{p}\sin\theta, \quad \hat{F}^\dagger(\theta)\hat{p}\hat{F}(\theta) = \hat{x}\sin\theta + \hat{p}\cos\theta, \tag{3.120}$$

or equivalently,

$$\begin{pmatrix} \hat{x}' \\ \hat{p}' \end{pmatrix} = \begin{pmatrix} \cos\theta & -\sin\theta \\ \sin\theta & \cos\theta \end{pmatrix} \begin{pmatrix} \hat{x} \\ \hat{p} \end{pmatrix}. \tag{3.121}$$

We next define $\hat{a}$ and $\hat{a}^\dagger$ as the annihilation and creation operators of the mode on which the operator works. The transformation by the rotation operator $\hat{F}(\theta)$ in the Heisenberg picture is given by

$$\hat{F}^\dagger(\theta)\hat{a}\hat{F}(\theta) = \hat{a}e^{i\theta}, \quad \hat{F}^\dagger(\theta)\hat{a}^\dagger\hat{F}(\theta) = \hat{a}^\dagger e^{-i\theta}. \tag{3.122}$$

3.5.3.3 Transformation of Nullifier

The transformation rule for nullifiers is given by

$$\hat{F}(\theta)\hat{x}\hat{F}^{\dagger}(\theta) = \hat{x}\cos\theta + \hat{p}\sin\theta, \quad \hat{F}(\theta)\hat{p}\hat{F}^{\dagger}(\theta) = -\hat{x}\sin\theta + \hat{p}\cos\theta, \tag{3.123}$$

$$\hat{F}(\theta)\hat{a}\hat{F}^{\dagger}(\theta) = \hat{a}e^{-i\theta}, \quad \hat{F}(\theta)\hat{a}^{\dagger}\hat{F}^{\dagger}(\theta) = \hat{a}^{\dagger}e^{i\theta}. \tag{3.124}$$

3.5.4 Fourier Operator

3.5.4.1 Definition

The Fourier operator $\hat{F}$ is defined as

$$\hat{F} = \hat{F}\left(\frac{\pi}{2}\right) = e^{\frac{i\pi}{4\hbar}(\hat{x}^2+\hat{p}^2)}. \tag{3.125}$$

It rotates a state in phase space by $\theta = \frac{\pi}{2}$ in a counter-clockwise direction. In this thesis, the notation $\hat{F}$ without argument represents the Fourier operator.

3.5.4.2 Schrödinger Picture

By applying the Fourier operator $\hat{F}$ on eigenstates of the position operator $|x = a\rangle$ and the momentum operator $|p = b\rangle$, we get

$$\hat{F}|x = a\rangle = |p = a\rangle, \quad \hat{F}|p = b\rangle = |x = -b\rangle. \tag{3.126}$$

3.5.4.3 Heisenberg Picture

We define $\hat{x}$ and $\hat{p}$ as the position and momentum operators of the mode on which the operator works. The transformation by the Fourier operator $\hat{F}$ in the Heisenberg picture is given by

$$\hat{F}^{\dagger}\hat{x}\hat{F} = -\hat{p}, \quad \hat{F}^{\dagger}\hat{p}\hat{F} = \hat{x}, \tag{3.127}$$

or equivalently,

$$\begin{pmatrix}\hat{x}'\\ \hat{p}'\end{pmatrix} = \begin{pmatrix}0 & -1\\ 1 & 0\end{pmatrix}\begin{pmatrix}\hat{x}\\ \hat{p}\end{pmatrix}. \tag{3.128}$$

We next define $\hat{a}$ and $\hat{a}^\dagger$ as the annihilation and creation operators of the mode on which the operator works. The transformation by the Fourier operator $\hat{F}$ in the Heisenberg picture is given by

$$\hat{F}^\dagger \hat{a} \hat{F} = i\hat{a}, \quad \hat{F}^\dagger \hat{a}^\dagger \hat{F} = -i\hat{a}^\dagger. \tag{3.129}$$

3.5.4.4 Transformation of Nullifier

The transformation rule for nullifiers is given by

$$\hat{F}\hat{x}\hat{F}^\dagger = \hat{p}, \quad \hat{F}\hat{p}\hat{F}^\dagger = -\hat{x}, \tag{3.130}$$

$$\hat{F}\hat{a}\hat{F}^\dagger = -i\hat{a}, \quad \hat{F}\hat{a}^\dagger\hat{F}^\dagger = i\hat{a}^\dagger. \tag{3.131}$$

3.5.5 *Quadratic Phase Operator*

3.5.5.1 Definition

The quadratic phase operator (shearing operator) $\hat{D}(\kappa)$ is defined as

$$\hat{D}(\kappa) = e^{\frac{i}{2\hbar}\kappa\hat{x}^2}, \quad \kappa \in \mathbb{R}. \tag{3.132}$$

It shears a state in phase space with respect to the x axis by a gradient of κ.

3.5.5.2 Heisenberg Picture

We define $\hat{x}$ and $\hat{p}$ as the position and momentum operators of the mode on which the operator works. The transformation by the quadratic phase operator $\hat{D}(\kappa)$ in the Heisenberg picture is given by

$$\hat{D}^\dagger(\kappa)\hat{x}\hat{D}(\kappa) = \hat{x}, \quad \hat{D}^\dagger(\kappa)\hat{p}\hat{D}(\kappa) = \hat{p} + \kappa\hat{x}, \tag{3.133}$$

or equivalently,

$$\begin{pmatrix} \hat{x}' \\ \hat{p}' \end{pmatrix} = \begin{pmatrix} 1 & 0 \\ \kappa & 1 \end{pmatrix} \begin{pmatrix} \hat{x} \\ \hat{p} \end{pmatrix}. \tag{3.134}$$

3.5.5.3 Transformation of Nullifier

The transformation rule for nullifiers is given by

$$\hat{D}(\kappa)\hat{x}\hat{D}^{\dagger}(\kappa) = \hat{x}, \quad \hat{D}(\kappa)\hat{p}\hat{D}^{\dagger}(\kappa) = \hat{p} - \kappa\hat{x}. \tag{3.135}$$

3.5.6 Two-Mode Interaction Operator

3.5.6.1 Definition

We define the following operator $\hat{Q}(g, \theta, \phi)$:

$$\hat{Q}(g, \theta, \phi) = e^{-\frac{i}{\hbar}g(\hat{x}_1 \cos\theta + \hat{p}_1 \sin\theta)(\hat{x}_2 \cos\phi + \hat{p}_2 \sin\phi)}, \quad g \in \mathbb{R}, 0 \leq \theta < 2\pi, 0 \leq \phi < 2\pi. \tag{3.136}$$

It shows a two-mode interaction operator with interaction gain g.

3.5.6.2 Heisenberg Picture

We define $\hat{x}_i$ and $\hat{p}_i$ as the position and momentum operators of mode i on which the operator works. The transformation by the two-mode interaction operator $\hat{Q}(g, \theta, \phi)$ in the Heisenberg picture is given by

$$\begin{pmatrix} \hat{x}_1' \\ \hat{p}_1' \\ \hat{x}_2' \\ \hat{p}_2' \end{pmatrix} = \begin{pmatrix} 1 & 0 & g\sin\theta\cos\phi & g\sin\theta\sin\phi \\ 0 & 1 & -g\cos\theta\cos\phi & -g\cos\theta\sin\phi \\ g\sin\phi\cos\theta & g\sin\phi\sin\theta & 1 & 0 \\ -g\cos\phi\cos\theta & -g\cos\phi\sin\theta & 0 & 1 \end{pmatrix} \begin{pmatrix} \hat{x}_1 \\ \hat{p}_1 \\ \hat{x}_2 \\ \hat{p}_2 \end{pmatrix}, \tag{3.137}$$

where we have utilized *xpxp* notation (Sect. 3.2.5).

3.5.7 Controlled-Z Operator

3.5.7.1 Definition

The controlled-Z operator $\hat{C}_Z(g)$ with gain g, sometimes called the weighted controlled-Z operator, is defined as the two-mode interaction operator $\hat{Q}(g, \theta, \phi)$ with $\theta = 0$ and $\phi = \pi$:

$$\hat{C}_Z(g) = \hat{Q}(g, 0, \pi) = e^{\frac{i}{\hbar}g\hat{x}_1\hat{x}_2}. \tag{3.138}$$

In the special case of unity gain $g = 1$, it is simply called the controlled-Z operator, which is expressed by $\hat{C}_Z$.

3.5.7.2 Heisenberg Picture

We define $\hat{x}_i$ and $\hat{p}_i$ as the position and momentum operators of mode i on which the operator works. The transformation by the controlled-Z operator $\hat{C}_Z(g)$ with gain g in the Heisenberg picture is given by

$$\hat{C}_Z^\dagger(g)\hat{x}_1\hat{C}_Z(g) = \hat{x}_1, \quad \hat{C}_Z^\dagger(g)\hat{p}_1\hat{C}_Z(g) = \hat{p}_1 + g\hat{x}_2, \tag{3.139}$$

$$\hat{C}_Z^\dagger(g)\hat{x}_2\hat{C}_Z(g) = \hat{x}_2, \quad \hat{C}_Z^\dagger(g)\hat{p}_2\hat{C}_Z(g) = \hat{p}_2 + g\hat{x}_1, \tag{3.140}$$

or equivalently,

$$\begin{pmatrix} \hat{x}'_1 \\ \hat{p}'_1 \\ \hat{x}'_2 \\ \hat{p}'_2 \end{pmatrix} = \begin{pmatrix} 1 & 0 & 0 & 0 \\ 0 & 1 & g & 0 \\ 0 & 0 & 1 & 0 \\ g & 0 & 0 & 1 \end{pmatrix} \begin{pmatrix} \hat{x}_1 \\ \hat{p}_1 \\ \hat{x}_2 \\ \hat{p}_2 \end{pmatrix}. \tag{3.141}$$

3.5.7.3 Transformation of Nullifier

The transformation rule for nullifiers is given by

$$\hat{C}_Z(g)\hat{x}_1\hat{C}_Z^\dagger(g) = \hat{x}_1, \quad \hat{C}_Z(g)\hat{p}_1\hat{C}_Z^\dagger(g) = \hat{p}_1 - g\hat{x}_2, \tag{3.142}$$

$$\hat{C}_Z(g)\hat{x}_2\hat{C}_Z^\dagger(g) = \hat{x}_2, \quad \hat{C}_Z(g)\hat{p}_2\hat{C}_Z^\dagger(g) = \hat{p}_2 - g\hat{x}_1. \tag{3.143}$$

3.5.8 Beam Splitter Operator

3.5.8.1 Definition in Heisenberg Picture

The transformation by a beam splitter (including phase space rotation) is a two-mode unitary transformation in which the number of photons is preserved. It is formulated as a linear transformation of annihilation operators $\hat{a}_1$ and $\hat{a}_2$:

$$\begin{pmatrix} \hat{a}'_1 \\ \hat{a}'_2 \end{pmatrix} = \begin{pmatrix} B_{11} & B_{12} \\ B_{21} & B_{22} \end{pmatrix} \begin{pmatrix} \hat{a}_1 \\ \hat{a}_2 \end{pmatrix}. \tag{3.144}$$

In order to preserve the number of photons (the law of conservation of energy), the following condition must be satisfied:

$$\hat{a}_1^\dagger \hat{a}_1 + \hat{a}_2^\dagger \hat{a}_2 = \hat{a}_1'^\dagger \hat{a}_1' + \hat{a}_2'^\dagger \hat{a}_2', \tag{3.145}$$

leading to

$$|B_{11}|^2 + |B_{21}|^2 = |B_{12}|^2 + |B_{22}|^2 = 1, \quad B_{11}^* B_{12} + B_{21}^* B_{22} = 0. \tag{3.146}$$

It shows that the matrix B is a 2×2 unitary matrix. Note that elements of B may not be real numbers.

In general, an arbitrary 2×2 unitary matrix B can be decomposed into the following four components [1]:

$$B = e^{i\frac{\Lambda}{2}} \begin{pmatrix} e^{i\frac{\Psi}{2}} & 0 \\ 0 & e^{-i\frac{\Psi}{2}} \end{pmatrix} \begin{pmatrix} \cos\frac{\Theta}{2} & \sin\frac{\Theta}{2} \\ -\sin\frac{\Theta}{2} & \cos\frac{\Theta}{2} \end{pmatrix} \begin{pmatrix} e^{i\frac{\Phi}{2}} & 0 \\ 0 & e^{-i\frac{\Phi}{2}} \end{pmatrix}, \quad \Lambda, \Psi, \Phi \in \mathbb{R}, 0 \leq \Theta \leq 2\pi. \tag{3.147}$$

3.5.8.2 Operator

The operator $\hat{B}$ which satisfies

$$\begin{pmatrix} \hat{a}_1' \\ \hat{a}_2' \end{pmatrix} = \hat{B}^\dagger \begin{pmatrix} \hat{a}_1 \\ \hat{a}_2 \end{pmatrix} \hat{B} \tag{3.148}$$

is given by

$$\hat{B} = e^{i\Lambda\hat{L}_0} e^{i\Psi\hat{L}_3} e^{i\Theta\hat{L}_2} e^{i\Phi\hat{L}_3}. \tag{3.149}$$

Here, we have defined the following four operators:

$$\hat{L}_0 = \frac{1}{2}(\hat{a}_1^\dagger \hat{a}_1 + \hat{a}_2^\dagger \hat{a}_2), \qquad \hat{L}_1 = \frac{1}{2}(\hat{a}_1^\dagger \hat{a}_2 + \hat{a}_1 \hat{a}_2^\dagger),$$
$$\hat{L}_2 = \frac{1}{2i}(\hat{a}_1^\dagger \hat{a}_2 - \hat{a}_1 \hat{a}_2^\dagger), \qquad \hat{L}_3 = \frac{1}{2}(\hat{a}_1^\dagger \hat{a}_1 - \hat{a}_2^\dagger \hat{a}_2), \tag{3.150}$$

or equivalently,

$$\hat{L}_0 = \frac{1}{4\hbar}(\hat{x}_1^2 + \hat{p}_1^2 + \hat{x}_2^2 + \hat{p}_2^2 - 2\hbar), \qquad \hat{L}_1 = \frac{1}{2\hbar}(\hat{x}_1\hat{x}_2 + \hat{p}_1\hat{p}_2),$$
$$\hat{L}_2 = \frac{1}{2\hbar}(\hat{x}_1\hat{p}_2 - \hat{p}_1\hat{x}_2), \qquad \hat{L}_3 = \frac{1}{4\hbar}(\hat{x}_1^2 + \hat{p}_1^2 - \hat{x}_2^2 - \hat{p}_2^2). \tag{3.151}$$

$\hat{L}_0$ commutes with $\hat{L}_1$, $\hat{L}_2$, and $\hat{L}_3$, and they satisfy $\hat{L}_0(\hat{L}_0 + 1) = \hat{L}_1^2 + \hat{L}_2^2 + \hat{L}_3^2$.

By using

$$e^{i\Lambda\hat{L}_0} = e^{-i\frac{\Lambda}{2}}\hat{F}_1\left(\frac{\Lambda}{2}\right)\hat{F}_2\left(\frac{\Lambda}{2}\right), \tag{3.152}$$

$$e^{i\Psi\hat{L}_3} = \hat{F}_1\left(\frac{\Psi}{2}\right)\hat{F}_2\left(-\frac{\Psi}{2}\right), \tag{3.153}$$

$$e^{i\Theta\hat{L}_2} = e^{\frac{i\Theta}{2\hbar}(\hat{x}_1\hat{p}_2-\hat{p}_1\hat{x}_2)}, \tag{3.154}$$

$\hat{B}$ is given by

$$\hat{B} = e^{-i\frac{\Lambda}{2}}\hat{F}_1\left(\frac{\Lambda}{2}\right)\hat{F}_2\left(\frac{\Lambda}{2}\right)\hat{F}_1\left(\frac{\Psi}{2}\right)\hat{F}_2\left(-\frac{\Psi}{2}\right)e^{\frac{i\Theta}{2\hbar}(\hat{x}_1\hat{p}_2-\hat{p}_1\hat{x}_2)}\hat{F}_1\left(\frac{\Phi}{2}\right)\hat{F}_2\left(-\frac{\Phi}{2}\right). \tag{3.155}$$

It is the operator representation of Eq. (3.147). The first term $e^{-i\frac{\Lambda}{2}}$ shows a global phase, which does not affect the operation. The terms $\hat{F}_i(*)$ show rotations in phase space. The term $e^{\frac{i\Theta}{2\hbar}(\hat{x}_1\hat{p}_2-\hat{p}_1\hat{x}_2)}$ shows a phase-free beam splitter. Λ, Ψ, Θ, Φ correspond to those in Eq. (3.147).

3.5.8.3 Four-Types of Beam Splitter Matrices and Thier Operators

As beam splitter matrices B for linear transformations of annihilation operators:

$$\begin{pmatrix}\hat{a}'_i\\ \hat{a}'_j\end{pmatrix} = B\begin{pmatrix}\hat{a}_i\\ \hat{a}_j\end{pmatrix}, \tag{3.156}$$

we define the following four unitary matrices[4]:

$$B^{(1)}_{ij}(\sqrt{R}) = \begin{pmatrix}-\sqrt{R} & \sqrt{T}\\ \sqrt{T} & \sqrt{R}\end{pmatrix},\quad B^{(2)}_{ij}(\sqrt{R}) = \begin{pmatrix}\sqrt{T} & -\sqrt{R}\\ \sqrt{R} & \sqrt{T}\end{pmatrix}, \tag{3.157}$$

$$B^{(3)}_{ij}(\sqrt{R}) = \begin{pmatrix}\sqrt{T} & \sqrt{R}\\ -\sqrt{R} & \sqrt{T}\end{pmatrix},\quad B^{(4)}_{ij}(\sqrt{R}) = \begin{pmatrix}\sqrt{R} & \sqrt{T}\\ \sqrt{T} & -\sqrt{R}\end{pmatrix}. \tag{3.158}$$

We define the following four beam splitter operators which correspond to the beam splitter matrices above:

$$\hat{B}^{(1)}_{ij}(\sqrt{R}),\quad \hat{B}^{(2)}_{ij}(\sqrt{R}),\quad \hat{B}^{(3)}_{ij}(\sqrt{R}),\quad \hat{B}^{(4)}_{ij}(\sqrt{R}). \tag{3.159}$$

---

[4] Beam splitter matrices and operators are usually defined just as the author pleases in each paper. In some cases, the definitions of beam splitter matrices are not mentioned explicitly. Such ambiguity of their definition might confuse the readers. In this thesis, we define the four-types of beam splitters for convenience. By explicitly declaring the type of its matrix when we use a beam splitter in theory, we can avoid confusion derived from the ambiguity of the beam splitter matrix.

Here, R and T represent the energy reflectivity and transmissivity of the beam splitter, where $T + R = 1$. In general, beam splitter operators and matrices depend on the order of modes on which they operate. Thus, we show the order of modes in the subscript. On the other hand, the superscript shows the entry of the matrix which is equal to $-\sqrt{R}$. That is,

1. In the case of $B_{ij}^{(1)}(\sqrt{R})$, the matrix element from mode i to mode i is $-\sqrt{R}$. ((1, 1) entry of the matrix when the order of modes is i, j.)
2. In the case of $B_{ij}^{(2)}(\sqrt{R})$, the matrix element from mode j to mode i is $-\sqrt{R}$. ((1, 2) entry of the matrix when the order of modes is i, j.)
3. In the case of $B_{ij}^{(3)}(\sqrt{R})$, the matrix element from mode i to mode j is $-\sqrt{R}$. ((2, 1) entry of the matrix when the order of modes is i, j.)
4. In the case of $B_{ij}^{(4)}(\sqrt{R})$, the matrix element from mode j to mode j is $-\sqrt{R}$. ((2, 2) entry of the matrix when the order of modes is i, j.)

If the order of modes is j, i, the matrices are given by

$$\hat{B}_{ji}^{(1)}(\sqrt{R}) = \hat{B}_{ij}^{(4)}(\sqrt{R}), \quad \hat{B}_{ji}^{(2)}(\sqrt{R}) = \hat{B}_{ij}^{(3)}(\sqrt{R}), \quad \hat{B}_{ji}^{(3)}(\sqrt{R}) = \hat{B}_{ij}^{(2)}(\sqrt{R}), \quad \hat{B}_{ji}^{(4)}(\sqrt{R}) = \hat{B}_{ij}^{(1)}(\sqrt{R}). \tag{3.160}$$

3.5.8.4 Transformation of Nullifier

Consider an operation which can be formulated as a linear transformation of annihilation operators (transformation by beam splitters with phase rotations):

$$\hat{\boldsymbol{a}}' = \hat{U}^{\dagger}\hat{\boldsymbol{a}}\hat{U} = U\hat{\boldsymbol{a}}, \tag{3.161}$$

where $\hat{\boldsymbol{a}}$ shows a vector of annihilation operators, while $\hat{U}$ and U show a beam splitter operator and a beam splitter matrix, respectively. Its inverse transformation is

$$\hat{U}\hat{\boldsymbol{a}}'\hat{U}^{\dagger} = \hat{\boldsymbol{a}} = U^{-1}\hat{\boldsymbol{a}}'. \tag{3.162}$$

Thus, it can be described as a linear transformation of annihilation operators by the unitary matrix U^{-1}. In the case of the four beam splitter matrices above, they are given by

$$B_{ij}^{(1)^{-1}}(\sqrt{R}) = B_{ij}^{(1)}(\sqrt{R}), \quad B_{ij}^{(2)^{-1}}(\sqrt{R}) = B_{ij}^{(3)}(\sqrt{R}), \tag{3.163}$$

$$B_{ij}^{(3)^{-1}}(\sqrt{R}) = B_{ij}^{(2)}(\sqrt{R}), \quad B_{ij}^{(4)^{-1}}(\sqrt{R}) = B_{ij}^{(4)}(\sqrt{R}). \tag{3.164}$$

It leads to the following transformation rule for nullifiers.

In the case of $\hat{B}_{ij}^{(1)}(\sqrt{R})$, it is given by

$$\hat{x}_i \to -\sqrt{R}\hat{x}_i + \sqrt{1-R}\hat{x}_j, \quad \hat{p}_i \to -\sqrt{R}\hat{p}_i + \sqrt{1-R}\hat{p}_j, \tag{3.165}$$

$$\hat{x}_j \to \sqrt{1-R}\hat{x}_i + \sqrt{R}\hat{x}_j, \quad \hat{p}_j \to \sqrt{1-R}\hat{p}_i + \sqrt{R}\hat{p}_j. \tag{3.166}$$

In the case of $\hat{B}_{ij}^{(2)}(\sqrt{R})$, it is given by

$$\hat{x}_i \to \sqrt{1-R}\hat{x}_i + \sqrt{R}\hat{x}_j, \quad \hat{p}_i \to \sqrt{1-R}\hat{p}_i + \sqrt{R}\hat{p}_j, \tag{3.167}$$

$$\hat{x}_j \to -\sqrt{R}\hat{x}_i + \sqrt{1-R}\hat{x}_j, \quad \hat{p}_j \to -\sqrt{R}\hat{p}_i + \sqrt{1-R}\hat{p}_j. \tag{3.168}$$

In the case of $\hat{B}_{ij}^{(3)}(\sqrt{R})$, it is given by

$$\hat{x}_i \to \sqrt{1-R}\hat{x}_i - \sqrt{R}\hat{x}_j, \quad \hat{p}_i \to \sqrt{1-R}\hat{p}_i - \sqrt{R}\hat{p}_j, \tag{3.169}$$

$$\hat{x}_j \to \sqrt{R}\hat{x}_i + \sqrt{1-R}\hat{x}_j, \quad \hat{p}_j \to \sqrt{R}\hat{p}_i + \sqrt{1-R}\hat{p}_j. \tag{3.170}$$

In the case of $\hat{B}_{ij}^{(4)}(\sqrt{R})$, it is given by

$$\hat{x}_i \to \sqrt{R}\hat{x}_i + \sqrt{1-R}\hat{x}_j, \quad \hat{p}_i \to \sqrt{R}\hat{p}_i + \sqrt{1-R}\hat{p}_j, \tag{3.171}$$

$$\hat{x}_j \to \sqrt{1-R}\hat{x}_i - \sqrt{R}\hat{x}_j, \quad \hat{p}_j \to \sqrt{1-R}\hat{p}_i - \sqrt{R}\hat{p}_j. \tag{3.172}$$

3.5.8.5 Multi-Mode Beam Splitter Network

Consider an N-mode operation which can be formulated as a linear transformation of annihilation operators

$$\hat{a}_i' = \sum_{j=1}^{N} U_{ij}\hat{a}_j, \tag{3.173}$$

or equivalently,

$$\begin{pmatrix} \hat{a}_1' \\ \vdots \\ \hat{a}_N' \end{pmatrix} = \begin{pmatrix} U_{11} & \cdots & U_{1N} \\ \vdots & \ddots & \vdots \\ U_{N1} & \cdots & U_{NN} \end{pmatrix} \begin{pmatrix} \hat{a}_1 \\ \vdots \\ \hat{a}_N \end{pmatrix}, \tag{3.174}$$

where U is a unitary matrix. This is a photon-number-preserving operation, thus it can be achieved by a network of beam splitters with phase rotations in phase space. In general, it can be implemented by using at most $\frac{N(N-1)}{2}$ beam splitters [7].

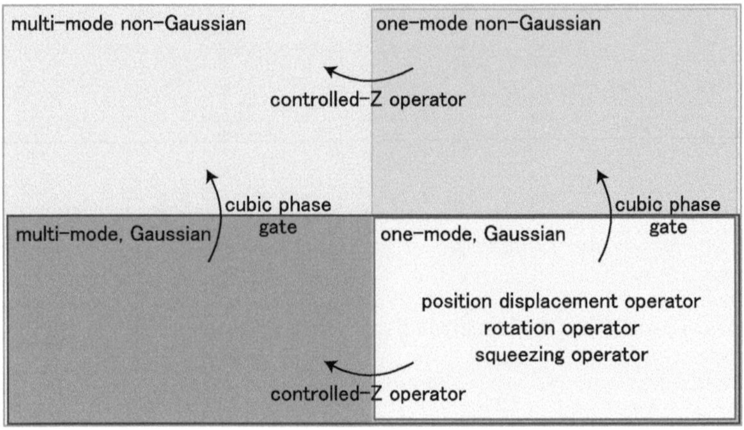

Fig. 3.1 Classification of unitary operators

3.6 Universality and Classical Simulation

3.6.1 Universality

Consider a set S of operators and its subset $T = \{\hat{b}_1, \hat{b}_2, \ldots \hat{b}_n\}$ with a finite number n of elements. If an arbitrary element $\hat{b} \in S$ can be approximated by a product of a finite number of $\hat{b}_k \in T$ with arbitrary precision, the set T is termed universal for the set S.

Figure 3.1 shows classification of unitary operators based on the order of Hamiltonians (Gaussian/Unitary with non-Gaussian) and the number of modes (one-mode/multi-mode). Universal quantum computation corresponds to multi-mode unitary operations [8].

3.6.1.1 One-Mode Gaussian Operation

The simplest set is the one that consists of one-mode Gaussian operations. Here, a one-mode Gaussian operation corresponds to a Clifford group operator whose Hamiltonian is an inhomogeneous quadratic polynomial in the quadrature operators of one mode (Sect. 3.4.3).

It is known that an arbitrary one-mode Gaussian operation can be achieved by combining the position displacement operator $\hat{X}(s)$, rotation operator $\hat{F}(\theta)$, and squeezing operator $\hat{S}(r)$.

3.6.1.2 One-Mode Unitary Operation

The second group consists of one-mode unitary operations whose Hamiltonians are arbitrary-order polynomials in the quadrature operators of one mode. It includes one-mode non-Gaussian operations whose Hamiltonians are more than or equal to third-order polynomials.

It is known that an arbitrary one-mode unitary operation can be achieved by combining the position displacement operator $\hat{X}(s)$, rotation operator $\hat{F}(\theta)$, squeezing operator $\hat{S}(r)$, and cubic phase operator $\hat{N}(\eta) = e^{i\eta\hat{x}^3}$.

3.6.1.3 Multi-Mode Gaussian Operation

The third group consists of multi-mode Gaussian operations. Here, a multi-mode Gaussian operation corresponds to a Clifford group operator whose Hamiltonian is an inhomogeneous quadratic polynomial in the quadrature operators of multiple modes (Sect. 3.4.3).

It is known that an arbitrary multi-mode Gaussian operation can be achieved by combining the position displacement operator $\hat{X}(s)$, rotation operator $\hat{F}(\theta)$, squeezing operator $\hat{S}(r)$, and controlled-Z operator $\hat{C}_Z(g)$.

3.6.1.4 Multi-Mode Unitary Operation

The last group consists of multi-mode unitary operations whose Hamiltonians are arbitrary-order polynomials in the quadrature operators of multiple modes. It includes multi-mode non-Gaussian operations whose Hamiltonians are more than or equal to third-order polynomials. This is the general case, and corresponds to the universal quantum computation.

It is known that an arbitrary multi-mode unitary operation can be achieved by combining the position displacement operator $\hat{X}(s)$, rotation operator $\hat{F}(\theta)$, squeezing operator $\hat{S}(r)$, controlled-Z operator $\hat{C}_Z(g)$, and cubic phase operator $\hat{N}(\eta) = e^{i\eta\hat{x}^3}$.

3.6.2 *Efficient Classical Simulation*

It was reported that quantum computations in some subgroup of the unitary group can be simulated efficiently by classical computers. Although it was initially proven in DV systems [1, 9], it was later extended to CV systems [10].

The subgroup of CV quantum computation which can be efficiently simulated by classical computers is summarized as follows. (1) Initial state is a Gaussian state. (2) The operations are composed of the following components: (2.a) Gaussian operations, (2.b) losses, (2.c) homodyne measurements, and (2.d) Gaussian operations based on homodyne measurements. It shows that non-Gaussian operations must be involved so that quantum computers stand at advantage over classical computers.

An important thing is that it does not mean Gaussian operations in quantum computations can be processed by assistant classical computers, instead of main quantum computers. This is because measurements, which destroy superpositions of quantum states, cannot be avoided before Gaussian operations are implemented by classical computers, since classical computers cannot process quantum states. Therefore, Gaussian operations should be implemented by quantum computers as well.

Experimental demonstrations in this thesis satisfy all conditions above, thus they can be efficiently simulated by classical computers. We can consider that the computer program "Quantum Computation Builder" in Sect. B.2 is an example of classical simulation of quantum computation.

3.7 Entangled States and Entanglement Criteria

3.7.1 Entangled States

Consider an n-mode quantum state with a density operator $\hat{\rho}$ [11]. We label each mode by $j = 1, 2, \ldots, n$. We define the set of all modes to be B. If the density operator $\hat{\rho}$ can be decomposed into a direct product of density operators of its l subsystems:

$$\hat{\rho} = \sum_i \lambda_i \hat{\rho}_{iB_1} \otimes \cdots \otimes \hat{\rho}_{iB_l}, \tag{3.175}$$

the quantum state is referred to be separable into $\{B_k\}$. Here, B_k shows the set of modes in k-th subsystem, while $\hat{\rho}_{iB_k}$ shows the density operator of k-th subsystem consisting of modes in B_k. Note that intersections of any B_j and B_k for $j \neq k$ are empty sets, while the union of all B_k for $k = 1, \ldots, l$ is B. λ_i shows the probability of the i-th state with the density operator $\hat{\rho}_i = \hat{\rho}_{iB_1} \otimes \cdots \otimes \hat{\rho}_{iB_l}$, where $\sum_i \lambda_i = 1$. The quantum state is a fully inseparable state, or genuine entangled state, if and only if its density operator cannot be decomposed into Eq. (3.175) with any set of B_k. In this section, we don't impose the condition that the subsystem composed of modes in B_k is inseparable. That is, it may be separable into sub-subsystems.

Although quantum entanglement was originally proposed as a paradox which showed imperfection of quantum dynamics [12], it was later demonstrated experimentally [13, 14]. It is now understood not only as a distinctive property of quantum dynamics, but the central key element of expanding its applications, such as quantum computations and quantum communications.

3.7.2 Entanglement Criteria

A condition which is used to distinguish entangled states from separable states is called *entanglement criterion*, or *inseparability criterion*.

In this thesis, we utilize two types of entanglement criteria. One is based on variances of quadrature operators (Sect. 3.7.4). The other is based on the partial transpose of density operators (Sect. 3.7.5). We choose an appropriate entanglement criterion for each experiment because each has its own advantages and disadvantages (Sect. 3.7.7).

3.7.3 Duan Criterion

The Duan entanglement criterion, which was theoretically proven by Duan in 2000 [15], is one of two major entanglement criteria for two-mode continuous variables. It is based on the total variance of quadrature operators.

We assume that $\hat{x}_i$ and $\hat{p}_i$ for $i = 1, 2$ represents quadrature operators of a two-mode continuous-variable system. If the quantum state is a separable state, it satisfies

$$\left\langle \Delta^2\left(|a|\hat{x}_1 + \frac{1}{a}\hat{x}_2\right)\right\rangle + \left\langle \Delta^2\left(|a|\hat{p}_1 - \frac{1}{a}\hat{p}_2\right)\right\rangle \geqq \left(a^2 + \frac{1}{a^2}\right)\hbar, \quad a \in \mathbb{R} \setminus \{0\}. \tag{3.176}$$

As the contraposition of this statement, we get the following sufficient condition for entanglement:

$$\exists a \in \mathbb{R} \setminus \{0\},\ \left\langle \Delta^2\left(|a|\hat{x}_1 + \frac{1}{a}\hat{x}_2\right)\right\rangle + \left\langle \Delta^2\left(|a|\hat{p}_1 - \frac{1}{a}\hat{p}_2\right)\right\rangle < \left(a^2 + \frac{1}{a^2}\right)\hbar$$
$$\Longrightarrow \text{inseparable into subsystems 1 and 2.} \tag{3.177}$$

Note that Eq. (3.177) does not provide a necessary and sufficient condition, but only a sufficient condition for entanglement. However, as for any Gaussian state, it is proven that we can get a necessary and sufficient condition for entanglement by performing standardization of its covariance matrix, which corresponds to virtual local Gaussian operations on the quantum state in question (see Ref. [15]).

3.7.4 Extension of van Loock-Furusawa Criterion

3.7.4.1 Background

Several years after the proposal of Duan entanglement criterion for two-mode continuous-variable states, van Loock and Furusawa showed its extension to multimode states [11]. Similar to the original, sufficiency for inseparability is proven by evaluating variances of linear combinations of n-mode position or momentum operators: $\hat{u} = \sum_j h_j\hat{x}_j$ and $\hat{v} = \sum_j g_j\hat{p}_j$, where $h_j, g_j \in \mathbb{R}$ for all j. When the density operator is decomposed into Eq. (3.175), the variances of these linear combinations satisfy

$$\langle \Delta^2 \hat{u} \rangle_\rho + \langle \Delta^2 \hat{v} \rangle_\rho \geqq \frac{1}{2} \sum_{k=1}^{l} \Big| \sum_{j \in B_k} h_j g_j \Big|. \tag{3.178}$$

Violation of Eq. (3.178) guarantees that the density operator of the quantum state cannot be decomposed into Eq. (3.175). For some n-mode quantum states, it can be proven that they are fully inseparable states, by showing that $(n-1)$ Eq. (3.178) are violated with an appropriate set of $(\hat{u}_j, \hat{v}_j, \{B_k\}_j)$ for $j = 1, \ldots, n-1$. The sets of modes $\{B_k\}_j$ should be carefully chosen so that every separable case is negated. Note that van Loock-Furusawa criterion gives us a sufficient condition for inseparability. On the contrary, necessity does not hold.

van Loock-Furusawa criterion works well in some specific cases. Several distinct examples are verifications of an EPR state and n-mode GHZ states. In these cases, $(\hat{u}_j, \hat{v}_j)$ are chosen from the set of their nullifiers: $\hat{u} = \hat{x}_1 - \hat{x}_2$ and $\hat{v} = \hat{p}_1 + \hat{p}_2$ for the EPR state, while $\hat{u}_j = \hat{x}_j - \hat{x}_n$ and $\hat{v}_j = \sum_{m=1}^{n} \hat{p}_m$ for the n-mode GHZ state. In the former case, B_k are set to be $B_1 = \{1\}$ and $B_2 = \{2\}$, while $\{B_1, B_2\}_j = \{\{j\}, B \setminus \{j\}\}$ in the latter case, where $B = \{1, \ldots, n\}$.

However, the selection of $(\hat{u}_j, \hat{v}_j)$ becomes problematic in many cases. It derives from the strong condition that $\hat{u}_j$ consists *only* of position operators, while $\hat{v}_j$ consists *only* of momentum operators. Therefore, van Loock-Furusawa entanglement criterion cannot be directly applied to entanglement detections of cluster states since their nullifiers consist of both position and momentum operators.

A well-used evasion scheme is, similar to the Duan criterion, to transform the separability criterion Eq. (3.178) by applying virtual local Gaussian operations, especially rotations in phase space, which effectively mix position and momentum operators in $\hat{u}$ and $\hat{v}$. It is valid since local operations do not affect separability of quantum states. By applying this scheme, we can expand the group of quantum states whose entanglement can be detected by the *extended* van Loock-Furusawa criteria. It includes n-mode linear cluster states and n-mode star cluster states.

In spite of the extension above, it does not provide entanglement criteria for all shapes of cluster states. For example, entanglement of the three-mode triangle cluster state with nullifiers

$$\hat{p}_1 - \hat{x}_2 - \hat{x}_3, \quad \hat{p}_2 - \hat{x}_1 - \hat{x}_3, \quad \hat{p}_3 - \hat{x}_1 - \hat{x}_2 \tag{3.179}$$

cannot be detected. In general, the extension above fails for any cluster state with a triangle structure.

Here we present a further extension of van Loock-Furusawa criteria, where the assumption on $\hat{u}$ and $\hat{v}$ are removed, except that they are observables. By this extension, all linear combinations of position and momentum operators can be directly used for detection of inseparability, without any virtual local unitary operations. In addition, nonlinear combinations are also acceptable.

3.7.4.2 Extension of van Loock-Furusawa Criterion

Consider that an n-mode quantum state with the density operator $\hat{\rho}$ can be decomposed into l subsystems in the form of Eq. (3.175). We define

$$\hat{h}_{B_k} \equiv h_{B_k}(\hat{x}_j, \hat{p}_j | j \in B_k), \tag{3.180}$$

$$\hat{g}_{B_k} \equiv g_{B_k}(\hat{x}_j, \hat{p}_j | j \in B_k), \tag{3.181}$$

where $h_{B_k}(\hat{x}_j, \hat{p}_j | j \in B_k)$ and $g_{B_k}(\hat{x}_j, \hat{p}_j | j \in B_k)$ are arbitrary Hermitian functions of $\hat{x}_j$ and $\hat{p}_j$ within the k-th subsystem B_k. We also define the summations of these operators:

$$\hat{\xi} = \sum_{k=1}^{l} \hat{h}_{B_k}, \qquad \hat{\zeta} = \sum_{k=1}^{l} \hat{g}_{B_k}. \tag{3.182}$$

By definition, the variance of the operator $\hat{\xi}$ is given by

$$\langle \Delta^2 \hat{\xi} \rangle = \sum_i \lambda_i \langle \hat{\xi}^2 \rangle_i - \left(\sum_i \lambda_i \langle \hat{\xi} \rangle_i \right)^2, \tag{3.183}$$

leading to

$$\langle \Delta^2 \hat{\xi} \rangle = \sum_i \lambda_i \left(\langle \Delta^2 \hat{\xi} \rangle_i + \langle \hat{\xi} \rangle_i^2 \right) - \left(\sum_i \lambda_i \langle \hat{\xi} \rangle_i \right)^2 \tag{3.184}$$

$$\geq \sum_i \lambda_i \langle \Delta^2 \hat{\xi} \rangle_i = \sum_i \lambda_i \left(\langle \hat{\xi}^2 \rangle_i - \langle \hat{\xi} \rangle_i^2 \right). \tag{3.185}$$

Here $\langle \hat{\xi} \rangle_i = \mathrm{tr} \hat{\rho}_i \hat{\xi}$ and $\langle \Delta^2 \hat{\xi} \rangle_i = \mathrm{tr} \hat{\rho}_i (\hat{\xi} - \langle \hat{\xi} \rangle_i)^2$ represent expectation values and variances of the i-th quantum state with $\hat{\rho}_i = \hat{\rho}_{iB_1} \otimes \cdots \otimes \hat{\rho}_{iB_l}$. We used the Cauchy-Schwarz inequality

$$1 \cdot \sum_i \lambda_i \langle \hat{\xi} \rangle_i^2 = \left(\sum_i \lambda_i \right) \left(\sum_i \lambda_i \langle \hat{\xi}^2 \rangle_i \right) \tag{3.186}$$

$$\geq \left(\sum_i \sqrt{\lambda_i} \times \sqrt{\lambda_i} \langle \hat{\xi} \rangle_i \right)^2, \tag{3.187}$$

with $\sum_i \lambda_i = 1$. By using $\left\langle \hat{h}_{B_k} \hat{h}_{B_{k'}} \right\rangle_i = \left\langle \hat{h}_{B_k} \right\rangle_i \left\langle \hat{h}_{B_{k'}} \right\rangle_i$ for $k \neq k'$, we get

$$\langle\hat{\xi}^2\rangle_i - \langle\hat{\xi}\rangle_i^2 = \left(\sum_{k=1}^{l}\left\langle\hat{h}_{B_k}^2\right\rangle_i + \sum_{k,k'=1}^{l}\left\langle\hat{h}_{B_k}\right\rangle_i\left\langle\hat{h}_{B_{k'}}\right\rangle_i\right)$$
$$-\left(\sum_{k=1}^{l}\left\langle\hat{h}_{B_k}\right\rangle_i^2 + \sum_{k,k'=1}^{l}\left\langle\hat{h}_{B_k}\right\rangle_i\left\langle\hat{h}_{B_{k'}}\right\rangle_i\right) \tag{3.188}$$
$$= \sum_{k=1}^{l}\left\langle\Delta^2\hat{h}_{B_k}\right\rangle_i, \tag{3.189}$$

and thus

$$\langle\Delta^2\hat{\xi}\rangle_\rho \geq \sum_i \lambda_i \sum_{k=1}^{l}\left\langle\Delta^2\hat{h}_{B_k}\right\rangle_i. \tag{3.190}$$

In a similar manner, we also get

$$\langle\Delta^2\hat{\zeta}\rangle_\rho \geq \sum_i \lambda_i \sum_{k=1}^{l}\left\langle\Delta^2\hat{g}_{B_k}\right\rangle_i. \tag{3.191}$$

By using these inequalities, the sum of these variances satisfies

$$\langle\Delta^2\hat{\xi}\rangle_\rho + \langle\Delta^2\hat{\zeta}\rangle_\rho \geq \sum_{k=1}^{l}\left|\sum_i \lambda_i\left\langle[\hat{h}_{B_k}, \hat{g}_{B_k}]\right\rangle_i\right| = \sum_{k=1}^{l}\left|\left\langle[\hat{h}_{B_k}, \hat{g}_{B_k}]\right\rangle\right|. \tag{3.192}$$

This is the final form of the necessary condition for separability. Here we have used the uncertainty principle

$$\left\langle\Delta^2\hat{A}\right\rangle + \left\langle\Delta^2\hat{B}\right\rangle \geq \left|\left\langle[\hat{A}, \hat{B}]\right\rangle\right|, \tag{3.193}$$

for Hermitian operators $\hat{A}$ and $\hat{B}$, and the triangle inequality

$$\sum_i |a_i| \geq \left|\sum_i a_i\right|. \tag{3.194}$$

for $a_i \in \mathbb{C}$.

In the same manner, the product of two variances satisfies

$$\langle\Delta^2\hat{\xi}\rangle_\rho\langle\Delta^2\hat{\zeta}\rangle_\rho \geq= \frac{1}{4}\sum_{k=1}^{l}\left|\left\langle[\hat{h}_{B_k}, \hat{g}_{B_k}]\right\rangle\right|^2. \tag{3.195}$$

The sufficient conditions for inseparability are given by taking the contrapositions of these statements. By choosing Eq. (3.192), we get the sufficient condition for inseparability:

$$\left\langle \Delta^2 \hat{\xi} \right\rangle_\rho + \left\langle \Delta^2 \hat{\zeta} \right\rangle_\rho < \sum_{k=1}^{l} \left| \left\langle \left[\hat{h}_{B_k}, \hat{g}_{B_k} \right] \right\rangle \right| \Longrightarrow \text{inseparable into } \{B_k\}. \tag{3.196}$$

Similar to the original van Loock-Furusawa criteria, Eq. (3.196) only negates the possibility that the quantum state can be separable into $\{B_k\}$. Full inseparability can be proven by showing that $(n-1)$ Eq. (3.196) hold with an appropriate set of $(\hat{u}_j, \hat{v}_j, \{B_k\}_j)$ for $j = 1, \ldots, n-1$. The sets of modes $\{B_k\}_j$ should be carefully chosen so that every separable case is negated.

3.7.5 Entanglement Criteria via Partial Transpose of Density Operators

3.7.5.1 Peres Criterion

Although Peres criterion was initially proposed as an entanglement criterion for low-dimensional systems by Peres and Horodecki [16, 17], it was later extended to continuous-variable systems.

Consider that a quantum state with a density operator $\hat{\rho}$ is separable into two subsystems A and B. In this case, the density operator is given by

$$\hat{\rho} = \sum_i \eta_i \hat{\rho}_{iA} \otimes \hat{\rho}_{iB}, \tag{3.197}$$

where $\hat{\rho}_{iA}$ and $\hat{\rho}_{iB}$ represent density operators in two subsystems A and B. We don't impose the condition that each subsystem A and B is inseparable. That is, each subsystem may be separable into sub-subsystems.

Consider the operator which is given by taking the partial transpose of the density operator with respect to the subsystem A:

$$\hat{\rho}^{T_A} = \sum_i \eta_i \hat{\rho}_{iA}^{T} \otimes \hat{\rho}_{iB}. \tag{3.198}$$

If the density operator $\hat{\rho}$ represents a quantum state which is separable into two subsystems A and B, $\hat{\rho}^{T_A}$ satisfies

$$\mathrm{tr}\hat{\rho}^{T_A} = 1, \quad \hat{\rho}^{T_A} \geq 0, \tag{3.199}$$

thus $\hat{\rho}^{T_A}$ represents a density operator of another physical state.

As the contraposition of this statement, we get the following sufficient condition for entanglement:

$$\hat{\rho}^{T_A} \text{ does not represent a physical state} \Longrightarrow \hat{\rho} \text{ is inseparable into } A \text{ and } B. \tag{3.200}$$

Furthermore, we get

$$\hat{\rho}^{T_A}\text{does not represent a physical state for an arbitrary set } \{A, B\} \Longrightarrow \hat{\rho}\text{ represents a fully inseparable state.} \tag{3.201}$$

3.7.5.2 Simon Criterion

By applying the Peres criterion, Simon found an inseparable criterion on covariance matrices of two-mode continuous-variable states [18].

We assume that the covariance matrix of a two-mode state is given by V. We label these two modes as A and B. We here utilize *xpxp* notation for covariance matrices (Sect. 3.2.5). Since the partial transpose with respect to the subsystem A corresponds to

$$\begin{pmatrix} \hat{x}_A \\ \hat{p}_A \\ \hat{x}_B \\ \hat{p}_B \end{pmatrix} \to \Lambda \begin{pmatrix} \hat{x}_A \\ \hat{p}_A \\ \hat{x}_B \\ \hat{p}_B \end{pmatrix}, \quad \Lambda = \begin{pmatrix} 1 & & & \\ & -1 & & \\ & & 1 & \\ & & & 1 \end{pmatrix}, \tag{3.202}$$

the covariance matrix V is transformed to

$$V \to \tilde{V} = \Lambda V \Lambda. \tag{3.203}$$

The Peres criterion shows that, if V is a covariance matrix of a separable state, $\tilde{V}$ represents a covariance matrix of a physical state. Thus, it satisfies

$$\tilde{V} + \frac{i\hbar}{2}\Omega \geq 0. \tag{3.204}$$

We define three 2×2 matrices A, B, and C as

$$V = \begin{pmatrix} A & C \\ C^T & B \end{pmatrix}. \tag{3.205}$$

If the covariance matrix V represents a separable state, it satisfies

$$\det A \det B + \left(\frac{\hbar^2}{4} - |\det C| \right)^2 - \text{tr}[AJCJBJC^T J] \geq \frac{\hbar^2}{4}(\det A + \det B), \quad (3.206)$$

where

$$J = \begin{pmatrix} 0 & 1 \\ -1 & 0 \end{pmatrix}. \quad (3.207)$$

By taking the contraposition of this statement, we get the following sufficient condition for entanglement:

For a covariance matrix V of a physical state,

$$\det A \det B + \left(\frac{\hbar^2}{4} - |\det C| \right)^2 - \text{tr}[AJCJBJC^T J] < \frac{\hbar^2}{4}(\det A + \det B)$$
$$\Longrightarrow V \text{ represents a covariance matrix of an entangled state.} \quad (3.208)$$

If V represents a covariance matrix of a Gaussian state, it leads to the following necessary and sufficient condition for entanglement:

For a covariance matrix V of a physical Gaussian state,

$$\det A \det B + \left(\frac{\hbar^2}{4} - |\det C| \right)^2 - \text{tr}[AJCJBJC^T J] < \frac{\hbar^2}{4}(\det A + \det B)$$
$$\Longleftrightarrow V \text{ represents a covariance matrix of an entangled state.} \quad (3.209)$$

3.7.5.3 Condition on PT Symplectic Eigenvalues

We assume that the covariance matrix of an n-mode state is given by V. In the CV system, partial transpose with respect to a subsystem B_1 corresponds to sign flips of momentum operators in B_1:

$$\begin{pmatrix} \hat{x}_1 \\ \hat{p}_1 \\ \vdots \\ \hat{x}_n \\ \hat{p}_n \end{pmatrix} \to \Lambda \begin{pmatrix} \hat{x}_1 \\ \hat{p}_1 \\ \vdots \\ \hat{x}_n \\ \hat{p}_n \end{pmatrix}, \quad \Lambda = \begin{pmatrix} 1 & & & & \\ & s(1) & & & \\ & & \ddots & & \\ & & & 1 & \\ & & & & s(n) \end{pmatrix}, \quad (3.210)$$

where

$$s(k) = \begin{cases} 1 & (k \in B_1) \\ -1 & (k \notin B_1) \end{cases}. \quad (3.211)$$

By applying the partial transpose, the covariance matrix V becomes

$$V \rightarrow \tilde{V} = \Lambda V \Lambda. \tag{3.212}$$

If the quantum state given by V is separable into two subsystems B_1 and B_2, the matrix $\tilde{V}$ represents a covariance matrix of some physical n-mode state. We define the partial transpose (PT) symplectic eigenvalues $\tilde{\nu}$ to be the symplectic eigenvalues of $\tilde{V}$. In the special case of two-mode states, we define $\tilde{\nu}_{\pm}$ ($\tilde{\nu}_+ \geq \tilde{\nu}_-$) as the two PT symplectic eigenvalues, where $\tilde{\nu}_+$ is necessarily $\tilde{\nu}_+ \geq \frac{\hbar}{2}$. As a result, we get

$$\min \tilde{\nu}_k \geq \frac{\hbar}{2}. \tag{3.213}$$

By taking the contraposition of this statement, we get the following sufficient condition for inseparability:

For a covariance matrix V of a physical Gaussian state,

$$\min \tilde{\nu}_k < \frac{\hbar}{2} \implies \text{not separable into } B_1 \text{ and } B_2. \tag{3.214}$$

Although we have assumed that V represents a covariance matrix of a *physical* state, it is not guaranteed that symmetric matrices which are acquired in experiments show covariance matrices of *physical* states because of experimental errors. Therefore, we have to verify that the experimental symmetric matrix surely shows a physical state as well:

$$\begin{aligned}&\text{A real symmetric matrix } V \text{ satisfies } V > 0 \text{ and } \min \nu_k \geq \frac{\hbar}{2} \text{ and } \min \tilde{\nu}_k < \frac{\hbar}{2}\\ &\qquad \implies V \text{ represents a physical state which is not separable into } B_1 \text{ and } B_2.\end{aligned} \tag{3.215}$$

Note that it does not give us necessary condition for inseparability in general. For some specific cases, such as $1 \times n$ Gaussian states, it is proven that it is a necessary and sufficient condition for inseparability. However, its necessity is violated even though V is a covariance matrix of a Gaussian state. The minimum case of its violation is the 2×2 Gaussian system [19].

3.7.6 Logarithmic Negativity

Although several measures of entanglement for mixed states are proposed [20–24], we utilize the logarithmic negativity, which can be easily acquired in both theory and experiment [25].

We assume that the density operator of a quantum state is given by $\hat{\rho}$. We define $\hat{\rho}^{T_A}$ to be the partial transpose operator of $\hat{\rho}$ with respect to the subsystem A. The logarithmic negativity $E_N(\hat{\rho})$ is defined as

$$E_N(\hat{\rho}) = \log \|\hat{\rho}^{T_A}\|_1, \tag{3.216}$$

where

$$\|\hat{U}\|_1 = \mathrm{tr}\sqrt{\hat{U}^\dagger \hat{U}} \tag{3.217}$$

is the trace norm of $\hat{U}$, which is equal to the sum of the singular values of $\hat{U}$, or equivalent to $\sum_k \sqrt{\lambda_k}$, where λ_k represent eigenvalues of $\hat{U}^\dagger \hat{U}$. If the operator $\hat{U}$ is a Hermitian operator, including density operators, singular values are equal to the absolute values of eigenvalues of $\hat{U}$.

It is known that the logarithmic negativity does not increase by local operations and classical communications (LOCC). Although it is entanglement monotone, it is not convex [25, 26].

We finally assume that the density operator $\hat{\rho}$ represents a Gaussian state. The logarithmic negativity is given by

$$E_N(\hat{\rho}) = \sum_k \max\left[0, -\log\left(\frac{2\tilde{\nu}_k}{\hbar}\right)\right]. \tag{3.218}$$

In the special case of a two-mode Gaussian state, it is given by

$$E_N(\hat{\rho}) = \max\left[0, -\log\left(\frac{2\tilde{\nu}_-}{\hbar}\right)\right], \tag{3.219}$$

since $\tilde{\nu}_+ \geq \frac{\hbar}{2}$.

3.7.7 Comparison of Entanglement Criteria

We discuss advantages and disadvantages of two different entanglement criteria (one is based on quadrature variances, while the other is based on partial transpose of density operators). In the following, we assume that the quantum state in question is a two-mode Gaussian state, which is the case with experiments in this thesis.

3.7.7.1 Disadvantage of Entanglement Criteria Based on Quadrature Variances

Entanglement criteria based on variances of *specific* functions of quadrature operators (Duan criterion and van Loock-Furusawa criterion) provide sufficient conditions for entanglement, not necessary and sufficient conditions, even though it is known in advance that the quantum states to be evaluated are Gaussian states. Although the standardization of its covariance matrix provides a necessary and sufficient conditions for entanglement, we have to change the quadrature operator functions. Thus we regard the revised criterion as another one. For each quantum state, there exist appropriate functions of quadrature operators with which its entanglement can be efficiently detected.

The fact that they only provide sufficient conditions imposes a strong requirement on us that we have to choose appropriate functions of quadrature operators. It means that we have to have some knowledge about the state in advance. Therefore, these entanglement criteria work well when quantum states to be checked are known in advance. One good example is verification of cluster state generation. It works well too when we verify output states of known quantum gates with known input states. On the contrary, they might not be suitable for verification of unknown quantum states.

We show another example where they are not suitable. Consider a two-mode Gaussian quantum state. We assume that a position operator $\hat{x}_i$ of mode i has correlation with momentum operator $\hat{p}_i$ of the same mode i. The standardization in Duan criterion leads to mixture of position and momentum operators $(\hat{x}_i, \hat{p}_i)$ of single mode. Although it is not problematic in theory, it complicates experimental setup because we have to measure variances of linear combinations of position and momentum operators. An example of this will be given in Chap. 9.

3.7.7.2 Advantage of Entanglement Criteria Based on Partial Transpose of Density Operators

Entanglement criteria based on partial transpose of density operators give us necessary and sufficient conditions for entanglement. Once the density matrix or covariance matrix is acquired, inseparability criteria can be directly evaluated. We do not have to *choose* an *appropriate* formulae, such as functions of quadrature operators in Duan criterion and van Loock-Furusawa criterion.

3.7.7.3 Disadvantage of Entanglement Criteria Based on Partial Transpose of Density Operators

In order to evaluate inseparability, we have to know all the elements of the covariance matrix. They includes excess elements which might not be needed for verification of entanglement.

3.7.7.4 Advantage of Entanglement Criteria Based on Quadrature Variances

Inseparability can be evaluated by acquiring the minimum set of variables (such as several elements of the covariance matrix) which are needed for verification of entanglement. It simplifies experimental demonstrations of entanglement detections.

References

1. Nielsen, M.A., Chuang, I. L.: Quantum Computation and Quantum Information. Cambridge University Press, Cambridge (2000)
2. van Loock, P., Weedbrook, C.: Building Gaussian cluster states by linear optics. Phys. Rev. A **76**, 032321 (2007)
3. Furusawa, A., van Loock, P.: Quantum Teleportation and Entanglement. Wiley-VCH, New York (2011)
4. Folland, G.B.: Harmonic Analysis in Phase Space, Annals of Mathematics Studies, vol. 112. Princeton University Press, Princeton (1989)
5. Marsden, J.E., Ratiu, T.S.: Introduction to Mechanics and Symmetry. Springer, Berlin (2002)
6. Leonhardt, U.: Measuring the Quantum State of Light. Cambridge University Press, Cambridge (1997)
7. Reck, M., Zeilinger, A., Bernstein, H.J., Bertani, P.: Experimental realization of any discrete unitary operator. Phys. Rev. Lett. **73**, 58 (1994)
8. Lloyd, S., Braunstein, S.L.: Quantum computation over continuous variables. Phys. Rev. Lett. **82**, 1784 (1999)
9. Corney, S.P., Delbourgo, R., Jarvis, P.D.: Proceedings of the XXII International Colloquium on Group Theoretical Methods in Physics. International Press, Cambridge (1999)
10. Bartlett, S.D., Sanders, B.C., Braunstein, S.L., Nemoto, K.: Efficient classical simulation of continuous variable quantum information processes. Phys. Rev. Lett. **88**, 097904 (2002)
11. van Loock, P., Furusawa, A.: Detecting genuine multipartite continuous-variable entanglement. Phys. Rev. A **67**, 052315 (2003)
12. Einstein, A., Podolsky, B., Rosen, N.: Can quantum-mechanical description of physical reality be considered complete? Phys. Rev. **47**, 777 (1935)
13. Aspect, A., Grangier, P., Roger, G.: Experimental tests of realistic local theories via Bell's theorem. Phys. Rev. Lett. **47**, 460 (1981)
14. Aspect, A., Dalibard, J., Roger, G.: Experimental test of Bell's inequalities using time-varying analyzers. Phys. Rev. Lett. **49**, 1804 (1982)
15. Duan, L.-M., Giedke, G., Cirac, J.I., Zoller1, P.: Inseparability criterion for continuous variable systems. Phys. Rev. Lett. **84**, 2722 (2000)
16. Peres, A.: Separability criterion for density matrices. Phys. Rev. Lett. **77**, 1413 (1996)
17. Horodecki, M., Horodecki, P., Horodecki, R.: Separability of mixed states: necessary and sufficient conditions. Phys. Lett. A **25**, 223 (1996)
18. Simon, R.: Peres-Horodecki separability criterion for continuous variable systems. Phys. Rev. Lett. **84**, 2726 (2000)
19. Werner, R.F., Wolf, M.M.: Bound entangled Gaussian states. Phys. Rev. Lett. **86**, 3658 (2001)
20. Bennett, C.H., DiVincenzo, D.P., Smolin, J.A., Wootters, W.K.: Mixed-state entanglement and quantum error correction. Phys. Rev. A **54**, 3824 (1996)
21. Vedral, V., Plenio, M.B.: Entanglement measures and purification procedures. Phys. Rev. A **57**, 1619 (1998)
22. Horodecki, M., Horodecki, P., Horodecki, R.: Limits for entanglement measures. Phys. Rev. Lett. **84**, 2014 (2000)
23. Henderson, L., Vedral, V.: Information, relative entropy of entanglement, and irreversibility. Phys. Rev. Lett. **84**, 2263 (2000)
24. Hayden, P.M., Horodecki, M., Terhal, B.M.: The asymptotic entanglement cost of preparing a quantum state. J. Phys. A: Math. Gen. **34**, 6891 (2001)
25. Vidal, G., Werner, R.F.: Computable measure of entanglement. Phys. Rev. A **65**, 032314 (2002)
26. Plenio, M.B.: Logarithmic negativity: a full entanglement monotone that is not convex. Phys. Rev. Lett. **95**, 090503 (2005)

Chapter 4
Offline Scheme and One-Way Quantum Computation

4.1 Quantum Teleportation and Gate Teleportation

4.1.1 Quantum Teleportation

Quantum teleportation is a protocol in which one can transmit an unknown quantum state to a receiver at a distance. The sender and the receiver of the quantum state are usually named "Alice" and "Bob", respectively. In order to distinguish quantum teleportation from one-mode teleportation (half-teleportation, Sect. 4.2), we sometimes refer to it as *ordinary quantum teleportation* or *full quantum teleportation*.

We consider that Alice possesses a one-mode quantum state $|\psi\rangle$. Alice will transmit it to Bob via a quantum teleportation.

For that purpose, Alice and Bob share an EPR state in advance. Alice has a half of the EPR state, while Bob has the other half of it. The procedure of quantum teleportation is as follows (see Figs. 4.1 and 4.2). First, Alice entangles the quantum state to be transmitted and a half of the EPR state which belongs to Alice. Alice measures the two outcomes in an appropriate measurement basis. The measurement results are sent to Bob through classical channels. By performing correction operations on the other half of the EPR state, Bob can reconstruct the quantum state which was initially owned by Alice. Note that the quantum state $|\psi\rangle$ does not pass through any quantum channels with which quantum states can be directly transmitted.

In Figs. 4.1 and 4.2, mode 1 and mode 2 correspond to an EPR state. Mode 1 is owned by Alice, while mode 2 is owned by Bob. In Fig. 4.2, we show the procedure for generation of the EPR state as well: it can be generated by combining two squeezed states on a beam splitter (we will show this later).

In the following, we formulate the process of quantum computation (Fig. 4.2) by using quadrature operators which specify particular quantum states (Sect. 3.2.8).

We assume that the initial states in modes 1 and 2 are vacuum states. We describe these modes as

$$\begin{cases} \hat{x}_1^{(0)} + i\hat{p}_1^{(0)} \\ \hat{x}_2^{(0)} + i\hat{p}_2^{(0)} \end{cases}, \tag{4.1}$$

R. Ukai, *Multi-Step Multi-Input One-Way Quantum Information Processing with Spatial and Temporal Modes of Light*, Springer Theses, DOI 10.1007/978-4-431-55019-8_4

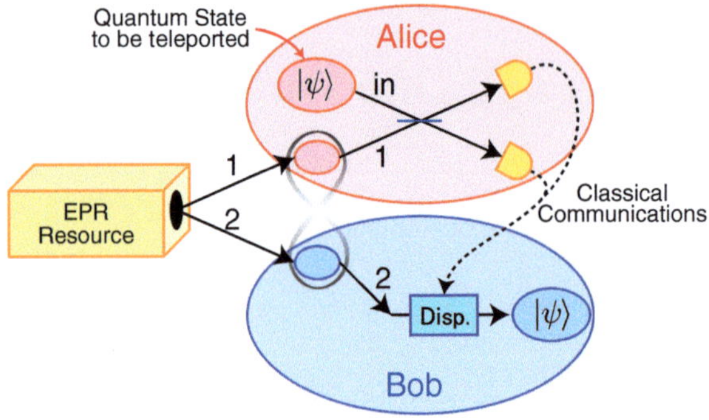

Fig. 4.1 Abstract illustration of quantum teleportation

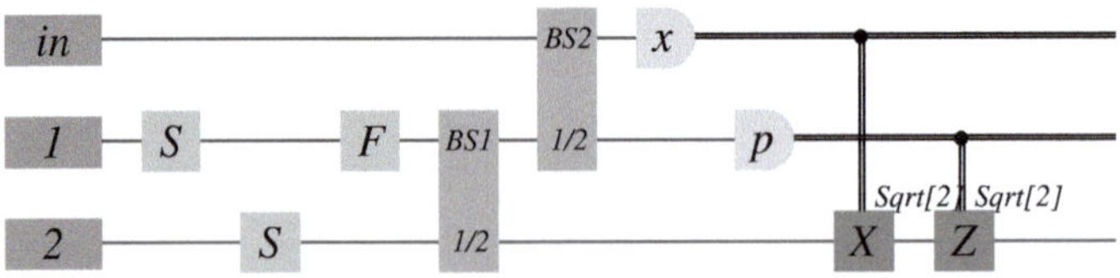

Fig. 4.2 Quantum circuit of quantum teleportation

where the subscript $k = 1, 2$ denotes the mode number, while the superscripts (0) show that these modes are in vacuum states.

As a next step, we apply p-squeezing operators on both modes. We can describe these two squeezed states as

$$\begin{cases} \mathrm{e}^{r_1}\hat{x}_1^{(0)} + i\mathrm{e}^{-r_1}\hat{p}_1^{(0)} \\ \mathrm{e}^{r_2}\hat{x}_2^{(0)} + i\mathrm{e}^{-r_2}\hat{p}_2^{(0)} \end{cases}, \tag{4.2}$$

where r_k represents the squeezing parameter for mode k.

We then perform a Fourier operation on mode 1, leading to

$$\begin{cases} -\mathrm{e}^{-r_1}\hat{p}_1^{(0)} + i\mathrm{e}^{r_1}\hat{x}_1^{(0)} \\ \mathrm{e}^{r_2}\hat{x}_2^{(0)} + i\mathrm{e}^{-r_2}\hat{p}_2^{(0)} \end{cases}. \tag{4.3}$$

The next step is an application of a beam splitter operator $\hat{B}_{12}^{(1)}(\frac{1}{\sqrt{2}})$ on mode 1 and mode 2. Note that the argument in the operator is the amplitude reflectivity, while that in the quantum circuit shown in Fig. 4.2 is the energy reflectivity. As a result, mode 1 and mode 2 become

$$\begin{cases} \hat{x}_1^{(e)} + i\hat{p}_1^{(e)} = \frac{1}{\sqrt{2}}\left(\mathrm{e}^{-r_1}\hat{p}_1^{(0)} + \mathrm{e}^{r_2}\hat{x}_2^{(0)}\right) + i\frac{1}{\sqrt{2}}\left(-\mathrm{e}^{r_1}\hat{x}_1^{(0)} + \mathrm{e}^{-r_2}\hat{p}_2^{(0)}\right) \\ \hat{x}_2^{(e)} + i\hat{p}_2^{(e)} = \frac{1}{\sqrt{2}}\left(-\mathrm{e}^{-r_1}\hat{p}_1^{(0)} + \mathrm{e}^{r_2}\hat{x}_2^{(0)}\right) + i\frac{1}{\sqrt{2}}\left(\mathrm{e}^{r_1}\hat{x}_1^{(0)} + \mathrm{e}^{-r_2}\hat{p}_2^{(0)}\right) \end{cases}. \tag{4.4}$$

By using these equations, we get

$$\hat{x}_1^{(e)} - \hat{x}_2^{(e)} = \sqrt{2}\mathrm{e}^{-r_1}\hat{p}_1^{(0)}, \quad \hat{p}_1^{(e)} + \hat{p}_2^{(e)} = \sqrt{2}\mathrm{e}^{-r_2}\hat{p}_2^{(0)}. \tag{4.5}$$

In the limit of infinite squeezing ($r_k \to \infty$), we get

$$\hat{x}_1^{(e)} - \hat{x}_2^{(e)} = 0, \quad \hat{p}_1^{(e)} + \hat{p}_2^{(e)} = 0. \tag{4.6}$$

Therefore, mode 1 and mode 2 represented by $\hat{x}_k^{(e)} + i\hat{p}_k^{(e)}$ $(k = 1, 2)$ are in an EPR state.[1]

We define the quadrature operators of the input mode as

$$\hat{x}_{in} + i\hat{p}_{in}. \tag{4.7}$$

We apply the beam splitter operator $\hat{B}_{in,1}^{(2)}(\frac{1}{\sqrt{2}})$ on mode in and mode 1, leading to

$$\begin{cases} \left[\frac{1}{\sqrt{2}}\hat{x}_{in} - \frac{1}{2}\left(\mathrm{e}^{-r_1}\hat{p}_1^{(0)} + \mathrm{e}^{r_2}\hat{x}_2^{(0)}\right)\right] + i\left[\frac{1}{\sqrt{2}}\hat{p}_{in} - \frac{1}{2}\left(-\mathrm{e}^{r_1}\hat{x}_1^{(0)} + \mathrm{e}^{-r_2}\hat{p}_2^{(0)}\right)\right] \\ \left[\frac{1}{\sqrt{2}}\hat{x}_{in} + \frac{1}{2}\left(\mathrm{e}^{-r_1}\hat{p}_1^{(0)} + \mathrm{e}^{r_2}\hat{x}_2^{(0)}\right)\right] + i\left[\frac{1}{\sqrt{2}}\hat{p}_{in} + \frac{1}{2}\left(-\mathrm{e}^{r_1}\hat{x}_1^{(0)} + \mathrm{e}^{-r_2}\hat{p}_2^{(0)}\right)\right] \\ \left[\frac{1}{\sqrt{2}}\left(-\mathrm{e}^{-r_1}\hat{p}_1^{(0)} + \mathrm{e}^{r_2}\hat{x}_2^{(0)}\right)\right] + i\left[\frac{1}{\sqrt{2}}\left(\mathrm{e}^{r_1}\hat{x}_1^{(0)} + \mathrm{e}^{-r_2}\hat{p}_2^{(0)}\right)\right] \end{cases}. \tag{4.8}$$

We perform measurements (homodyne measurements) on modes in and 1. We measure $\hat{x}$ in mode in, while $\hat{p}$ in mode 1. The measurement observables $\hat{s}_{in}$, $\hat{s}_1$ for the measurements on modes in and 1 are given by

$$\hat{s}_{in} = \frac{1}{\sqrt{2}}\hat{x}_{in} - \frac{1}{2}\left(\mathrm{e}^{-r_1}\hat{p}_1^{(0)} + \mathrm{e}^{r_2}\hat{x}_2^{(0)}\right), \quad \hat{s}_1 = \frac{1}{\sqrt{2}}\hat{p}_{in} + \frac{1}{2}\left(-\mathrm{e}^{r_1}\hat{x}_1^{(0)} + \mathrm{e}^{-r_2}\hat{p}_2^{(0)}\right). \tag{4.9}$$

Without using the position operators $\mathrm{e}^{r_k}\hat{x}_k^{(0)}$ for the squeezed-state modes $k = 1, 2$, the quadrature operators for the output mode $k = 2$ can be reformulated as

$$\left[\hat{x}_{in} - \sqrt{2}\mathrm{e}^{-r_1}\hat{p}_1^{(0)} - \sqrt{2}\hat{s}_{in}\right] + i\left[\hat{p}_{in} + \sqrt{2}\mathrm{e}^{-r_2}\hat{p}_2^{(0)} - \sqrt{2}\hat{s}_1\right]. \tag{4.10}$$

Note that the reformulation above holds before the measurements are carried out. On the contrary, when the measurements on modes in and 1 have been carried out indeed, the quadratures of the output modes become

[1] To be precise, mode 1 and mode 2 represented by $\hat{x}_k^{(e)} + i\hat{p}_k^{(e)}$ $(k = 1, 2)$ are in an EPR state in the limit of infinite squeezing. In the case of finite squeezing, $\hat{x}_1^{(e)} - \hat{x}_2^{(e)}$ and $\hat{p}_1^{(e)} + \hat{p}_2^{(e)}$ have nonzero variances. Thus, these two modes are in an approximate EPR state. We sometimes refer to this finitely correlated state as a two-mode squeezed state.

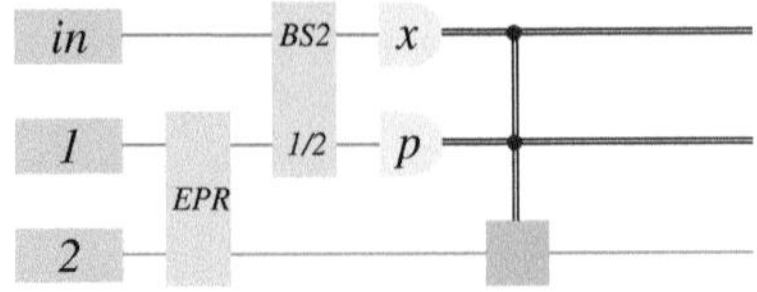

Fig. 4.3 Quantum circuit of quantum teleportation

$$\left[\hat{x}_{in} - \sqrt{2}e^{-r_1}\hat{p}_1^{(0)} - \sqrt{2}s_{in}\right] + i\left[\hat{p}_{in} + \sqrt{2}e^{-r_2}\hat{p}_2^{(0)} - \sqrt{2}s_1\right], \tag{4.11}$$

where s_{in} and s_1 represent measurement results of measurement variables $\hat{s}_{in}$ and $\hat{s}_1$, respectively. By performing feed-forwards (displacement operations) based on the measurement results:

$$\hat{X}_2\left(\sqrt{2}s_{in})\hat{Z}_2(\sqrt{2}s_1\right), \tag{4.12}$$

we get

$$\left[\hat{x}_{in} - \sqrt{2}e^{-r_1}\hat{p}_1^{(0)}\right] + i\left[\hat{p}_{in} + \sqrt{2}e^{-r_2}\hat{p}_2^{(0)}\right]. \tag{4.13}$$

In the limit of infinite squeezing $r_k \to \infty$, we get

$$\hat{x}_{in} + i\hat{p}_{in}, \tag{4.14}$$

which is identical to the quadrature operators of the input mode in. Therefore, teleportation of the quantum state is finished successfully.

In the following, we substitute Fig. 4.3 for Fig. 4.2, for simplicity.

4.1.2 Gate Teleportation

Although quantum teleportation was initially proposed as a protocol to transmit a quantum state, it was later found that it can be applied to implement quantum computation. In this subsection, we show an example which is called the "teleportation of quantum gate", "gate teleportation", or "offline scheme".

4.1.2.1 Gate Teleportation

Consider that we perform a unitary operator $\hat{U}$ on the output state of a quantum teleportation (Fig. 4.4).

We next consider a change of the order of the unitary operation by $\hat{U}$ and the feed-forward operation. Here, the feed-forward operator is given by $\hat{P}_A = \hat{X}_2(\sqrt{2}s_{in})\hat{Z}_2(\sqrt{2}s_1)$, where s_{in} and s_1 represent measurement results. In general, the unitary operator $\hat{U}$ and the displacement operator $\hat{P}_A$ do not commute with each

Fig. 4.4 Gate teleportation 1

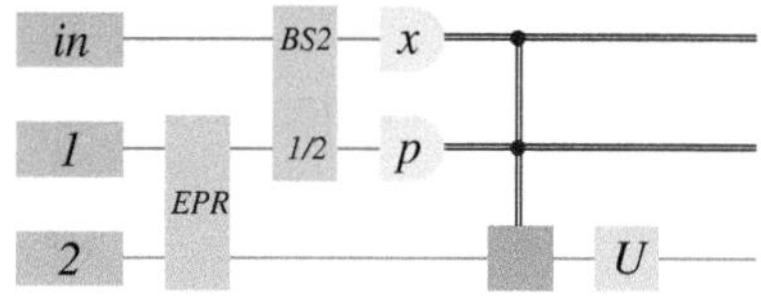

other. We here define another feed-forward unitary operator $\hat{N}$:

$$\hat{N} = \hat{U}\hat{P}_A\hat{U}^\dagger, \tag{4.15}$$

which is determined by measurement results s_{in}, s_1 and the unitary operator $\hat{U}$. We then have

$$\hat{U}\hat{P}_A = \hat{N}\hat{U}. \tag{4.16}$$

Therefore, we can change the order of the unitary operation by $\hat{U}$ and the feed-forward operation without changing the unitary operator $\hat{U}$.

4.1.2.2 Classification by Order of Unitary Operator

We consider the new feed-forward unitary operator $\hat{N}$. The original feed-forward operator $\hat{P}_A$ is a member of the Pauli group (Sect. 3.4.1). In general, the unitary operator $\hat{U}$ is formulated by a polynomial $f(\hat{x}, \hat{p})$ of quadrature operators $\hat{x}, \hat{p}$:

$$\hat{U} = e^{if(\hat{x},\hat{p})}. \tag{4.17}$$

If the order of $f(\hat{x}, \hat{p})$ is less than or equal to two, the unitary operator $\hat{U}$ belongs to the Clifford group (Sect. 3.4.3). If the order of $f(\hat{x}, \hat{p})$ is larger than or equal to three, it does not belong to the Clifford group.

We first consider that the operator $\hat{U}$ is a member of the Clifford group. In general, for an arbitrary Clifford operator $\hat{C}$ and Pauli operator $\hat{P}_A$, there exists a Pauli operator $\hat{P}_B$ which satisfies (Sect. 3.4.4)

$$\hat{C}\hat{P}_A\hat{C}^\dagger = \hat{P}_B. \tag{4.18}$$

As a result, we get $\hat{U}\hat{P}_A = \hat{P}_B\hat{U}$, thus the revised feed-forward operator $\hat{P}_B$ is a member of the Pauli group (displacement operator) as well.

We then consider the case where the order of the polynomial $f(\hat{x}, \hat{p})$ is more than or equal to three. The new feed-forward unitary operator $\hat{N}$ is not a member of the Pauli group, thus it is not a displacement operation. We define a group of unitary operators C_3:

$$C_3 = \left\{\hat{U}|\hat{U}P\hat{U}^\dagger \subseteqq C\right\}, \tag{4.19}$$

where P and C represent the Pauli group and the Clifford group. The cubic phase operator $\hat{V}(\gamma)$ is a member of the group C_3:

$$\hat{V}(\gamma) = e^{i\gamma\hat{x}^3} \in C_3. \tag{4.20}$$

If the unitary operator $\hat{U}$ is a member of the group C_3, there exists a Clifford group operator $\hat{C}$ which satisfies $\hat{U}\hat{P}_A = \hat{C}\hat{U}$, where $\hat{C}$ is determined by the measurement results s_{in}, s_1 and the unitary operator $\hat{U}$. Thus, the revised feed-forward operation is a Gaussian operation.

4.1.2.3 Gate Teleportation

In the original quantum circuit shown in Fig. 4.4, the unitary operator $\hat{U}$ is applied to the output state of the quantum teleportation. On the contrary, in the transformed version of the quantum circuit shown in Fig. 4.5, the unitary operator is applied to the EPR state before it is utilized as a resource for the quantum teleportation. This kind of transformation of quantum circuit is called *teleportation of quantum gate* or *gate teleportation.*

It is reported that gate teleportation can be applied to implement quantum computation (*offline scheme* of quantum computation). Consider that we perform a particular unitary operator $\hat{U}$ on an arbitrary quantum state $|\psi\rangle$ based on Fig. 4.4 or Fig. 4.5. In the original circuit (Fig. 4.4), the operator $\hat{U}$ is applied to the quantum state $|\psi\rangle$. Thus, we have to be able to apply $\hat{U}$ on an arbitrary quantum state. On the contrary, in the transformed version of quantum circuit (Fig. 4.5), the operator $\hat{U}$ is applied to the resource EPR state. Thus, we only have to be able to apply $\hat{U}$ on a particular quantum state, leading to a *revised resource state*. In addition, the revised resource state is also a known state. Therefore, if we can generate it by using a different way, we can replace the original method for generation of the revised state with the newer one.

It provides another advantage of gate teleportation. Consider that implementation of the operator $\hat{U}$ involves inevitable losses in experiment. If the loss is too large, the quantum state would be destroyed because of interaction with its environment (decoherence). This problem can be solved if we can generate the revised resource state by a less-loss process, instead of by performing the unitary operator $\hat{U}$ to the original resource EPR state.

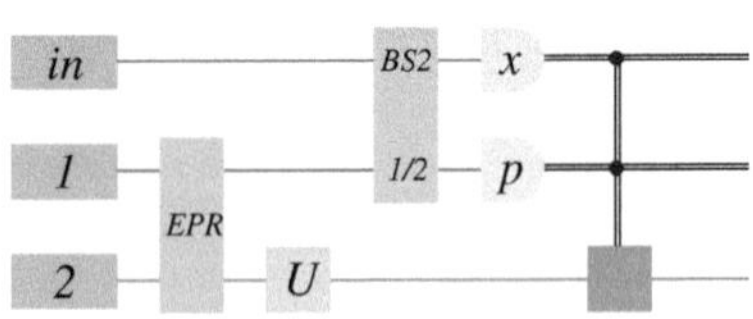

Fig. 4.5 Gate teleportation 2

Consider another case where implementation of the operator $\hat{U}$ cannot be achieved deterministically in experiment. The quantum state would be destroyed as well when it fails. This problem can be solved by applying the gate teleportation since the preparation of the revised resource state can be detached from the main (online) computation. That is, even if the revised resource state can be generated probabilistically, we can repeat generation of the revised resource state until we successfully generate it. The generated state is stored in a memory, and then utilized in the online computation. In addition, we can also apply the distillation [1, 2] in order to purify the revised resource state.

We briefly mention universality of quantum computation with gate teleportation. It is known that an arbitrary unitary operation can be implemented by combining Gaussian operations and a one-mode non-Gaussian operation (Sect. 3.6). Remember that a one-mode non-Gaussian operation with the cubic phase operator $\hat{V}(\gamma)$ can be implemented with Gaussian feed-forward operations by using the gate teleportation. Therefore, universality is achieved by combining Gaussian operations and preparation of the revised resource state for the cubic phase operator $\hat{V}(\gamma)$.

4.2 One-Mode Teleportation Circuit and Elementary Circuit for One-Way Quantum Computation

In this section, we introduce another application of quantum teleportation to quantum computation. It is the *one-way quantum computation*, which is the main theme in this thesis. It is also called *cluster model* of quantum computation, or *cluster computation*.

Since quantum teleportation circuit is not the minimum component for quantum computation, we first show the minimum circuit called the *one-mode teleportation circuit*. It is also called the *half-teleportation circuit*[2] or *elementary one-mode one-way quantum computation circuit*. We then refer to offline scheme and one-way quantum computation.

4.2.1 One-Mode Teleportation Circuit (Half-Teleportation Circuit, Elementary One-Mode One-Way Quantum Computation Circuit)

4.2.1.1 Quantum Circuit

Figure 4.6 shows the quantum circuit of the one-mode teleportation circuit.

Mode *in* represents the input mode, which can be in an arbitrary state. Mode 1 is the resource mode for the one-mode teleportation circuit. In the ideal case, it

[2] It is called the half-teleportation circuit since two one-mode teleportation circuits are equivalent to the ordinary quantum teleportation circuit.

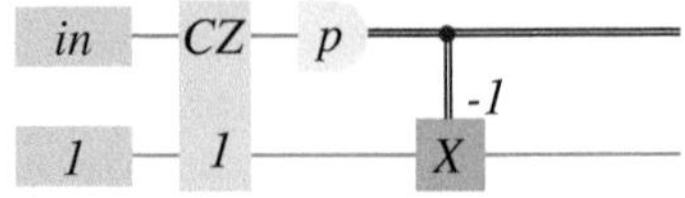

Fig. 4.6 One-mode teleportation circuit

is initially in the zero eigenstate $|p = 0\rangle$ of the momentum operator $\hat{p}$. It can be approximated by a p-squeezed state.

We perform a controlled-Z gate on mode in and mode 1 (Sect. 3.5.7). It is represented by an operator $\hat{C}_Z = \mathrm{e}^{\frac{i}{\hbar}\hat{x}_{in}\hat{x}_1}$. We then perform a measurement on mode in with measurement observable $\hat{p}$, followed by feed-forward operation to the position operator of mode 1.

4.2.1.2 Schrödinger Picture

We show the process of the one-mode teleportation circuit in the Schrödinger picture.

We define $|\psi\rangle_{in}$ to be the initial state in mode in, while the initial state in mode 1 is given by $|p = 0\rangle_1$. Thus, the initial state in two modes is

$$|\psi\rangle_{in}|p = 0\rangle_1. \tag{4.21}$$

We expand these two states as

$$|\psi\rangle_{in} = \int_{-\infty}^{\infty} \mathrm{d}s\ \psi(s)|x = s\rangle_{in}, \tag{4.22}$$

$$|p = 0\rangle_1 = \frac{1}{\sqrt{2\pi\hbar}} \int_{-\infty}^{\infty} \mathrm{d}t\ |x = t\rangle_1. \tag{4.23}$$

By performing the controlled-Z operator $\hat{C}_Z = \mathrm{e}^{\frac{i}{\hbar}\hat{x}_{in}\hat{x}_1}$, it becomes

$$|\Psi\rangle_{in1}^{(C)} = \hat{C}_Z|\psi\rangle_{in}|p = 0\rangle_1 = \frac{1}{\sqrt{2\pi\hbar}} \int_{-\infty}^{\infty} \mathrm{d}s \int_{-\infty}^{\infty} \mathrm{d}t\ \psi(s)\mathrm{e}^{\frac{i}{\hbar}st}|x = s\rangle_{in}|x = t\rangle_1. \tag{4.24}$$

Next, we perform a projection measurement on mode in with the set of projection operators $\{\hat{E}(m)\}$:

$$\{\hat{E}(m)\} = \big\{|p = m\rangle\langle p = m|\big|m \in \mathbb{R}\big\}. \tag{4.25}$$

When the measurement result is m, the quantum state after the measurement is given by

$$|\Psi\rangle_{in1}^{(M)} \propto \hat{E}(m)|\Psi\rangle_{in1}^{(C)}. \tag{4.26}$$

We have

$$\hat{E}(m)|\Psi\rangle_{in1}^{(C)} = |p=m\rangle_{in}\frac{1}{2\pi\hbar}\int_{-\infty}^{\infty} \mathrm{d}s \int_{-\infty}^{\infty} \mathrm{d}t\ \psi(s)\mathrm{e}^{\frac{i}{\hbar}st}\mathrm{e}^{-\frac{i}{\hbar}sm}|x=t\rangle_1 \tag{4.27}$$

$$= |p=m\rangle_{in}\frac{1}{\sqrt{2\pi\hbar}}\int_{-\infty}^{\infty} \mathrm{d}s\ \psi(s)\hat{X}_1(m)|p=s\rangle_1 \tag{4.28}$$

$$= |p=m\rangle_{in}\frac{1}{\sqrt{2\pi\hbar}}\hat{X}_1(m)\int_{-\infty}^{\infty} \mathrm{d}s\ \psi(s)\hat{F}_1|x=s\rangle_1 \tag{4.29}$$

$$= |p=m\rangle_{in}\frac{1}{\sqrt{2\pi\hbar}}\hat{X}_1(m)\hat{F}_1|\psi\rangle_1, \tag{4.30}$$

where

$$\hat{X}(m) = \mathrm{e}^{-\frac{i}{\hbar}m\hat{p}}, \quad \hat{F} = \mathrm{e}^{\frac{i\pi}{4\hbar}(\hat{x}^2+\hat{p}^2)} \tag{4.31}$$

are the position displacement operator and the Fourier operator, respectively. By extracting mode 1 with an appropriate normalization factor, we get

$$|\Psi\rangle_1^{(M)} = \hat{X}_1(m)\hat{F}_1|\psi\rangle_1. \tag{4.32}$$

By performing a feed-forward operation $\hat{X}(-m)$, it becomes

$$|\Psi\rangle_1^{(out)} = \hat{F}_1|\psi\rangle_1. \tag{4.33}$$

Therefore, the input state $|\psi\rangle$ is transmitted from mode in to mode 1 with the Fourier operator $\hat{F}$ applied.

4.2.1.3 Heisenberg Picture

Next, we show the process of one-mode teleportation circuit in the Heisenberg picture.

We define the initial quadrature operators to be

$$\begin{cases} \hat{x}_{in} + i\hat{p}_{in} \\ \hat{x}_1 + i\hat{p}_1 \end{cases}. \tag{4.34}$$

By performing the controlled-Z operator on mode in and mode 1, they become

$$\begin{cases} \hat{x}_{in} + i(\hat{p}_{in} + \hat{x}_1) \\ \hat{x}_1 + i(\hat{p}_1 + \hat{x}_{in}) \end{cases}. \tag{4.35}$$

Next, we perform a projection measurement on mode in. The measurement observable $\hat{s}_{in}$ is $\hat{p}$, thus

$$\hat{s}_{in} = \hat{p}_{in} + \hat{x}_1. \tag{4.36}$$

Without using the position operator $\hat{x}_1$ for the resource squeezed-state mode $k = 1$, the quadrature operators for the output mode $k = 1$ can be reformulated as

$$\hat{s}_{in} - \hat{p}_{in} + i(\hat{p}_1 + \hat{x}_{in}). \tag{4.37}$$

Note that the reformulation above holds before the measurement is carried out. On the contrary, when the measurement on mode in has been carried out indeed, the quadratures of the output mode become

$$s_{in} - \hat{p}_{in} + i(\hat{p}_1 + \hat{x}_{in}), \tag{4.38}$$

where s_{in} represents the measurement result of the measurement variable $\hat{s}_{in}$. By performing the feed-forward operation (displacement operation) based on the measurement result:

$$\hat{X}_1(-s_{in}), \tag{4.39}$$

we get

$$\hat{x}_{out} + i\hat{p}_{out} = -\hat{p}_{in} + i(\hat{p}_1 + \hat{x}_{in}). \tag{4.40}$$

Note that the input-output relation above holds independently of the initial states in mode in and mode 1.

We consider the case where the initial state in the resource mode 1 is the zero eigenstate $|p = 0\rangle$ of the momentum operator $\hat{p}$, whose nullifier is $\hat{p}_1 = 0$. The input-output relation becomes

$$\hat{x}_{out} + i\hat{p}_{out} = -\hat{p}_{in} + i\hat{x}_{in}, \tag{4.41}$$

or equivalently,

$$\begin{pmatrix} \hat{x}_{out} \\ \hat{p}_{out} \end{pmatrix} = \begin{pmatrix} 0 & -1 \\ 1 & 0 \end{pmatrix} \begin{pmatrix} \hat{x}_{in} \\ \hat{p}_{in} \end{pmatrix}. \tag{4.42}$$

Thus, we find that the Fourier transformation on the input mode is performed.

We next consider another case where the initial state in the resource mode 1 is a p-squeezed state. We use quadrature operators which specify particular quantum states (Sect. 3.2.8):

$$\hat{x}_1 + i\hat{p}_1 = \mathrm{e}^{r_1}\hat{x}_1^{(0)} + i\mathrm{e}^{-r_1}\hat{p}_1^{(0)}. \tag{4.43}$$

The input-output relation becomes

$$-\hat{p}_{in} + i\left(\mathrm{e}^{-r_1}\hat{p}_1^{(0)} + \hat{x}_{in}\right), \tag{4.44}$$

or equivalently,

$$\begin{pmatrix}\hat{x}_{out}\\ \hat{p}_{out}\end{pmatrix} = \begin{pmatrix}0 & -1\\ 1 & 0\end{pmatrix}\begin{pmatrix}\hat{x}_{in}\\ \hat{p}_{in}\end{pmatrix} + \begin{pmatrix}0\\ \mathrm{e}^{-r_1}\hat{p}_1^{(0)}\end{pmatrix}. \tag{4.45}$$

In the limit of infinite squeezing $r_1 \to \infty$, it becomes identical to the ideal case. On the contrary, when the squeezing parameter r_1 is finite, the additional term $\mathrm{e}^{-r_1}\hat{p}_1^{(0)}$ does not vanish. It leads to error of one-way quantum computation which derives from the finite level of resource squeezing.

4.2.1.4 Controlled-Z Gate with Gain g

Although we usually use the unity-gain controlled-Z gate for the one-mode teleportation circuit, we sometimes use a controlled-Z gate with gain g (Fig. 4.7). In this case, the feed-forward operation should be changed to $\hat{X}_1(-\frac{1}{g}s_{in})$. The input-output relationship becomes

$$\hat{x}_{out} + i\hat{p}_{out} = -\frac{1}{g}\hat{p}_{in} + i(\hat{p}_1 + g\hat{x}_{in}). \tag{4.46}$$

Note that the input-output relation above holds independently of the initial states in mode in and mode 1.

In the ideal case where the resource mode is initially in $|p = 0\rangle$ or p-squeezed state with $r \to \infty$, the additional term $\hat{p}_1$ vanishes, and an ideal operation can be achieved. The input-output relation becomes

$$\begin{pmatrix}\hat{x}_{out}\\ \hat{p}_{out}\end{pmatrix} = \begin{pmatrix}0 & -\frac{1}{g}\\ g & 0\end{pmatrix}\begin{pmatrix}\hat{x}_{in}\\ \hat{p}_{in}\end{pmatrix} = \begin{pmatrix}1/g & 0\\ 0 & g\end{pmatrix}\begin{pmatrix}0 & -1\\ 1 & 0\end{pmatrix}\begin{pmatrix}\hat{x}_{in}\\ \hat{p}_{in}\end{pmatrix} = \begin{pmatrix}0 & -1\\ 1 & 0\end{pmatrix}\begin{pmatrix}g & 0\\ 0 & 1/g\end{pmatrix}\begin{pmatrix}\hat{x}_{in}\\ \hat{p}_{in}\end{pmatrix}. \tag{4.47}$$

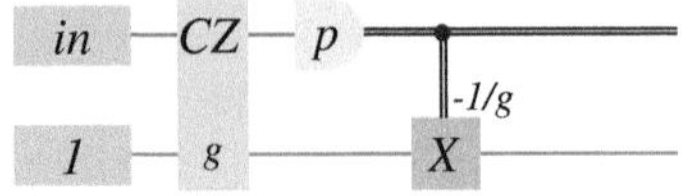

Fig. 4.7 One-mode teleportation gate with non-unity-gain controlled-Z gate

4.2.2 Offline Scheme

In order to compare one-way quantum computation with offline scheme, we develop the offline scheme based on the one-mode teleportation circuit. Although the essence in this subsection is similar to that of the gate teleportation (Sect. 4.1.2), it is not an example of the gate teleportation in a strict sense because the following discussion does not utilize the ordinary quantum teleportation.

Consider that we perform a unitary operator $\hat{D}$ on the output state of a one-mode teleportation circuit (Fig. 4.8).

We assume that the Hamiltonian of the operator $\hat{D}$ is a polynomial $f(\hat{x})$ of the position operator $\hat{x}$:

$$\hat{D}(\hat{x}) = e^{\frac{i}{\hbar} f(\hat{x})}. \tag{4.48}$$

It is obvious that the output state is given by

$$|\Psi\rangle_1^{(out)} = \hat{D}_1(\hat{x})\hat{F}_1|\psi\rangle_1. \tag{4.49}$$

We consider a change of the order of the displacement operation with $\hat{X}(-s)$ and the unitary gate with the operator $\hat{D}(\hat{x})$ (Fig. 4.9). We define an operator $\hat{M}(s, f)$ as

$$\hat{M}(s, f) = \hat{D}(\hat{x})\hat{X}(-s)\hat{D}^{\dagger}(\hat{x}), \tag{4.50}$$

which is determined by the measurement result s and the function f. Since $\hat{M}(s, f)\hat{D}(\hat{x}) = \hat{D}(\hat{x})\hat{X}(-s)$, we find that these two operators can be swapped by changing the feed-forward operation. Note that the unitary operator $\hat{D}(\hat{x})$ is preserved.

The original feed-forward operator $\hat{X}(-s)$ is a member of the Pauli group. On the contrary, the new feed-forward operator is a unitary operator $\hat{M}(s, f)$. If the order of the polynomial $f(\hat{x})$ is less than or equal to two, the unitary operator $\hat{D}(\hat{x})$ belongs to the Clifford group (Sect. 3.4.3), thus $\hat{M}(s, f)$ belongs to the Pauli group (Sect. 4.1.2). If the unitary operator $\hat{D}(\hat{x})$ is a member of the group C_3 Eq. (4.19), $\hat{M}(s, f)$ belongs to the Clifford group, thus the new feed-forward operation is a Gaussian operation.

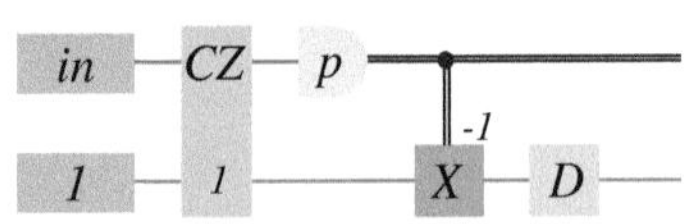

Fig. 4.8 Offline scheme 1

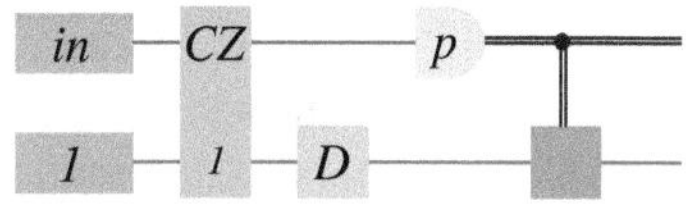

Fig. 4.9 Offline scheme 2

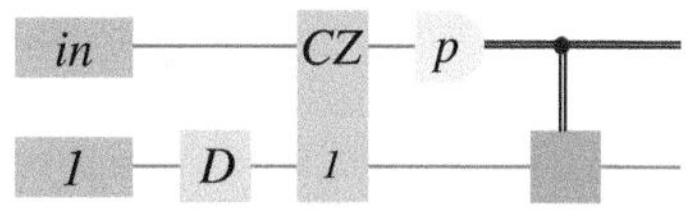

Fig. 4.10 Offline scheme 3

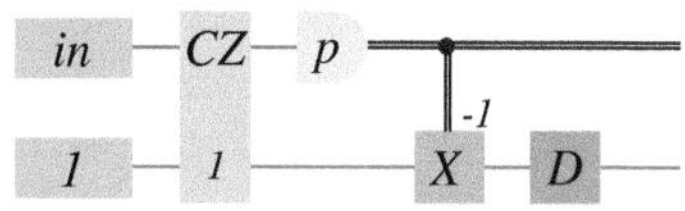

Fig. 4.11 Elementary circuit for one-way quantum computation 1

We then consider a change of the order of the controlled-Z gate and the unitary gate with the operator $\hat{D}(\hat{x})$ (Fig. 4.10). Since their Hamiltonians are diagonal to $\hat{x}_i$, they commute with each other.

Figure 4.10 shows a circuit where the unitary operator $\hat{D}(\hat{x})$ is applied to the resource state in mode 1 in advance, and it acts indirectly on the input state via the one-mode teleportation circuit. Therefore, we can consider that it is an example of the offline scheme.

4.2.3 Elementary Circuit for One-Way Quantum Computation

4.2.3.1 Transformation of Quantum Circuit

Consider that we perform a unitary operator $\hat{D}$ on the output state of a one-mode teleportation circuit (Fig. 4.11).

Different from Sect. 4.2.2,[3] We assume that the Hamiltonian of the operator $\hat{D}$ is a polynomial $f(\hat{p})$ of the momentum operator $\hat{p}$:

$$\hat{D}(\hat{p}) = \mathrm{e}^{\frac{i}{\hbar} f(\hat{p})}. \tag{4.51}$$

It is obvious that the output state is given by

$$|\Psi\rangle_1^{(out)} = \hat{D}_1(\hat{p})\hat{F}_1|\psi\rangle_1. \tag{4.52}$$

By using $\hat{F}^\dagger \hat{p} \hat{F} = \hat{x}$ and $\hat{F}^\dagger \hat{D}(\hat{p}) \hat{F} = \hat{D}(\hat{x})$, it is equivalent to

$$|\Psi\rangle_1^{(out)} = \hat{F}_1 \hat{D}_1(\hat{x})|\psi\rangle_1. \tag{4.53}$$

[3] In Sect. 4.2.2, we have assumed that $\hat{D}(\hat{x}) = \mathrm{e}^{\frac{i}{\hbar} f(\hat{x})}$. Although we have changed the assumption for simplicity, the essence is the same.

Remember that the output state of a one-mode teleportation circuit is given by $\hat{F}_1|\phi\rangle_1$ when the input state is $|\phi\rangle_{in}$. Thus, the state Eq. (4.53) can be considered as the output state of the one-mode teleportation circuit with the input state $|\phi\rangle_{in} = \hat{D}_{in}(\hat{x})|\psi\rangle_{in}$. It is also equivalent to the output state of the quantum circuit where the unitary operator $\hat{D}_{in}(\hat{x})$ is first applied to the input state $|\psi\rangle_{in}$, and then the state is transmitted to mode 1 via the one-mode teleportation circuit (Fig. 4.12).

We then consider a change of the order of the controlled-Z gate and the unitary gate with the operator $\hat{D}(\hat{x})$ (Fig. 4.13). Since their Hamiltonians are diagonal to $\hat{x}_i$, they commute with each other.

Here, the measurement is a projection measurement with the set of projection operators $\{\hat{E}(m)\} = \{|p = m\rangle\langle p = m| \big| m \in \mathbb{R}\}$. Note that it is carried out after the operator $\hat{D}(\hat{x})$ is performed. The combination of the operator and the measurement is equivalent to another measurement with the set of projection operators

$$\{\hat{E}'(m)\} = \{\hat{D}^\dagger(\hat{x})|p = m\rangle\langle p = m|\hat{D}(\hat{x}) \big| m \in \mathbb{R}\}. \tag{4.54}$$

Its measurement basis is $\hat{D}^\dagger(\hat{x})|p = m\rangle$, while observable is $\hat{p}' = \hat{D}^\dagger(\hat{x})\hat{p}\hat{D}(\hat{x})$ (Fig. 4.14).

Different from the offline scheme, the feed-forward operation is not changed during transformation of the quantum circuit above.

In summary, by changing the measurement basis from $|p = m\rangle$ to $\hat{D}^\dagger(\hat{x})|p = m\rangle$, we can perform a unitary operator $\hat{D}(\hat{x})$ which is determined by the measurement basis. The key of *one-way quantum computation* is that operations are controlled by measurement bases.

4.2.3.2 Observable

In the case of the offline scheme, the observable is not changed from the momentum operator $\hat{p}$. On the contrary, it is changed to $\hat{p}' = \hat{D}^\dagger(\hat{x})\hat{p}\hat{D}(\hat{x})$ in one-way quantum computation. We assume that the Hamiltonian of the operator $\hat{D}(\hat{x})$ is a polynomial of the position operator $\hat{x}$:

$$\hat{D}(\hat{x}) = \mathrm{e}^{\frac{i}{\hbar} f(\hat{x})} = \mathrm{e}^{\frac{i}{\hbar} \sum_n a_n \hat{x}^n}. \tag{4.55}$$

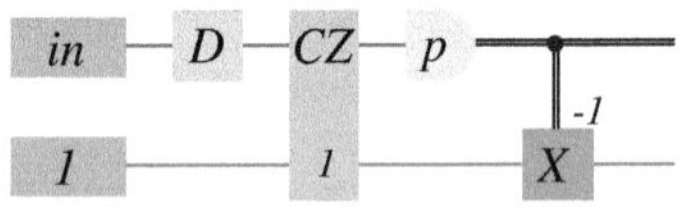

Fig. 4.12 Elementary circuit for one-way quantum computation 2

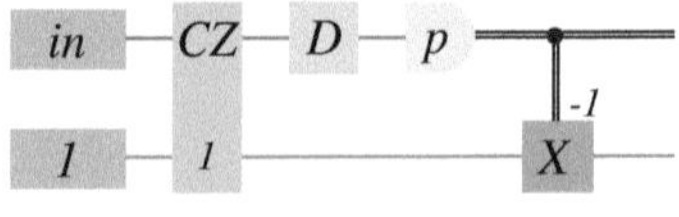

Fig. 4.13 Elementary circuit for one-way quantum computation 3

Fig. 4.14 Elementary circuit for one-way quantum computation 4

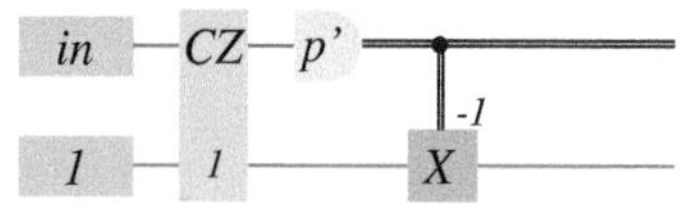

By using $[\hat{x}^n, \hat{p}] = i\hbar n\hat{x}^{n-1}$, we get

$$\hat{p}' = \hat{D}^\dagger(\hat{x})\hat{p}\hat{D}(\hat{x}) = \hat{p} + f'(\hat{x}), \tag{4.56}$$

where $f'(x)$ is the first derivative of $f(x)$ with respect to x: $f'(x) = \frac{\mathrm{d}f(x)}{\mathrm{d}x}$. When $f(\hat{x})$ is a quadratic function of $\hat{x}$, $\hat{p}'$ becomes a linear combination of quadrature operators $\hat{x}$ and $\hat{p}$. It can be achieved by a homodyne measurement. When the order of $f(\hat{x})$ is larger than or equal to three, non-linear measurement is required.

4.2.3.3 Heisenberg Picture

Since four figures (Figs. 4.11, 4.12, 4.13 and 4.14) are equivalent to each other, we choose Fig. 4.12 in order to acquire the input-output relation in the Heisenberg picture for simplicity.

We define that the input mode is $\hat{x}_{in} + i\hat{p}_{in}$. By performing the unitary operator $\hat{D}(\hat{x})$, the position and momentum operators become

$$\hat{x}' + i\hat{p}' = \hat{D}^\dagger(\hat{x}_{in})(\hat{x}_{in} + i\hat{p}_{in})\hat{D}(\hat{x}_{in}) = \hat{x}_{in} + i(\hat{p}_{in} + f'(\hat{x}_{in})). \tag{4.57}$$

Since the input-output relation of the one-mode teleportation circuit in the Heisenberg picture is given by Eq. (4.40), the input-output relation of the elementary circuit for the one-way quantum computation is

$$\hat{x}_{out} + i\hat{p}_{out} = -(\hat{p}_{in} + f'(\hat{x}_{in})) + i(\hat{p}_1 + \hat{x}_{in}). \tag{4.58}$$

Note that it holds independently of the initial states in mode in and mode 1. In the ideal case where the resource mode is initially in $|p = 0\rangle$ or p-squeezed state with $r \to \infty$, the additional term $\hat{p}_1$ vanishes, and an ideal operation can be achieved.

4.2.3.4 Controlled-Z Gate with Gain g

Although we usually use the unity-gain controlled-Z gate for the elementary circuit, we also use a controlled-Z gate with gain g on rare occasions (Fig. 4.15). In this case, the feed-forward operation should be changed to $\hat{X}_1(-\frac{1}{g}s_{in})$. The input-output relationship becomes

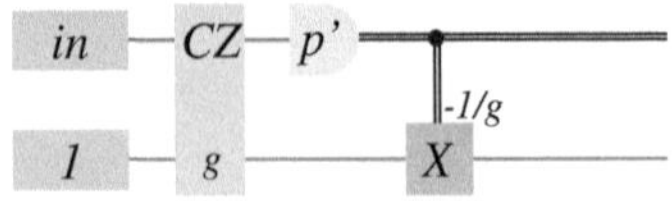

Fig. 4.15 Elementary circuit for one-way quantum computation with non-unity-gain controlled-Z gate

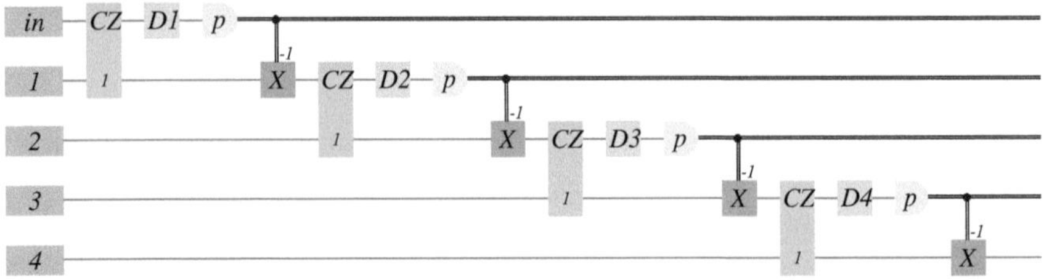

Fig. 4.16 Four-step one-mode teleportation circuit 1

$$\hat{x}_{out} + i\hat{p}_{out} = -\frac{1}{g}(\hat{p}_{in} + f'(\hat{x}_{in})) + i(\hat{p}_1 + g\hat{x}_{in}). \tag{4.59}$$

Note that it holds independently of the initial states in mode in and mode 1. In the ideal case where the resource mode is initially in $|p = 0\rangle$ or p-squeezed state with $r \to \infty$, the additional term $\hat{p}_1$ vanishes, and an ideal operation can be achieved.

4.2.4 Multi-Step One-Mode Teleportation Circuit and Cluster States

In general, an arbitrary quantum computation can be achieved by combining several elementary gates (Sect. 3.6.1). In this subsection, we show how multi-step quantum computation can be implemented in one-way quantum computation.

4.2.4.1 Multi-Step One-Mode Teleportation Circuit and Its Output State

Figure 4.16 shows a four-step one-mode teleportation circuit. Although we will show a four-step case as an example, the following discussion can be easily extended to the general case.

It is obvious that the output state is given by

$$|\Psi\rangle^{(out)} = \hat{D}_4(\hat{p})\hat{F}\hat{D}_3(\hat{p})\hat{F}\hat{D}_2(\hat{p})\hat{F}\hat{D}_1(\hat{p})\hat{F}|\psi\rangle \tag{4.60}$$

$$= \hat{F}\hat{D}_4(\hat{x})\hat{F}\hat{D}_3(\hat{x})\hat{F}\hat{D}_2(\hat{x})\hat{F}\hat{D}_1(\hat{x})|\psi\rangle. \tag{4.61}$$

Note that the subscripts are utilized to discriminate operators. Equation (4.61) can be transformed into

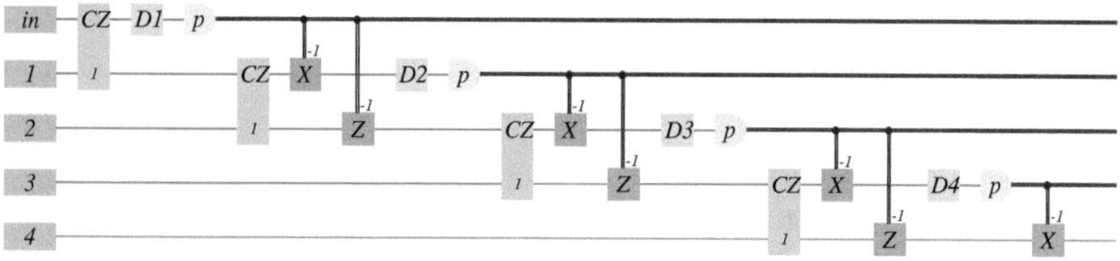

Fig. 4.17 Four-step one-mode teleportation circuit 2

$$\hat{F}\hat{D}_4(\hat{x})\hat{F}\hat{D}_3(\hat{x})\hat{F}\hat{D}_2(\hat{x})\hat{F}\hat{D}_1(\hat{x}) = \hat{F}\hat{D}_4(\hat{x})\hat{F}\hat{D}_3(\hat{x})\hat{F}^2\hat{F}^\dagger\hat{D}_2(\hat{x})\hat{F}\hat{D}_1(\hat{x}) \quad (4.62)$$

$$= \hat{F}\hat{D}_4(\hat{x})\hat{F}\hat{F}^2\hat{F}^{\dagger 2}\hat{D}_3(\hat{x})\hat{F}^2\hat{D}_2(-\hat{p})\hat{D}_1(\hat{x}) \quad (4.63)$$

$$= \hat{F}\hat{F}^3\hat{F}^{\dagger 3}\hat{D}_4(\hat{x})\hat{F}^3\hat{D}_3(-\hat{x})\hat{D}_2(-\hat{p})\hat{D}_1(\hat{x}) \quad (4.64)$$

$$= \hat{F}^4\hat{D}_4(\hat{p})\hat{D}_3(-\hat{x})\hat{D}_2(-\hat{p})\hat{D}_1(\hat{x}). \quad (4.65)$$

Thus, operators $\hat{D}_k(\hat{x})$ and $\hat{D}_k(\hat{p})$ can be performed by turns. In general, the output state becomes

$$\hat{F}\hat{D}_n(\hat{x})\cdots\hat{F}\hat{D}_2(\hat{x})\hat{F}\hat{D}_1(\hat{x}) = \begin{cases} \hat{F}^n\hat{D}_n(\hat{x})\cdots\hat{D}_2(-\hat{p})\hat{D}_1(\hat{x}), \ (n = 4m+1) \\ \hat{F}^n\hat{D}_n(-\hat{p})\cdots\hat{D}_2(-\hat{p})\hat{D}_1(\hat{x}), \ (n = 4m+2) \\ \hat{F}^n\hat{D}_n(-\hat{x})\cdots\hat{D}_2(-\hat{p})\hat{D}_1(\hat{x}), \ (n = 4m+3) \\ \hat{F}^n\hat{D}_n(\hat{p})\cdots\hat{D}_2(-\hat{p})\hat{D}_1(\hat{x}), \ (n = 4m) \end{cases}. \quad (4.66)$$

4.2.4.2 Transformation and Cluster State

We consider transformation of Fig. 4.16. By using

$$\hat{C}_{Zij}\hat{X}_k(s) = \hat{X}_k(s)\hat{Z}_{k'}(s)\hat{C}_{Zij}, \quad (k, k') = (i, j), (j, i) \quad (4.67)$$

for a controlled-Z operator $\hat{C}_{Zij}$ and position and momentum displacement operators $\hat{X}_k(s)$ and $\hat{Z}_{k'}(s)$, we can transform Fig. 4.16 into Fig. 4.17.

Next, since $\hat{C}_{Zij}$ commutes with both $\hat{Z}_k(s)$ and $\hat{C}_{Zmn}$, we can transform Fig. 4.17 into Fig. 4.18.

The procedure in Fig. 4.18 is summarized as follows:

- A resource state generated by combining four zero-eigenstates of momentum operators ($|p = 0\rangle$) via three controlled-Z gates is prepared in advance (modes 1 to 4).

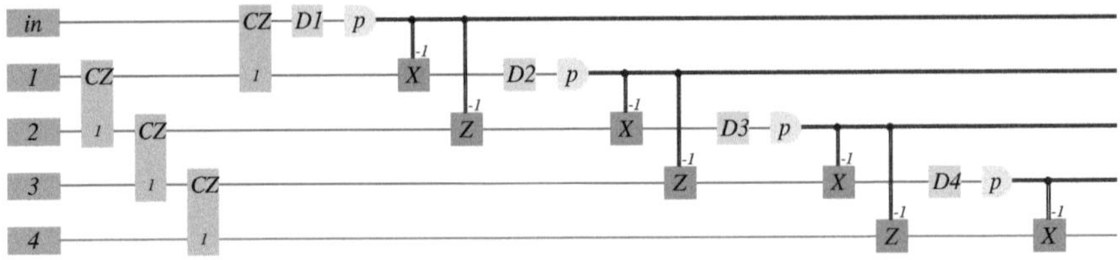

Fig. 4.18 Four-step one-mode teleportation circuit 3

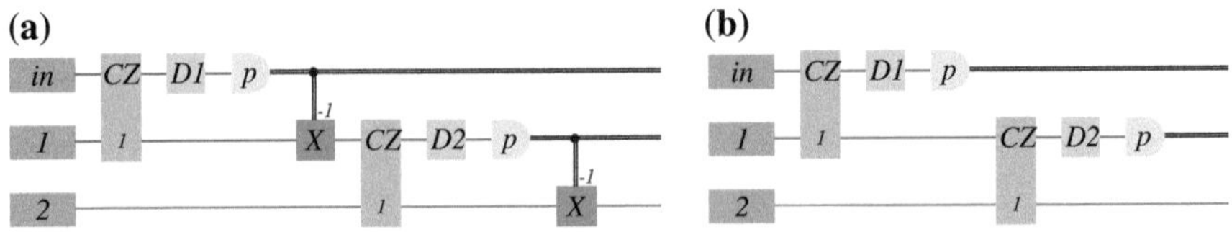

Fig. 4.19 Postponement of feed-forward operations. **a** Postponement of feed-forward operations 1. **b** Postponement of feed-forward operations 2

- The input state is coupled with the resource state via a controlled-Z gate.
- Measurements and feed-forward operations are performed.
- The output state is given by Eq. (4.61).

The resource state generated by combining zero eigenstates of momentum operators ($|p = 0\rangle$) via controlled-Z gates is called a *cluster state*. Many shapes of cluster states can be defined based on the network of controlled-Z gates. In this example, the cluster state is called the *four-mode linear cluster state* since four $|p = 0\rangle$ are entangled linearly.

4.2.4.3 Postponement of Feed-Forward Operations

We then consider a change of the order of the feed-forward operations and measurements. For this purpose, we start with the two-step circuit shown in Fig. 4.19a. We next consider a circuit where the feed-forward operations are removed (Fig. 4.19b). We then acquire the feed-forward operator so that the output state does not depend on measurement results.

It is obvious that the output state in Fig. 4.19b is given by

$$|\Psi\rangle^{(out)} = \hat{X}(s_1)\hat{F}\hat{D}_2(\hat{x})\hat{X}(s_{in})\hat{F}\hat{D}_1(\hat{x})|\psi\rangle. \tag{4.68}$$

Note that it depends on the measurement results s_{in}, s_1 since the feed-forward operations are removed. We omit $(\hat{x})$ from $\hat{D}_k(\hat{x})$, for simplicity.

In the following, we classify $\hat{D}_2$ into the following two cases: the case where $\hat{D}_2$ is a member of the Clifford group $C(1)$; and the case where $\hat{D}_2$ is not a member of $C(1)$.

$$\hat{D}_2 \in C(1).$$

When $\hat{D}_2$ is a member of the Clifford group, $\hat{F}\hat{D}_2$ is also a member of the Clifford group. Thus, the operator $\hat{P}(s_{in})$:

$$\hat{F}\hat{D}_2\hat{X}(s_{in})(\hat{F}\hat{D}_2)^\dagger = \hat{P}(s_{in}) \tag{4.69}$$

becomes a member of the Pauli group (Sect. 3.4.4). As a result, the output state is equivalent to

$$\hat{X}(s_1)\hat{P}(s_{in})\hat{F}\hat{D}_2\hat{F}\hat{D}_1|\psi\rangle. \tag{4.70}$$

It shows that the original operation $(\hat{F}\hat{D}_2\hat{F}\hat{D}_1)$ before the transformation (Fig. 4.19a) can be achieved by using the same measurement bases and another feed-forward operation $\hat{P}^\dagger(s_{in})\hat{X}^\dagger(s_1)$ in the Pauli group.

Similarly, the output state of the N-step case is given by

$$\hat{X}(s_{N-1})\hat{F}\hat{D}_N \cdots \hat{X}(s_{in})\hat{F}\hat{D}_1|\psi\rangle = \hat{P}(s_{in}, \ldots, s_{N-1})\hat{F}\hat{D}_N \cdots \hat{F}\hat{D}_2\hat{F}\hat{D}_1|\psi\rangle. \tag{4.71}$$

Therefore, all measurements can be performed simultaneously, where measurement bases are not changed during transformation of the circuit. After all measurements have been finished, all feed-forward operations can be performed simultaneously. Although the feed-forward operations are changed during the transformation of the circuit, they are still members of the Pauli group. These properties are called the *Gaussian parallelism*. We will show an example of it in Sect. 9.2.7.

$$\hat{D}_2 \notin C(1).$$

When $\hat{D}_2$ is not a member of the Clifford group, $\hat{F}\hat{D}_2$ is not a member of the Clifford group as well. Thus, the operator $\hat{U}(s_{in})$:

$$\hat{F}\hat{D}_2\hat{X}(s_{in})(\hat{F}\hat{D}_2)^\dagger = \hat{U}(s_{in}) \tag{4.72}$$

is not a member of the Pauli group. By using $\hat{U}(s_{in})$, the output state becomes

$$\hat{X}(s_1)\hat{U}(s_{in})\hat{F}\hat{D}_2\hat{F}\hat{D}_1|\psi\rangle. \tag{4.73}$$

It shows that the operation $\hat{F}\hat{D}_2\hat{F}\hat{D}_1$ which is the same to that before the transformation is achieved by using the same measurement bases and different feed-forward operation $\hat{U}^\dagger(s_{in})\hat{X}^\dagger(s_1)$. However, the feed-forward operation is no longer a member of the Pauli group. Although $\hat{U}(s_{in})$ is in the Clifford group if $\hat{D}_2$ is in the group C_3 Eq. (4.19), $\hat{U}(s_{in})$ is not in the Clifford group if $\hat{D}_2$ is not in the group C_3.

We consider another transformation of the equation. We assume that $\hat{F}\hat{D}_2$ is not a member of the Clifford group. By using $\hat{V}(s_{in})$:

$$\hat{F}^{\dagger}\hat{X}^{\dagger}(s_{in})\hat{F}\hat{D}_2\hat{X}(s_{in}) = \hat{V}(s_{in}), \tag{4.74}$$

the output of the second step is given by

$$\hat{X}(s_1)\hat{X}(s_{in})\hat{F}\hat{V}(s_{in})\hat{F}\hat{D}_1|\psi\rangle, \tag{4.75}$$

where $\hat{V}(s_{in})$ is an operator which is not a member of the Clifford group. The revised feed-forward operator $\hat{X}^{\dagger}(s_{in})\hat{X}^{\dagger}(s_1)$ is a member of the Pauli group in this case. However, in order to achieve the same operation $\hat{F}\hat{D}_2\hat{F}\hat{D}_1$, we have to choose a different measurement basis which is determined by $\hat{V}(s_{in})$ at the second step. It also depends on the measurement result s_{in} in the first step.

By performing the feed-forward operation in each step, we can determine measurement basis independently of the other steps. In addition, all feed-forward operations are members of the Pauli group. Therefore, it is preferred to perform feed-forward operations in each step when we perform non-Clifford operations.

References

1. Bennett, C.H., Bernstein, H.J., Popescu, S., Schumacher, B.: Concentrating partial entanglement by local operations. Phys. Rev. A **53**, 2046 (1996)
2. Bennett, C.H., DiVincenzo, D.P., Smolin, J.A., Wootters, W.K.: Mixed-state entanglement and quantum error correction. Phys. Rev. A **54**, 3824 (1996)

Chapter 5
Cluster States and One-Way Quantum Computation

5.1 Cluster States

5.1.1 Definition by State Vector

Consider an n-vertex weighted and undirected graph without self-loops. Each node is labeled by integers $1, \ldots, n$. We assume that the weight of edge between two modes j and k is real and equal to g_{jk}. We define E to be the set of all edges. The weighted cluster state associated with this graph is defined to be

$$|\Psi_G\rangle = \hat{C}_{Ztotal}|p = 0\rangle^{\otimes n}, \tag{5.1}$$

where $\hat{C}_{Ztotal}$ is given by

$$\hat{C}_{Ztotal} = \prod_{\{j,k\}\in E} \hat{C}_{Z\,jk}(g_{jk}) = \prod_{\{j,k\}\in E} e^{\frac{i}{\hbar} g_{jk}\hat{x}_j\hat{x}_k}. \tag{5.2}$$

It is called an unweighted cluster state if all g_{jk} satisfy $g_{jk} = 1$. Since all controlled-Z gates commute, we do not have to specify the order of controlled-Z gates.

In general, a graph is represented by its adjacency matrix A. In this case, A is an $n \times n$ real symmetric matrix with $A_{jk} = g_{jk}$. Since the graph does not have self-loops, all diagonal elements A_{jj} satisfy $A_{jj} = 0$. By using the adjacency matrix, $\hat{C}_{Ztotal}$ is given by

$$\hat{C}_{Ztotal} = \prod_{\{j,k\}\in G} \hat{C}_{Zjk}(g_{jk}) = \exp\left[\frac{i}{2\hbar}\hat{\boldsymbol{x}}^T A\hat{\boldsymbol{x}}\right], \quad \hat{\boldsymbol{x}} = (\hat{x}_1, \ldots, \hat{x}_n)^T. \tag{5.3}$$

5.1.2 Stabilizers and Nullifiers

Since the stabilizer of the zero eigenstate $|p = 0\rangle$ of the momentum operator $\hat{p}$ is given by $\{\hat{X}(s) = e^{-\frac{i}{\hbar}s\hat{p}}\}$, the stabilizer of $|p = 0\rangle^{\otimes n}$ is given by $\{\hat{X}_a(s_a) =$

R. Ukai, *Multi-Step Multi-Input One-Way Quantum Information Processing with Spatial and Temporal Modes of Light*, Springer Theses,
DOI 10.1007/978-4-431-55019-8_5

$e^{-\frac{i}{\hbar}s_a\hat{p}_a}\}$. By performing $\hat{C}_{Ztotal}$ on $|p=0\rangle^{\otimes n}$, the stabilizer becomes

$$\{\hat{C}_{Ztotal}\hat{X}_a(s_a)\hat{C}^{\dagger}_{Ztotal}\}. \tag{5.4}$$

By using

$$\hat{C}_{Ztotal}\hat{p}_a\hat{C}^{\dagger}_{Ztotal} = \hat{p}_a - (A\hat{x})_a, \tag{5.5}$$

we get the stabilizer of the cluster state:

$$\{\hat{C}_{Ztotal}\hat{X}_a(s_a)\hat{C}^{\dagger}_{Ztotal}\} = \{e^{-\frac{i}{\hbar}s_a(\hat{p}_a-(A\hat{x})_a)}\} = \{\hat{X}_a(s_a)\prod_k \hat{Z}_k(s_a g_{ak})\}, \tag{5.6}$$

where we have utilized $A_{ak} = g_{ak}$.

Although the stabilizer is well formulated above, we sometimes rewrite it as

$$\{\hat{C}_{Ztotal}\hat{X}_a(s_a)\hat{C}^{\dagger}_{Ztotal}\} = \{\hat{X}_a(s_a)\prod_{k\in N(a)} \hat{Z}_k(s_a g_{ak})\}, \tag{5.7}$$

where $N(a)$ shows the set of vertices which are connected to the vertex a with non-zero weight ($g_{ak} \neq 0$) in the graph. We refer to it as the set of nearest-neighbor modes of mode a.

We next consider the nullifiers of the cluster state. Since the nullifiers of $|p=0\rangle^{\otimes n}$ is $\{\hat{p}_a\}$, the nullifiers of the cluster state is given by

$$\hat{\delta}_a = \hat{C}_{Ztotal}\hat{p}_a\hat{C}^{\dagger}_{Ztotal} = \hat{p}_a - (A\hat{x})_a = \hat{p}_a - \sum_k A_{ak}\hat{x}_k = \hat{p}_a - \sum_k g_{ak}\hat{x}_k = \hat{p}_a - \sum_{k\in N(a)} g_{ak}\hat{x}_k, \tag{5.8}$$

or equivalently,

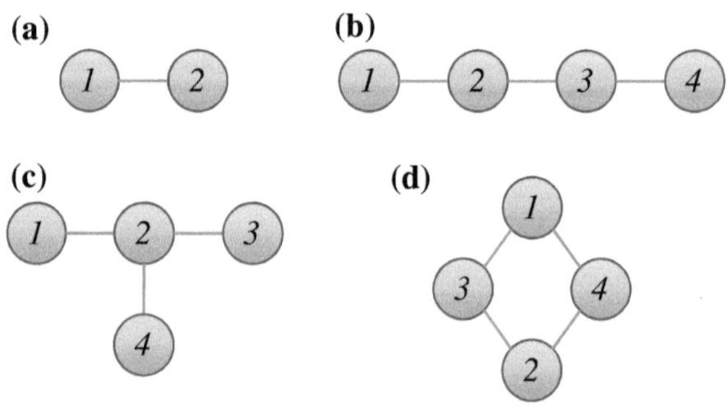

Fig. 5.1 Examples of cluster state. **a** Two-mode cluster state, **b** four-mode linear cluster state, **c** four-mode T-shaped cluster state, **d** four-mode diamond-shaped cluster state

$$\hat{\delta} = \hat{p} - A\hat{x}. \tag{5.9}$$

Equation (5.9) gives n independent nullifiers, each of which commutes with the others. The cluster state is the simultaneous zero eigenstate of these n nullifiers.

5.1.3 Examples of Cluster States

Figure 5.1 shows several graphs of unweighted cluster states.

5.1.3.1 Two-Mode Cluster State

The simplest cluster state is the two-mode cluster state $|\Psi_2\rangle$.

$$|\Psi_2\rangle = \hat{C}_{Z12}|p=0\rangle_1|p=0\rangle_2 = e^{\frac{i}{\hbar}\hat{x}_1\hat{x}_2}|p=0\rangle_1|p=0\rangle_2. \tag{5.10}$$

By using

$$|x\rangle = \frac{1}{\sqrt{2\pi\hbar}}\int_{-\infty}^{\infty} e^{-\frac{ixp}{\hbar}}|p\rangle \mathrm{d}p, \quad |p\rangle = \frac{1}{\sqrt{2\pi\hbar}}\int_{-\infty}^{\infty} e^{\frac{ixp}{\hbar}}|x\rangle \mathrm{d}x, \tag{5.11}$$

we get

$$|\Psi_2\rangle = \frac{1}{\sqrt{2\pi\hbar}}\int_{-\infty}^{\infty} \mathrm{d}a|x_1=a\rangle|p_2=a\rangle = \frac{1}{\sqrt{2\pi\hbar}}\int_{-\infty}^{\infty} \mathrm{d}b|p_1=b\rangle|x_2=b\rangle. \tag{5.12}$$

The nullifiers of the two-mode cluster state are given by

$$\hat{p}_1 - \hat{x}_2 = 0, \quad \hat{p}_2 - \hat{x}_1 = 0. \tag{5.13}$$

They satisfy

$$(\hat{p}_1 - \hat{x}_2)|\Psi_2\rangle = 0, \quad (\hat{p}_2 - \hat{x}_1)|\Psi_2\rangle = 0. \tag{5.14}$$

5.1.3.2 Three-Mode Linear Cluster State

The three-mode linear cluster state $|\Psi_{3L}\rangle$ is given by

$$|\Psi_{3L}\rangle = \hat{C}_{Z12}\hat{C}_{Z23}|p=0\rangle_1|p=0\rangle_2|p=0\rangle_3 \tag{5.15}$$

$$= \frac{1}{2\pi\hbar}\int_{-\infty}^{\infty} da \int_{-\infty}^{\infty} db|x_1 = a\rangle|p_2 = a+b\rangle|x_3 = b\rangle \tag{5.16}$$

$$= \frac{1}{\sqrt{2\pi\hbar}}\int_{-\infty}^{\infty} da|p_1 = a\rangle|x_2 = a\rangle|p_3 = a\rangle. \tag{5.17}$$

Its nullifiers are

$$\hat{p}_1 - \hat{x}_2 = 0, \quad \hat{p}_2 - \hat{x}_1 - \hat{x}_3 = 0, \quad \hat{p}_3 - \hat{x}_2 = 0. \tag{5.18}$$

5.1.3.3 Four-Mode Linear Cluster State

The four-mode linear cluster state $|\Psi_{4L}\rangle$ is given by

$$|\Psi_{4L}\rangle = \hat{C}_{Z12}\hat{C}_{Z23}\hat{C}_{Z34}|p=0\rangle_1|p=0\rangle_2|p=0\rangle_3|p=0\rangle_4 \tag{5.19}$$

$$= \frac{1}{2\pi\hbar}\int_{-\infty}^{\infty} db \int_{-\infty}^{\infty} dce^{\frac{i}{\hbar}bc}|p_1 = b\rangle|x_2 = b\rangle|x_3 = c\rangle|p_4 = c\rangle \tag{5.20}$$

$$= \frac{1}{\sqrt{2\pi\hbar}^3}\int_{-\infty}^{\infty} da \int_{-\infty}^{\infty} dc \int_{-\infty}^{\infty} dde^{\frac{i}{\hbar}cd}|x_1 = a\rangle|p_2 = a+c\rangle|x_3 = c\rangle|x_4 = d\rangle \tag{5.21}$$

$$= \frac{1}{\sqrt{2\pi\hbar}^3}\int_{-\infty}^{\infty} da \int_{-\infty}^{\infty} db \int_{-\infty}^{\infty} dde^{\frac{i}{\hbar}ab}|x_1 = a\rangle|x_2 = b\rangle|p_3 = b+d\rangle|x_4 = d\rangle. \tag{5.22}$$

Its nullifiers are

$$\hat{p}_1 - \hat{x}_2 = 0, \quad \hat{p}_2 - \hat{x}_1 - \hat{x}_3 = 0, \quad \hat{p}_3 - \hat{x}_2 - \hat{x}_4 = 0, \quad \hat{p}_4 - \hat{x}_3 = 0. \tag{5.23}$$

5.1.3.4 Four-Mode T-Shaped Cluster State

The nullifiers of the four-mode T-shaped cluster state are given by

$$\hat{p}_1 - \hat{x}_2 = 0, \quad \hat{p}_2 - \hat{x}_1 - \hat{x}_3 - \hat{x}_4 = 0, \quad \hat{p}_3 - \hat{x}_2 = 0, \quad \hat{p}_4 - \hat{x}_2 = 0. \tag{5.24}$$

5.1.3.5 Four-Mode Diamond-Shaped Cluster State

The nullifiers of the four-mode diamond-shaped cluster state are given by

$$\hat{p}_1 - \hat{x}_3 - \hat{x}_4 = 0, \quad \hat{p}_2 - \hat{x}_3 - \hat{x}_4 = 0, \quad \hat{p}_3 - \hat{x}_1 - \hat{x}_2 = 0, \quad \hat{p}_4 - \hat{x}_1 - \hat{x}_2 = 0. \tag{5.25}$$

5.1.4 Entanglement Criteria for Cluster States

5.1.4.1 General Theory

We consider entanglement criteria for n-mode cluster states. They are directly acquired as examples of the extended van Loock-Furusawa criteria (Sect. 3.7.4).

Consider an n-mode cluster state which is associated with an undirected and weighted graph described by an $n \times n$ adjacency matrix A. The nullifiers are given by $\{\boldsymbol{\delta}\} = \{\hat{\boldsymbol{p}} - A\hat{\boldsymbol{x}}\}$. We assume that i and j are labels of two vertices which are located next to each other. In general, two nullifiers δ_i and δ_j are reformed as

$$\hat{\delta}_i = \hat{p}_i - A_{ij}\hat{x}_j - \sum_m A_{im}\hat{x}_m - \sum_t A_{it}\hat{x}_t, \tag{5.26}$$

$$\hat{\delta}_j = \hat{p}_j - A_{ji}\hat{x}_i - \sum_n A_{jn}\hat{x}_n - \sum_t A_{jt}\hat{x}_t, \tag{5.27}$$

where $m \in N_i \cap \overline{N_i \cap N_j} \setminus j$, $n \in N_j \cap \overline{N_i \cap N_j} \setminus i$, and $t \in N_i \cap N_j$. We define $h_s(\hat{x}_s, \hat{p}_s)$ and $g_s(\hat{x}_s, \hat{p}_s)$ to be the terms consisting of mode s in δ_i and δ_j, respectively. Using $h_s(\hat{x}_s, \hat{p}_s)$ and $g_s(\hat{x}_s, \hat{p}_s)$, two nullifiers are given by $\hat{\delta}_i = \sum_{s\in B} h_s(\hat{x}_s, \hat{p}_s)$ and $\hat{\delta}_j = \sum_{s\in B} g_s(\hat{x}_s, \hat{p}_s)$. We apply the necessary condition for separability shown in Eq. (3.192) to the two nullifiers δ_i and δ_j. If a quantum state is separable into $\{B_k\}$, the inequality

$$\left\langle \Delta^2\hat{\delta}_i \right\rangle + \left\langle \Delta^2\hat{\delta}_j \right\rangle \geq \sum_{k=1}^{l} \left| \sum_{s\in B_k} [h_s(\hat{x}_s, \hat{p}_s), g_s(\hat{x}_s, \hat{p}_s)] \right| \tag{5.28}$$

$$= \begin{cases} 2\hbar|A_{ij}| & i \in B_k,\ j \in B_{k'} \\ 0 & i, j \in B_k \end{cases} \tag{5.29}$$

is satisfied. Here, we have used commutation relations between $h_s(\hat{x}_s, \hat{p}_s)$ and $g_s(\hat{x}_s, \hat{p}_s)$:

$$[h_s(\hat{x}_s, \hat{p}_s), g_s(\hat{x}_s, \hat{p}_s)] = \begin{cases} i\hbar A_{ji} & s = i \\ -i\hbar A_{ij} & s = j \\ 0 & s \neq i, j \end{cases} . \tag{5.30}$$

Note that this inequality holds for any mode set $\{B_k\}$. It follows that, if the sum of variances for a quantum state satisfies $\langle \Delta^2 \hat{\delta}_i \rangle + \langle \Delta^2 \hat{\delta}_j \rangle < 2\hbar |A_{ij}|$, two modes i and j are not separable ($i, j \in B_k$ for some k). We can derive the same sufficient condition for inseparability for each pair of nearest neighbor modes i and j. Therefore, we finally get the following sufficient condition for full inseparability of the cluster state.

Full inseparability of cluster state—Consider an n vertex undirected and weighted graph G with an adjacency matrix A. An n-mode quantum state is fully entangled in the form of the graph G if it satisfies $n-1$ inequalities $\langle \Delta^2 \hat{\delta}_i \rangle + \langle \Delta^2 \hat{\delta}_j \rangle < 2\hbar |A_{ij}|$ for $j \in N_i$, where $\hat{\boldsymbol{\delta}} = \hat{\boldsymbol{p}} - A\hat{\boldsymbol{x}}$ represent the nullifiers of the cluster state associated with G, while N_j represents the set of nearest neighbor modes of mode j. Note that the mode set $C_k = \{i, j\}_k$ for the k-th inequality should be chosen so that $|A_{ij}| \neq 0$ and the union of C_k for $k = 1, \ldots, n-1$ is equivalent to $B = \{1, \ldots, n\}$.

By using this statement, we can also acquire the following more severe condition for full inseparability.

Full inseparability of cluster state—An n-mode quantum state is fully entangled in the form of the graph G if it satisfies n inequalities $\langle \Delta^2 \hat{\delta}_a \rangle < \hbar \min_{b \in N_a} (|A_{ab}|)$ for all $a \in B$.

5.1.4.2 Examples

Four-Mode Linear Cluster State

We consider the (unweighted) four-mode linear cluster state. The adjacency matrix A is given by

$$A = \begin{pmatrix} 0 & 1 & 0 & 0 \\ 1 & 0 & 1 & 0 \\ 0 & 1 & 0 & 1 \\ 0 & 0 & 1 & 0 \end{pmatrix}. \tag{5.31}$$

The nullifiers are

$$\hat{\delta}_1 = \hat{p}_1 - \hat{x}_2, \quad \hat{\delta}_2 = \hat{p}_2 - \hat{x}_1 - \hat{x}_3, \quad \hat{\delta}_3 = \hat{p}_3 - \hat{x}_2 - \hat{x}_4, \quad \hat{\delta}_4 = \hat{p}_4 - \hat{x}_3. \tag{5.32}$$

We choose the set of modes C_k as

$$C_1 = \{1, 2\}, \quad C_2 = \{2, 3\}, \quad C_3 = \{3, 4\}. \tag{5.33}$$

A quantum state is fully entangled in the form of the four-mode linear cluster state if it satisfies

$$\langle \Delta^2(\hat{p}_1 - \hat{x}_2) \rangle + \langle \Delta^2(\hat{p}_2 - \hat{x}_1 - \hat{x}_3) \rangle < 2\hbar, \tag{5.34}$$

$$\langle \Delta^2(\hat{p}_2 - \hat{x}_1 - \hat{x}_3) \rangle + \langle \Delta^2(\hat{p}_3 - \hat{x}_2 - \hat{x}_4) \rangle < 2\hbar, \tag{5.35}$$

$$\langle \Delta^2(\hat{p}_3 - \hat{x}_2 - \hat{x}_4) \rangle + \langle \Delta^2(\hat{p}_4 - \hat{x}_3) \rangle < 2\hbar. \tag{5.36}$$

Four-Mode T-Shaped Cluster State

We consider the (unweighted) four-mode T-shaped cluster state. The adjacency matrix A is given by

$$A = \begin{pmatrix} 0 & 1 & 0 & 0 \\ 1 & 0 & 1 & 1 \\ 0 & 1 & 0 & 0 \\ 0 & 1 & 0 & 0 \end{pmatrix}. \tag{5.37}$$

The nullifiers are

$$\hat{\delta}_1 = \hat{p}_1 - \hat{x}_2, \quad \hat{\delta}_2 = \hat{p}_2 - \hat{x}_1 - \hat{x}_3 - \hat{x}_4, \quad \hat{\delta}_3 = \hat{p}_3 - \hat{x}_2, \quad \hat{\delta}_4 = \hat{p}_4 - \hat{x}_2. \tag{5.38}$$

We choose the set of modes C_k as

$$C_1 = \{1, 2\}, \quad C_2 = \{3, 2\}, \quad C_3 = \{4, 2\}. \tag{5.39}$$

A quantum state is fully entangled in the form of the four-mode T-shaped cluster state if it satisfies

$$\langle \Delta^2(\hat{p}_1 - \hat{x}_2) \rangle + \langle \Delta^2(\hat{p}_2 - \hat{x}_1 - \hat{x}_3 - \hat{x}_4) \rangle < 2\hbar, \tag{5.40}$$

$$\langle \Delta^2(\hat{p}_3 - \hat{x}_2) \rangle + \langle \Delta^2(\hat{p}_2 - \hat{x}_1 - \hat{x}_3 - \hat{x}_4) \rangle < 2\hbar, \tag{5.41}$$

$$\langle \Delta^2(\hat{p}_4 - \hat{x}_2) \rangle + \langle \Delta^2(\hat{p}_2 - \hat{x}_1 - \hat{x}_3 - \hat{x}_4) \rangle < 2\hbar. \tag{5.42}$$

5.2 Generation of Cluster States

5.2.1 Canonical Cluster States

A straight-forward scheme to build cluster states is the canonical way [1, 2], which is directly given by the definition of cluster states.

Originally, a cluster state is defined to be the state which is generated by entangling n copies of $|p = 0\rangle$ with a network of C_Z gates. Since a $|p = 0\rangle$ requires infinite amount of energy and is an unphysical state, it is approximated by a p-squeezed state. We define

$$\hat{x}_a^{(r)} + i\hat{p}_a^{(r)} = e^{r_a}\hat{x}_a^{(0)} + ie^{-r_a}\hat{p}_a^{(0)}, \quad a = 1, \ldots, n, \tag{5.43}$$

as annihilation operators of n squeezed states. Here, $\hat{x}_a^{(0)} + i\hat{p}_a^{(0)}$ is the annihilation operator of the vacuum state in mode a, while r_a represents the squeezing parameter of mode a. An approximate cluster state is acquired by entangling them with the C_Z gate network. We assume that the ideal C_Z gate is applicable. As a result, the annihilation operators of the cluster state is given by

$$\hat{x}_a^{(c)} + i\hat{p}_a^{(c)} = e^{r_a}\hat{x}_a^{(0)} + i\Big[e^{-r_a}\hat{p}_a^{(0)} + \sum_k A_{ak}e^{r_k}\hat{x}_k^{(0)}\Big], \tag{5.44}$$

for $a = 1, \ldots, n$. The state generated by using the above procedure is called a canonical cluster state. With finite levels of squeezing, the nullifier terms:

$$\hat{\delta}_a = \hat{p}_a^{(c)} - \sum_k A_{ak}\hat{x}_k^{(c)} = e^{-r_a}\hat{p}_a^{(0)}, \tag{5.45}$$

have non-zero values. To the contrary, they vanish in the limit of $r_a \to \infty$, thus an ideal cluster state is acquirable.

By applying the Bloch-Messiah reduction, an arbitrary two-mode Gaussian operation, including the C_Z gate, is implemented by two single-mode squeezers sandwiched between two beam splitters [3, 4]. Since concatenation of Gaussian operations is also a Gaussian operation, the C_Z gate network which generate the cluster state is a member of Gaussian operations on n modes. Thus it can be achieved by an n-mode beam-splitter network, followed by n single-mode squeezers, and another n-mode beam-splitter network.

5.2.2 Gaussian Cluster States

5.2.2.1 Generation of Gaussian Cluster States

Shortly after the original proposal of CV cluster states, Peter et al. showed that an approximate cluster state, called the Gaussian cluster state, can be efficiently generated by combining n copies of $|p = 0\rangle$ by using a network of beam splitters [2]. The beam splitter network should be carefully chosen so that the nullifiers $\hat{\delta}_a$ of the cluster state become zero. As is the case with the canonical scheme, $|p = 0\rangle$ is approximated by a p-squeezed state. In this case, $\hat{\delta}_a$ consists only of momentum operators of the initial squeezed states.

We define

$$\hat{\boldsymbol{a}}^{(r)} = \hat{\boldsymbol{x}}^{(r)} + i\hat{\boldsymbol{p}}^{(r)} \tag{5.46}$$

as annihilation operators of n-mode p-squeezed states, where $\hat{\boldsymbol{x}}^{(r)} = (\hat{x}_1^{(r)}, \dots, \hat{x}_n^{(r)})^T$ and $\hat{\boldsymbol{p}}^{(r)} = (\hat{p}_1^{(r)}, \dots, \hat{p}_n^{(r)})^T$ are vectors of position and momentum operators, respectively. For simplicity, we assume that each squeezed state has the same squeezing parameter r, thus $\hat{\boldsymbol{x}}^{(r)} = e^r\hat{\boldsymbol{x}}^{(0)}$ and $\hat{\boldsymbol{p}}^{(r)} = e^{-r}\hat{\boldsymbol{p}}^{(0)}$, where $\hat{\boldsymbol{x}}^{(0)}$ and $\hat{\boldsymbol{p}}^{(0)}$ are vectors of position and momentum operators of n-mode vacuum states, respectively. In general, a beam splitter network on n-mode quantum state is represented as a linear transformation of n-mode annihilation operators. Thus the annihilation operators of the cluster state is given by

$$\hat{\boldsymbol{x}}^{(c)} + i\hat{\boldsymbol{p}}^{(c)} = U(\hat{\boldsymbol{x}}^{(r)} + i\hat{\boldsymbol{p}}^{(r)}) \tag{5.47}$$

$$= \Re U\hat{\boldsymbol{x}}^{(r)} - \Im U\hat{\boldsymbol{p}}^{(r)} + i\big(\Im U\hat{\boldsymbol{x}}^{(r)} + \Re U\hat{\boldsymbol{p}}^{(r)}\big). \tag{5.48}$$

In order to preserve the commutation relations, the $n \times n$ matrix U is a unitary matrix: $UU^\dagger = I$. The values of nullifiers become

$$\hat{\boldsymbol{p}}^{(c)} - A\hat{\boldsymbol{x}}^{(c)} = (\Im U - A\Re U)\hat{\boldsymbol{x}}^{(r)} + (\Re U + A\Im U)\hat{\boldsymbol{p}}^{(r)}. \tag{5.49}$$

The necessary and sufficient condition that they become zero in the limit of squeezing parameter $r \to \infty$ is

$$\Im U - A\Re U = O, \tag{5.50}$$

where O is the $n \times n$ zero matrix. In this case, the nullifiers become

$$\hat{\boldsymbol{\delta}} = \hat{\boldsymbol{p}}^{(c)} - A\hat{\boldsymbol{x}}^{(c)} = (\Re U + A\Im U)\hat{\boldsymbol{p}}^{(r)} \to 0, \quad r_a \to \infty. \tag{5.51}$$

We define an $n \times n$ matrix R and $1 \times n$ row vectors $\boldsymbol{\alpha}_j$ to be

$$\Re U = R = \begin{pmatrix} \boldsymbol{\alpha}_1 \\ \vdots \\ \boldsymbol{\alpha}_n \end{pmatrix}. \tag{5.52}$$

By using $\Im U - A\Re U = O$, we get

$$U = \Re U + i \Im U = (I + iA)R. \tag{5.53}$$

Since U is a unitary matrix, the condition

$$UU^\dagger = (I + iA)\ RR^T (I - iA) = (I + iA)S(I - iA) = I, \tag{5.54}$$

should be satisfied. Here,

$$S = RR^T = \begin{pmatrix} \alpha_1\alpha_1^T & \cdots & \alpha_1\alpha_n^T \\ \vdots & \ddots & \vdots \\ \alpha_n\alpha_1^T & \cdots & \alpha_n\alpha_n^T \end{pmatrix} \tag{5.55}$$

is a symmetric matrix, each member of which corresponds to inner product of $\boldsymbol{\alpha}$. Since S, A, and I are real matrices, both real and imaginary parts should be zero, thus

$$S + ASA = I, \quad SA - AS = O, \tag{5.56}$$

should be satisfied. This is the necessary and sufficient condition that the generated state become identical to the ideal cluster state in the limit of infinite squeezing. From these two equations, we get $(I + A^2)S = I$:

$$S + ASA = I \text{ and } SA - AS = O \implies (I + A^2)S = I \tag{5.57}$$

Since A is a real symmetric matrix, A^2 becomes a positive semi-definite matrix, thus $I + A^2$ is a positive definite matrix. Therefore, there exists the inverse matrix of $I + A^2$, which leads to

$$(I + A^2)S = I \iff S = (I + A^2)^{-1}. \tag{5.58}$$

Although the discussion above gives us a matrix $S_1 = (I + A^2)^{-1}$ which satisfies Eq. (5.56), uniqueness of S is not proven. However, if we assume that there exists another S_2 which satisfies Eq. (5.56), we get $(I + A^2)S_2 = I$ from Eq. (5.57), leading to $S_2 = (I + A^2)^{-1}$. It is a contradiction to the assumption that $S_1 \neq S_2$. Therefore, the unitary matrix S which satisfies Eq. (5.56) is uniquely determined by the adjacency matrix A which specifies the cluster state and Eq. (5.58).

In summary, an approximation of the ideal cluster state can be generated in the following procedure:

1. Get $S = (I + A^2)^{-1}$ for the adjacency matrix A.
2. Get $\boldsymbol{\alpha}_k$ from S, which shows the inner products of row vectors $\boldsymbol{\alpha}_k$.
3. R is determined by $\boldsymbol{\alpha}_k$, which leads to the unitary matrix $U = (I + iA)R$.
4. Linear transformations of n-mode annihilation operators represented by U is achieved by at most $\frac{n(n-1)}{2}$ beam splitters [3].

In the procedure above, the matrix S is uniquely determined by the matrix A. At the second step, $\boldsymbol{\alpha}_k$ is necessarily acquired from S. However, its uniqueness is not held. At the third step, R and U are uniquely determined by $\boldsymbol{\alpha}_k$. At the final step, U is necessarily decomposed into a beam splitter network, though its uniqueness is not held. Therefore, one necessarily acquire a nonunique unitary matrix U and a beam splitter network for any adjacency matrix A. It means that any cluster state can be generated in this scheme with several unitary transformations, or identically, beam splitter networks.

5.2.2.2 Covariance Matrices

We explore several properties of Gaussian cluster states. Since Gaussian cluster states are members of n-mode Gaussian states, they are uniquely described by their first-order moments and covariance matrices. In the procedure above, the initial resource state for a Gaussian cluster state is an n-mode squeezed state, whose first-order moments and covariance matrix are given by

$$\langle \hat{\boldsymbol{q}}^{(r)} \rangle = \mathbf{0}, \quad \Sigma^{(r)} = \frac{1}{4}\begin{pmatrix} e^{2r} I & O \\ O & e^{-2r} I \end{pmatrix}, \tag{5.59}$$

where $\mathbf{0}$ is the $2n$-dimensional zero vector, while I and O are the $n \times n$ identity and zero matrices, respectively. Note again that we have assumed that each squeezed state has the same squeezing parameter r. By applying the unitary transformation represented by a unitary matrix U on the vector of annihilation operators, the first-order moments and covariance matrix of the Gaussian cluster state become

$$\langle \hat{\boldsymbol{q}}^{(c)} \rangle = \mathbf{0}, \tag{5.60}$$

$$\begin{aligned} \Sigma^{(c)} &= U_{xxpp} \Sigma^{(r)} U_{xxpp}^T \\ &= \frac{1}{4}\begin{pmatrix} e^{-2r} I + 2S \sinh 2r & 2AS \sinh 2r \\ 2AS \sinh 2r & e^{2r} I - 2S \sinh 2r \end{pmatrix}, \end{aligned} \tag{5.61}$$

where $U_{xxpp} = \begin{pmatrix} R & -AR \\ AR & R \end{pmatrix}$ is the transformation matrix represented in $xxpp$ notation. Since $S = (I + A^2)^{-1}$ is uniquely determined by the adjacency matrix A, the first-order moments and covariance matrix of the Gaussian cluster state are also determined uniquely by A. Therefore, the property of Gaussian cluster states are uniquely determined by the shape of the graph represented by A, and the squeezing parameter r. It does not depend on how U is chosen, nor how U is decomposed into a beam splitter network. It follows that all Gaussian cluster states generated by different beam splitter networks for a specific graph are equivalent to each other. Thus they can be used in one-way quantum computations as resources without any discriminations.

Before we finish our discussion in this section, we refer to the covariance matrix of nullifiers of a Gaussian cluster state. Here, the nullifiers of the Gaussian cluster

Fig. 5.2 n-mode linear cluster state

state are given by

$$\hat{\boldsymbol{\delta}} = \hat{\boldsymbol{p}}^{(c)} - A\hat{\boldsymbol{x}}^{(c)} = (I + A^2)R\hat{\boldsymbol{p}}^{(r)}. \tag{5.62}$$

We define the covariance matrix of nullifiers to be $\langle \hat{\boldsymbol{\delta}}\hat{\boldsymbol{\delta}}^T \rangle$. By assuming that each squeezed state has the same squeezing parameter r, we get

$$\langle \hat{\boldsymbol{\delta}}\hat{\boldsymbol{\delta}}^T \rangle = (I + A^2)R\langle \hat{\boldsymbol{p}}^{(r)}\hat{\boldsymbol{p}}^{(r)T} \rangle R^T (I + A^2)^T \tag{5.63}$$

$$= \frac{1}{4} e^{-2r}(I + A^2). \tag{5.64}$$

Since the covariance matrix of the Gaussian cluster state is uniquely determined by its adjacency matrix A, and not by its U, nor its decomposition into beam splitter networks, the covariance matrix of nullifiers is also uniquely determined by A.

We consider a special case where the cluster state is unweighted ($A_{jk} = 0$ or 1 for all j and k). In general, each element $(A^n)_{jk}$ is equivalent to the number of paths from vertex j to k with the length of n. By defining $N(i)$ to be the set of nearest neighbor vertices of vertex i, $(A^2)_{jk}$ represents the number of elements in the intersection of two sets $N(j)$ and $N(k)$, which we describe by $M_{N(i)\cap N(j)}$. As a result, the covariance matrix of the nullifiers become

$$\langle \hat{\boldsymbol{\delta}}\hat{\boldsymbol{\delta}}^T \rangle_{jk} = \frac{\hbar}{2} e^{-2r}\left[\delta_{jk} + M_{N(j)\cap N(k)}\right]. \tag{5.65}$$

The variances of the nullifiers are acquired by choosing its diagonal elements:

$$\langle \hat{\delta}_i^2 \rangle = \frac{\hbar}{2} e^{-2r}\left[\delta_j + M_{N(j)}\right], \tag{5.66}$$

where $M_{N(j)}$ represents the number of nearest modes of mode j.

5.2.3 Examples of Gaussian Cluster States

In this section, we show two examples of Gaussian cluster states: the n-mode linear cluster state and the n-mode star cluster state.

5.2.3.1 Linear Cluster States

Consider an unweighted and undirected graph with n vertices shown in Fig. 5.2, where each vertex is connected linearly. The cluster state associated with this graph is called an n-mode linear cluster state. By definition, its adjacency matrix $A^{(L,n)}$ is an $n \times n$ real symmetric matrix with $A^{(L,n)}_{j,j+1} = A^{(L,n)}_{j+1,j} = 1$ for $j = 1, \dots, n-1$, while the other elements are zero. It gives us n independent nullifiers of the n-mode linear cluster state:

$$\hat{p}_1 - \hat{x}_2 = 0, \tag{5.67}$$

$$\hat{p}_k - \hat{x}_{k-1} - \hat{x}_{k+1} = 0 \; (k = 2, \dots, n-1), \tag{5.68}$$

$$\hat{p}_n - \hat{x}_{n-1} = 0. \tag{5.69}$$

The n-mode Gaussian cluster state associated with the adjacency matrix $A^{(L,n)}$ can be generated by combining n-mode p-squeezed states by using a beam splitter network represented by

$$U^{(L,n)} = B^{(L)}_{1,2}\big(\sqrt{R_1}\big) B^{(L)}_{2,3}\big(\sqrt{R_2}\big) \cdots B^{(L)}_{n-1,n}\big(\sqrt{R_{n-1}}\big), \tag{5.70}$$

where $B^{(L)}_{j,k}\big(\sqrt{R}\big)$ is an $n \times n$ identity matrix except for $(B^{(L)}_{j,k})_{jj} = (B^{(L)}_{j,k})_{kk} = \sqrt{R}$ and $(B^{(L)}_{j,k})_{jk} = (B^{(L)}_{j,k})_{kj} = i\sqrt{1-R}$. Their energy reflectivities R_k are given by

$$R_k = \frac{F_{k+1}}{F_{k+2}}, \tag{5.71}$$

where

$$F_k = \frac{1}{\sqrt{5}}\left[\left(\frac{1+\sqrt{5}}{2}\right)^k - \left(\frac{1-\sqrt{5}}{2}\right)^k\right] \tag{5.72}$$

represents the Fibonacci sequence defined by

$$F_0 = 0, \quad F_1 = 1, \quad F_{k+2} = F_{k+1} + F_k. \tag{5.73}$$

In general, linear transformations of n-mode annihilation operators are achieved by a beam splitter network consisting of at most $n(n-1)/2$ two-mode beam splitters. To the contrary, n-mode linear cluster state can be generated by using $n-1$ beam splitters as shown in Eq. (5.70). By considering that we need at least $n-1$ beam splitters in order to combine n modes without excess nor deficiency, we find that an arbitrary n-mode linear cluster state can be generated by using the least beam splitters.

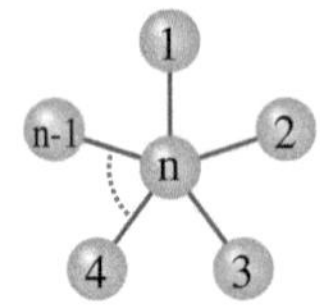

Fig. 5.3 n-mode star cluster state

5.2.3.2 Star Cluster States

Consider an unweighted and undirected graph with n vertices shown in Fig. 5.3, where each vertex except for vertex n is connected to the vertex n. The cluster state associated with this graph is called an n-mode star cluster state. By definition, its adjacency matrix $A^{(S,n)}$ is an $n \times n$ real symmetric matrix with $A^{(S,n)}_{j,n} = A^{(S,n)}_{n,j} = 1$ for $j = 1, \ldots, n-1$, while the other elements are zero. It gives us n independent nullifiers of the n-mode star cluster state:

$$\hat{p}_k - \hat{x}_n = 0 \ (k = 1, \ldots, n-1), \quad \hat{p}_n - \sum_{m=1}^{n-1} \hat{x}_m = 0. \tag{5.74}$$

Although a unitary matrix by which the Gaussian star cluster state is generated can be acquired from the general discussion in Sect. 5.2.2, we can find a solution from analogy between the n-mode star cluster state and the n-mode GHZ state. Here, the nullifiers of the n-mode GHZ state are

$$\hat{x}_k - \hat{x}_n = 0 \ (k = 1, \ldots, n-1), \quad \hat{p}_n + \sum_{m=1}^{n-1} \hat{p}_m = 0. \tag{5.75}$$

Thus, the n-mode star cluster state is acquired by applying Fourier transformations on through mode 1 to mode $n-1$ of the n-mode GHZ state. It is known that an n-mode GHZ state can be generated by combining n copies of p-squeezed states by using a beam splitter network described by an $n \times n$ unitary matrix

$$U^{(G,n)} = B^{(1)}_{1,2}\Big(\frac{1}{\sqrt{2}}\Big) B^{(1)}_{2,3}\Big(\frac{1}{\sqrt{3}}\Big) \cdots B^{(1)}_{n-1,n}\Big(\frac{1}{\sqrt{n}}\Big) F_1 \cdots F_{n-1}, \tag{5.76}$$

where F_k represents the Fourier transformation on mode k, while $B^{(1)}_{jk}(\sqrt{R})$ represents a beam splitter on modes j and k with energy reflectivity R. F_k is defined to be the $n \times n$ identity matrix except for $(F_k)_{kk} = i$, while $B^{(1)}_{jk}(\sqrt{R})$ is defined to be the $n \times n$ identity matrix except for $(B^{(1)}_{jk}(\sqrt{R}))_{jj} = -(B^{(1)}_{jk}(\sqrt{R}))_{kk} = -\sqrt{R}$ and $(B^{(1)}_{jk}(\sqrt{R}))_{jk} = (B^{(1)}_{jk}(\sqrt{R}))_{kj} = \sqrt{1-R}$. Therefore, an n-mode Gaussian star cluster state can be generated by combining n copies of p-squeezed states by using a beam splitter network described by an $n \times n$ unitary matrix

$$U^{(S,n)} = F_1 \cdots F_{n-1} U^{(G,n)}. \tag{5.77}$$

Similar to the case of the n-mode Gaussian linear cluster states, an arbitrary n-mode Gaussian star cluster state can be generated by using the least beam splitters.

5.3 Input Coupling

In general, quantum computation can be generously considered to be a unitary transformation on an input state. In the context of one-way quantum computation, once the input state is coupled with a cluster state, it is achieved by succession of single-mode measurements. The topic of this section is how to handle the input state, or equivalently, how to start the computation in the one-way scheme.

One idea is to generate the input state from the resource cluster state at the beginning of computation. Since the one-way scheme is a universal model of quantum computation, an arbitrary input state can be generated by measurements on the cluster state. In this settings, the input state is embedded in the cluster state at the beginning.

Another situation is when the input state is prepared independently of the cluster state. In this general case, the quantum computer opens its port for an arbitrary input state, and works as a unitary gate on it. It is also applied to the case where the total quantum computation is divided into sub-computations. In this case, the input state of the succeeding computation is the output of the earlier computation.

In this section, we show three types of input-coupling schemes for one-way quantum computation. The first scheme is a orthodox one which is based on a controlled-Z gate. It is a natural result of the discussion in Sect. 4.2.4. The main theme in this section is the other two schemes. One is based on quantum teleportation, while the other is based on a squeezer. In all schemes, an arbitrary input state can be connected with a cluster state. After the input coupling, one-way quantum computation can be achieved by performing measurements on the cluster state.

In the following discussion, we do not impose any condition on input mode *in*, thus it can be in an arbitrary state.

5.3.1 κ Representation, η Representation, and θ Representation

In this section, we introduce three types of notations for homodyne measurements: κ representation, η representation, and θ representation (Fig. 5.4).

5.3.1.1 κ Representation

Consider that the annihilation operator of the mode to be measured is given by $\hat{x}+i\hat{p}$. We perform an operator $\hat{D}_\kappa(\kappa) = e^{i\frac{\kappa}{2\hbar}\hat{x}^2}$ before the homodyne measurement with the observable of the momentum operator $\hat{p}$. This is the procedure of the homodyne measurement in the κ representation. It corresponds to Fig. 4.13, where the operator $\hat{D}$ in Fig. 4.13 is given by $\hat{D}_\kappa(\kappa)$. Since

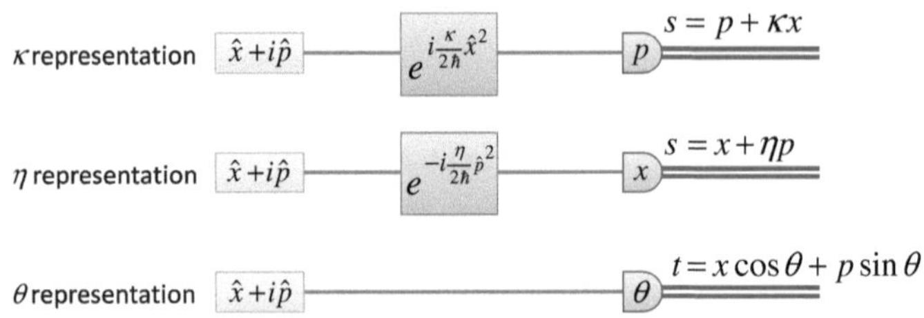

Fig. 5.4 κ representation, η representation, and θ representation

$$\hat{D}_\kappa^\dagger(\kappa)(\hat{x}+i\hat{p})\hat{D}_\kappa(\kappa)=\hat{x}+i(\hat{p}+\kappa\hat{x}), \tag{5.78}$$

the measurement with the observable $\hat{p}$ after the operator $\hat{D}_\kappa(\kappa)$ is applied to the mode to be measured is equivalent to a measurement with the observable $\hat{p}+\kappa\hat{x}$ without application of the operator.

5.3.1.2 η Representation

Consider that the annihilation operator of the mode to be measured is given by $\hat{x}+i\hat{p}$. We perform an operator $\hat{D}_\eta(\eta)=e^{-i\frac{\eta}{2\hbar}\hat{p}^2}$ before the homodyne measurement with the observable of the position operator $\hat{x}$. This is the procedure of the homodyne measurement in the η representation. Since

$$\hat{D}_\eta^\dagger(\eta)(\hat{x}+i\hat{p})\hat{D}_\eta(\eta)=\hat{x}+\eta\hat{p}+i\hat{p}, \tag{5.79}$$

the measurement with the observable $\hat{x}$ after the operator $\hat{D}_\eta(\eta)$ is applied to the mode to be measured is equivalent to a measurement with the observable $\hat{x}+\eta\hat{p}$ without application of the operator.

5.3.1.3 Comparison Between κ Representation and η Representation

Although the κ representation is almost equivalent to the η representation, there is a slight difference in measurement bases which cannot be formulated in these representations. In the κ representation, a homodyne measurement with the observable $\hat{p}$ can be formulated by $\kappa=0$, while that with the observable $\hat{x}$ cannot be formulated. On the other hand, in the η representation, a homodyne measurement with the observable $\hat{x}$ can be formulated by $\eta=0$, while that with the observable $\hat{p}$ cannot be formulated.

The common advantage in both representations is that calculations become simpler compared with the θ representation.

5.3.1.4 θ Representation

Consider that the annihilation operator of the mode to be measured is given by $\hat{x}+i\hat{p}$. We perform the homodyne measurement with the observable of $\hat{x}_\theta = \hat{x}\cos\theta + \hat{p}\sin\theta$. This is the homodyne measurement in the θ representation.

The advantage of this representation is that all homodyne measurement bases can be formulated. However, calculations become complicated compared with the other two representations.

5.3.1.5 κ Representation, η Representation, and θ Representation

The κ representation, η representation, and θ representation are equivalent to each other except that the observable $\hat{x}$ cannot be formulated in the κ representation, and the observable $\hat{p}$ cannot be formulated in the η representation.

The relationship between the κ representation and the θ representation is given by

$$s = \hat{p} + \kappa\hat{x} = \frac{1}{\sin\theta}(\hat{x}\cos\theta + \hat{p}\sin\theta) = \frac{t}{\sin\theta}, \quad s = \frac{t}{\sin\theta}, \quad \kappa = \frac{1}{\tan\theta}. \tag{5.80}$$

Similarly, the relationship between the η representation and the θ representation is given by

$$s = \hat{x} + \eta\hat{p} = \frac{1}{\cos\theta}(\hat{x}\cos\theta + \hat{p}\sin\theta) = \frac{t}{\cos\theta}, \quad s = \frac{t}{\cos\theta}, \quad \eta = \tan\theta. \tag{5.81}$$

These equations show that the homodyne measurement in the κ representation and the η representation can be decomposed into a homodyne measurement with a relative phase θ and an electric gain.

5.3.2 Controlled-Z-Based Input-Coupling Scheme

5.3.2.1 One-Way Quantum Computation Starting with C_Z-Based Input-Coupling Scheme

Although we have already mentioned the essence of the C_Z-based scheme in Sect. 4.2.3, we rewrite it in the context of input coupling, where the measurement observable is p and we continue one-way quantum computation after the input coupling.

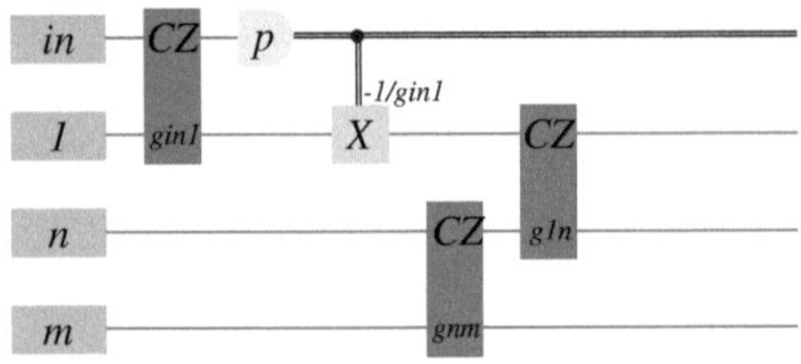

Fig. 5.5 One-way quantum computation starting with C_Z-based input-coupling scheme 1

Consider a quantum circuit described in Fig. 5.5. The label *in* shows a mode of an input state. It can be one mode of a multi-mode quantum state, in general. The initial state of mode 1 is a zero eigenstate of the momentum operator, or a p-squeezed state (which is infinitely squeezed in the ideal case). We assume that the gain of the controlled-Z gate between modes *in* and 1 is g_{in1}. The first three steps on modes *in* and 1 are equivalent to the elementary one-way gate. Here, we assume that the measurement variable $\hat{s}$ for mode *in* is $\hat{s} = \hat{D}^\dagger \hat{p} \hat{D} = \hat{p}$ with $\hat{D} = \hat{I}$. The labels of modes n and m show other resources, which are initially zero eigenstates of momentum operators, or p-squeezed states, as well. Mode n represents a nearest neighbor mode of mode 1, while mode m represents a nearest neighbor mode of mode n. Although modes n and m are symbolically described as single modes in Fig. 5.5, they can be generalized to the case where there are multiple nearest neighbor modes of mode 1 and mode n, respectively. In addition, other modes are also acceptable which are located at a distance of more than two from mode 1. We assume that the gain of the controlled-Z gate between modes 1 and n is g_{1n}, while that between modes n and m is g_{nm}. As a result, Fig. 5.5 shows the beginning of a one-way quantum computation, where the information in mode *in* is first teleported to mode 1, followed by the coupling between mode 1 and a cluster state consisting of modes n and m (and the other modes).

We consider a transformation of the circuit described in Fig. 5.5. By using

$$\hat{C}_{Z\,1n}(g)\hat{X}_1(-s) = \hat{X}_1(-s)\hat{Z}_n(-gs)\hat{C}_{Z\,1n}(g), \tag{5.82}$$

we can change the order of the feed-forward operator $\hat{X}_1$ and the controlled-Z gate $\hat{C}_{Z\,1n}(g_{1n})$ by adding another feed-forward to mode n (Fig. 5.6). It shows the beginning of a one-way quantum computation, where a cluster state consisting of modes 1, n, and m (and the other modes) is used as a resource, and the input mode is coupled with the cluster state using a controlled-Z gate.

The equivalence between Figs. 5.5 and 5.6 gives us the following statement on the C_Z-based input coupling scheme.

5.3.2.2 Summary of C_Z-Based Input-Coupling Scheme

Consider a weighted cluster state. We assume that mode 1 of it is the destination of the input coupling of mode *in* (Fig. 5.7). We describe the cluster state as

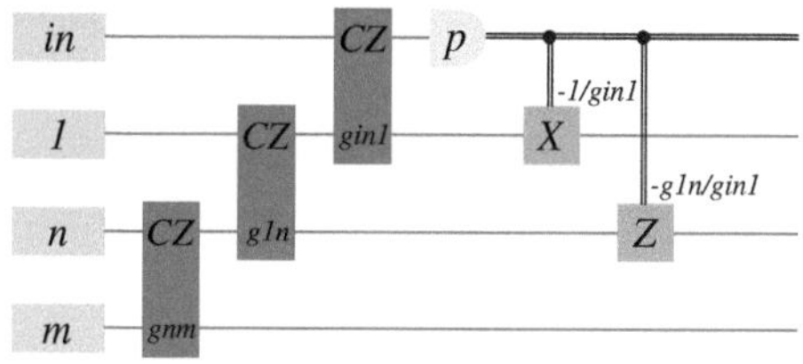

Fig. 5.6 One-way quantum computation starting with C_Z-based input-coupling scheme 2

$$\prod_{a\in N(1)} \hat{C}_{Z1a}(g_a)|p=0\rangle_1|\psi\rangle_C, \tag{5.83}$$

where $|\psi\rangle_C$ represents the cluster state without mode 1. We define $|\phi\rangle_{in,L}$ to be a k-mode input state, where the label *in* represents the mode to be coupled, while L represents the other modes. By performing the input coupling with the C_Z-based scheme (Fig. 5.6), the output state becomes

$$\prod_{a\in N(1)} \hat{C}_{Z\,in,a}(g_a)\hat{S}_{in}(\log g_{in1})\hat{F}_{in}|\phi\rangle_{in,L}|\psi\rangle_C, \tag{5.84}$$

which is acquired from the equivalence between Figs. 5.5 and 5.6. In the special case of $g_{in1}=1$, the term of squeezing operation $\hat{S}_{in}(\log g_{in1})$ vanishes.

5.3.3 Teleportation-Based Input-Coupling Scheme

5.3.3.1 Quantum Teleportation

The core of teleportation-based input coupling scheme is, literally, quantum teleportation, whose diagram is shown in Fig. 5.8. Mode *in* represents an input mode, while modes 1 and 2 represent two resource modes. In the following, we do not impose any conditions for these three modes, thus they can be in any state.

We define the annihilation operators for modes *in*, 1, and 2 at (A) as

$$\hat{x}_{in}+i\hat{p}_{in},\quad \hat{x}_1+i\hat{p}_1,\quad \hat{x}_2+i\hat{p}_2. \tag{5.85}$$

We perform a controlled-Z gate with gain g on mode 1 and mode 2. The annihilation operators for modes 1 and 2 at (B) become

$$\begin{pmatrix}\hat{x}_{1B}+i\hat{p}_{1B}\\ \hat{x}_{2B}+i\hat{p}_{2B}\end{pmatrix}=\begin{pmatrix}\hat{x}_1+i(\hat{p}_1+g\hat{x}_2)\\ \hat{x}_2+i(\hat{p}_2+g\hat{x}_1)\end{pmatrix}. \tag{5.86}$$

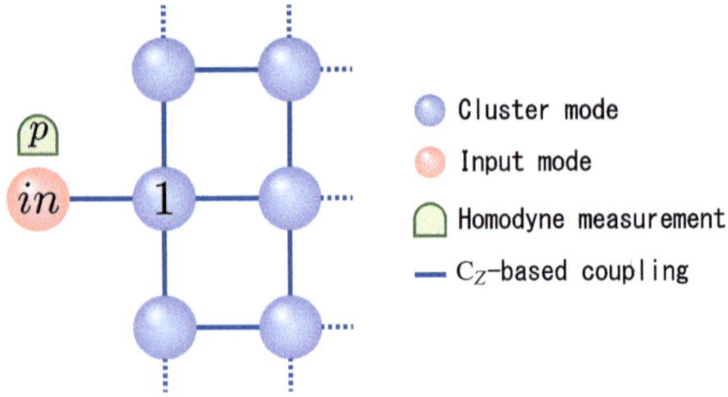

Fig. 5.7 C_Z-based input-coupling scheme

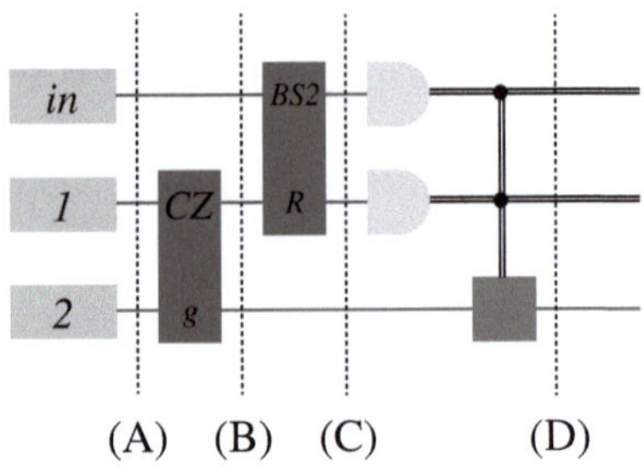

Fig. 5.8 Teleportation-based input-coupling scheme

The input mode *in* and the resource mode 1 are combined by using a beam splitter. We choose the linear transformation matrix to be $B_{in1}^{(2)}(\sqrt{R})$ with the energy reflectivity R. The linear transformation is given by

$$\begin{pmatrix} \hat{a}'_{in} \\ \hat{a}'_1 \end{pmatrix} = B_{in1}^{(2)}(\sqrt{R}) \begin{pmatrix} \hat{a}_{in} \\ \hat{a}_1 \end{pmatrix} = \begin{pmatrix} \sqrt{1-R} & -\sqrt{R} \\ \sqrt{R} & \sqrt{1-R} \end{pmatrix} \begin{pmatrix} \hat{a}_{in} \\ \hat{a}_1 \end{pmatrix}, \tag{5.87}$$

leading to the annihilation operators of modes *in* and 1 at (C):

$$\begin{pmatrix} \hat{x}_{inC} + i\hat{p}_{inC} \\ \hat{x}_{1C} + i\hat{p}_{1C} \end{pmatrix} = \begin{pmatrix} \sqrt{1-R}\hat{x}_{in} - \sqrt{R}\hat{x}_{1B} + i(\sqrt{1-R}\hat{p}_{in} - \sqrt{R}\hat{p}_{1B}) \\ \sqrt{R}\hat{x}_{in} + \sqrt{1-R}\hat{x}_{1B} + i(\sqrt{R}\hat{p}_{in} + \sqrt{1-R}\hat{p}_{1B}) \end{pmatrix}. \tag{5.88}$$

We perform homodyne measurements on modes *in* and 1. We utilize the θ *representation* for the measurements on modes *in* and 1 (Sect. 5.3.1). Therefore, the measurement observables $\hat{t}_{in}, \hat{t}_1$ on modes *in* and 1 are

$$\hat{t}_k = \hat{x}_k \cos\theta_k + \hat{p}_k \sin\theta_k, \quad k = in, 1. \tag{5.89}$$

Without using the position operators $\hat{x}_k$ for the resource modes $k = 1, 2$, the quadrature operators for the output mode $k = 2$ at (C), $\hat{x}_{2C} + i\hat{p}_{2C} = \hat{x}_{2B} + i\hat{p}_{2B}$,

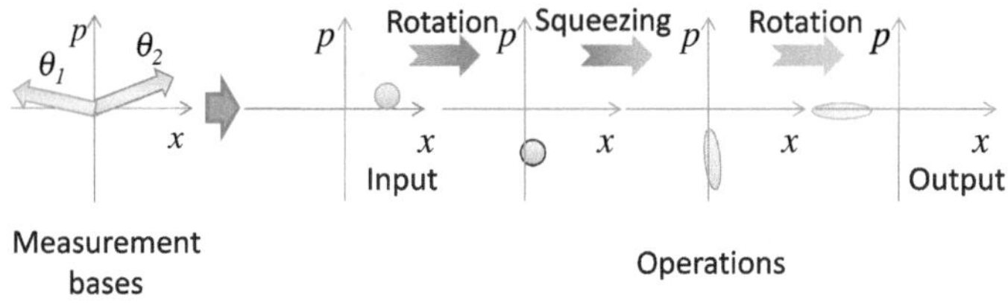

Fig. 5.9 Decomposition of operation

can be reformulated as

$$\begin{pmatrix}\hat{x}_{2C}\\ \hat{p}_{2C}\end{pmatrix} = M_{tele}\begin{pmatrix}\hat{x}_{in}\\ \hat{p}_{in}\end{pmatrix} + M_M\begin{pmatrix}\hat{t}_{in}\\ \hat{t}_1\end{pmatrix} + \begin{pmatrix}-1/g & 0\\ 0 & 1\end{pmatrix}\begin{pmatrix}\hat{p}_1\\ \hat{p}_2\end{pmatrix}, \tag{5.90}$$

where matrices M_{tele} and M_M are given by

$$M_{tele} = \frac{1}{2\sin\theta_-\sqrt{R(1-R)}}\begin{pmatrix}\frac{1}{g} & 0\\ 0 & g\end{pmatrix}\begin{pmatrix}\cos\theta_+ + \cos\theta_- & \sin\theta_+ + (1-2R)\sin\theta_-\\ -\sin\theta_+ + (1-2R)\sin\theta_- & \cos\theta_+ - \cos\theta_-\end{pmatrix}, \tag{5.91}$$

$$M_M = \frac{1}{\sin\theta_-\sqrt{R(1-R)}}\begin{pmatrix}\frac{1}{g} & 0\\ 0 & g\end{pmatrix}\begin{pmatrix}-\sqrt{1-R}\cos\theta_1 & -\sqrt{R}\cos\theta_{in}\\ \sqrt{1-R}\sin\theta_1 & \sqrt{R}\sin\theta_{in}\end{pmatrix}. \tag{5.92}$$

Here, we have defined $\theta_\pm = \theta_{in} \pm \theta_1$. Note that the reformulation above holds before the measurements are carried out.

Equation (5.90) consists of three terms. The first term represents the main operation on the input mode *in*, which is determined by the homodyne measurement angles θ_k. We cannot choose $\sin\theta_- = 0$ since denominators of M_{tele} and M_M become zero. The second term derives from the measurement variables $\hat{t}_k$. The last term consists of the momentum operators of mode 1 and mode 2 at (A) in Fig. 5.8.

When the measurements on modes *in* and 1 have been carried out, the quadratures of the output mode becomes

$$\begin{pmatrix}\hat{x}_{2C}\\ \hat{p}_{2C}\end{pmatrix} = M_{tele}\begin{pmatrix}\hat{x}_{in}\\ \hat{p}_{in}\end{pmatrix} + M_M\begin{pmatrix}t_{in}\\ t_1\end{pmatrix} + \begin{pmatrix}-1/g & 0\\ 0 & 1\end{pmatrix}\begin{pmatrix}\hat{p}_1\\ \hat{p}_2\end{pmatrix}, \tag{5.93}$$

where t_{in} and t_1 represent measurement results of the measurement variables $\hat{t}_{in}$ and $\hat{t}_1$, respectively. By performing feed-forward operations on mode 2 so that the random displacements in phase space are perfectly cancelled out, the quadrature operators of the output mode at (D) become

$$\begin{pmatrix}\hat{x}_{2D}\\ \hat{p}_{2D}\end{pmatrix} = M_{tele}\begin{pmatrix}\hat{x}_{in}\\ \hat{p}_{in}\end{pmatrix} + \begin{pmatrix}-1/g & 0\\ 0 & 1\end{pmatrix}\begin{pmatrix}\hat{p}_1\\ \hat{p}_2\end{pmatrix}. \tag{5.94}$$

This is the general form of the input-output relation for Fig. 5.8.

In the following, we consider a special case with $g = 1$ and $R = \frac{1}{2}$. M_{tele} and M_M become

$$M_{tele} = \frac{1}{\sin\theta_-}\begin{pmatrix}\cos\theta_+ + \cos\theta_- & \sin\theta_+ \\ -\sin\theta_+ & \cos\theta_+ - \cos\theta_-\end{pmatrix}, \tag{5.95}$$

$$M_M = \frac{\sqrt{2}}{\sin\theta_-}\begin{pmatrix}-\cos\theta_1 & -\cos\theta_{in} \\ \sin\theta_1 & \sin\theta_{in}\end{pmatrix}. \tag{5.96}$$

M_{tele} can be decomposed into

$$M_{tele} = R\left(-\frac{1}{2}\theta_+\right) S\left(\ln\tan\frac{1}{2}\theta_-\right) R\left(-\frac{1}{2}\theta_+\right) \tag{5.97}$$

$$= \begin{pmatrix}\cos\theta\cosh r - \sinh r & \cosh r\sin\theta \\ -\cosh r\sin\theta & \cos\theta\cosh r + \sinh r\end{pmatrix}, \tag{5.98}$$

where we have defined $r = \ln\tan\frac{1}{2}\theta_-, \theta = \theta_+$. Two matrices $R(\theta)$ and $S(r)$:

$$R(\theta) = \begin{pmatrix}\cos\theta & -\sin\theta \\ \sin\theta & \cos\theta\end{pmatrix}, \quad S(r) = \begin{pmatrix}e^{-r} & 0 \\ 0 & e^r\end{pmatrix} \tag{5.99}$$

represent a rotation operation and a squeezing operation, respectively. Therefore, the operation is a rotation with the angle $\theta = -\frac{1}{2}\theta_+$, followed by a squeezing operation with the squeezing parameter $r = \ln\tan\frac{1}{2}\theta_-$, and another rotation with the angle $\theta = -\frac{1}{2}\theta_+$, where $r \in \mathbb{R}$ and $0 \leq \theta < 2\pi$ (Fig. 5.9). Note again that these two parameters are determined by the homodyne measurement angles.

5.3.3.2 κ Representation

We assume that $g = 1$, $R = \frac{1}{2}$. The operation matrix M_{tele} can be reformulated in the κ representation:

$$M_{tele} = \frac{1}{\kappa_{in} - \kappa_1}\begin{pmatrix}-2\kappa_{in}\kappa_1 & -\kappa_{in} - \kappa_1 \\ \kappa_{in} + \kappa_1 & 2\end{pmatrix}. \tag{5.100}$$

Note that the measurement with the observable $\hat{x}$ cannot be formulated in the κ representation. The denominator $\kappa_{in} - \kappa_1 \neq 0$ corresponds to $\sin\theta_- = 0$ in the θ representation.

5.3.3.3 η, κ Representation

We assume that $g = 1$, $R = \frac{1}{2}$. We here use the η representation for mode *in*, while the κ representation for mode 1. The operation matrix M_{tele} can be reformulated as

$$M_{tele} = \frac{1}{1 - \kappa_1 \eta_{in}} \begin{pmatrix} -2\kappa_1 & -(1 + \kappa_1 \eta_{in}) \\ 1 + \kappa_1 \eta_{in} & 2\eta_{in} \end{pmatrix}. \tag{5.101}$$

Note that the measurements with the observables $\hat{p}_{in}$ and $\hat{x}_1$ cannot be formulated since we are now using the η representation and the κ representation for modes *in* and 1, respectively. The denominator $1 - \kappa_1 \eta_{in} \neq 0$ corresponds to $\sin \theta_- = 0$.

5.3.3.4 Case with p-Squeezed States in Modes 1 and 2

Although the discussion above is independent of the initial states in modes 1 and 2, we consider a special case where they are in p-squeezed states. In the following, we use quadrature operators which specify particular quantum states (Sect. 3.2.8).

We describe two p-squeezed-state modes as

$$\hat{x}_k + i\hat{p}_k = e^{r_k}\hat{x}_k^{(0)} + ie^{-r_k}\hat{p}_k^{(0)}, \quad k = 1, 2. \tag{5.102}$$

In this case, the quantum state at (B) in Fig. 5.8 is an approximate two-mode cluster state. The output-mode quadratures in Fig. 5.8 are given by

$$\begin{pmatrix} \hat{x}_{2D} \\ \hat{p}_{2D} \end{pmatrix} = M_{tele} \begin{pmatrix} \hat{x}_{in} \\ \hat{p}_{in} \end{pmatrix} + \begin{pmatrix} -1/g & 0 \\ 0 & 1 \end{pmatrix} \begin{pmatrix} e^{-r_1}\hat{p}_1^{(0)} \\ e^{-r_2}\hat{p}_2^{(0)} \end{pmatrix}, \tag{5.103}$$

where $e^{-r_k}\hat{p}_k^{(0)}$ represents the squeezing component of the resource mode k. In the limit of infinite squeezing $r_k \to \infty$, the ideal operation:

$$\begin{pmatrix} \hat{x}_{2D} \\ \hat{p}_{2D} \end{pmatrix} = M_{tele} \begin{pmatrix} \hat{x}_{in} \\ \hat{p}_{in} \end{pmatrix} \tag{5.104}$$

can be achieved.

5.3.3.5 One-Way Quantum Computation Starting with Teleportation-Based Input-Coupling Scheme

Next, we consider the teleportation-based input-coupling scheme for one-way quantum computation. It is given by replacing the C_Z gate for the input coupling in Fig. 5.5 with the teleportation. The fundamental circuit is described in Fig. 5.10.

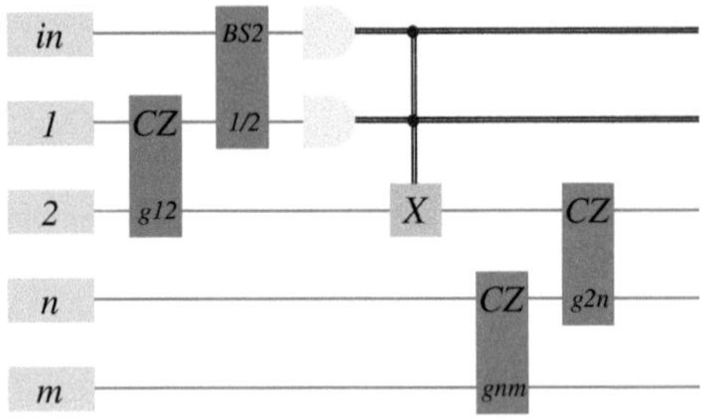

Fig. 5.10 One-way quantum computation starting with teleportation-based input-coupling scheme 1

The definition of mode labels is the same as that in Sect. 5.3.2, except that we have an additional mode labeled by "2", and mode n represents a nearest neighbor mode of mode 2. The initial states of modes $1, 2, n, m$ are zero eigenstates of the momentum operators, or p-squeezed states (which are infinitely squeezed in the ideal case). Note again that mode *in* can be one mode of a multi-mode input state. As a result, Fig. 5.10 shows the beginning of a one-way quantum computation, where the information in mode *in* is first transmitted to mode 2 with Gaussian operations applied on it through quantum teleportation, followed by coupling between mode 2 and a cluster state consisting of modes n and m (and the other modes).

We exchange the feed-forward and the C_Z gate on modes 2 and n by using

$$\hat{C}_{Z\,2n}\hat{X}_2\hat{Z}_2 = \hat{X}_2\hat{Z}_n\hat{C}_{Z\,2n}\hat{Z}_2 = \hat{X}_2\hat{Z}_2\hat{Z}_n\hat{C}_{Z\,2n}. \tag{5.105}$$

We find that Fig. 5.10 is equivalent to Fig. 5.11, which leads to the following statement on the teleportation-based input coupling scheme.

5.3.3.6 Summary of Teleportation-Based Input-Coupling Scheme

Consider a weighted cluster state. We assume that modes 1 and 2 of it are a resource and the destination of the input coupling of mode *in*, respectively (Fig. 5.12).

Fig. 5.11 One-way quantum computation starting with teleportation-based input-coupling scheme 2

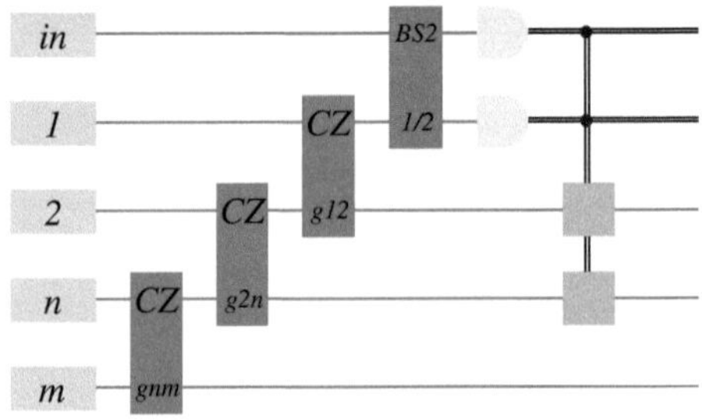

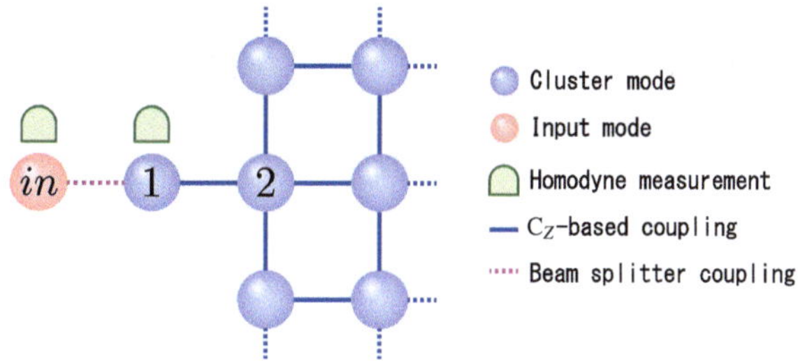

Fig. 5.12 Teleportation-based input-coupling scheme

We describe the cluster state as

$$\hat{C}_{Z12}(g_{12}) \prod_{a \in N(2)\backslash 1} \hat{C}_{Z2a}(g_{2a})|p=0\rangle_1|p=0\rangle_2|\psi\rangle_C, \tag{5.106}$$

where $|\psi\rangle_C$ represents the cluster state without modes 1 and 2. We define $|\phi\rangle_{in,L}$ to be a k-mode input state, where the label *in* represents the mode to be coupled, while L represents the other modes. By performing the input coupling with the teleportation-based scheme (Fig. 5.11), the output state becomes

$$\prod_{a \in N(2)\backslash 1} \hat{C}_{Z\,in,a}(g_{2a})\hat{M}_{tele\,in}|\phi\rangle_{in,L}|\psi\rangle_C, \tag{5.107}$$

which is acquired from the equivalence between Figs. 5.10 and 5.11.

In many cases, we usually choose $g_{12} = 1$ and $R = \frac{1}{2}$, for simplicity. The operation given by the matrix $\hat{M}_{tele\,in}$ can be controlled by the measurement bases at the homodyne detections. By choosing $\theta_- = \frac{\pi}{2}, \theta_+ = 0$, we can set it to the identity operator: $\hat{M}_{tele\,in} = \hat{I}$.

5.3.4 Squeezer-Based Input-Coupling Scheme

In the teleportation-based input-coupling scheme, two modes of a cluster state is consumed in order to couple one mode of an input state. This is a requirement derived from quantum teleportation, where a two-mode entangled state is used as a resource. One can easily imagine that it is not the minimum configuration.

In this section, we present a simpler input-coupling scheme composed of a one-mode resource state.

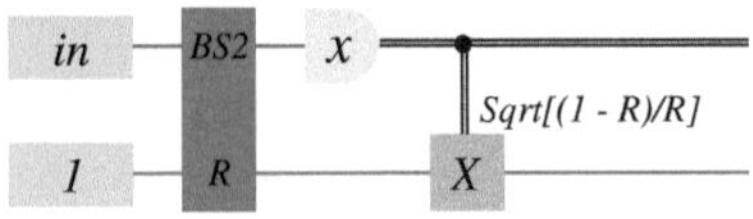

Fig. 5.13 Squeezer

5.3.4.1 Squeezer

Consider a quantum circuit described in Fig. 5.13, which is usually referred to as a single-mode squeezer.

Similar to the case of the teleportation-based scheme, mode *in* represents an input mode, while mode 1 represents a resource mode. In the following, we do not impose any conditions for these two modes, thus they can be in any state.

We define the annihilation operators at the beginning as

$$\hat{x}_{in} + i\hat{p}_{in}, \quad \hat{x}_1 + i\hat{p}_1. \tag{5.108}$$

At the first step, these two modes are combined by using a beam splitter. We choose the linear transformation matrix to be $B^{(2)}_{in1}(\sqrt{R})$ with the energy reflectivity R, which is identical to the teleportation-based scheme. The quadrature operators are transformed into

$$\left[\sqrt{1-R}\hat{x}_{in} - \sqrt{R}\hat{x}_1\right] + i\left[\sqrt{1-R}\hat{p}_{in} - \sqrt{R}\hat{p}_1\right], \quad \left[\sqrt{R}\hat{x}_{in} + \sqrt{1-R}\hat{x}_1\right] + i\left[\sqrt{R}\hat{p}_{in} + \sqrt{1-R}\hat{p}_1\right]. \tag{5.109}$$

Mode *in* is subject to be measured in the measurement basis $|x\rangle$. Thus the measurement variable $\hat{s}_{in}$ is given by

$$\hat{s}_{in} = \sqrt{1-R}\hat{x}_{in} - \sqrt{R}\hat{x}_1. \tag{5.110}$$

Without using the position operator $\hat{x}_1$ for the resource mode $k = 1$, the quadrature operators for the output mode $k = 1$ can be reformulated as

$$\left[\frac{1}{\sqrt{R}}\hat{x}_{in} - \sqrt{\frac{1-R}{R}}\hat{s}_{in}\right] + i\left[\sqrt{R}\hat{p}_{in} + \sqrt{1-R}\hat{p}_1\right]. \tag{5.111}$$

Note that the reformulation above holds before the measurement is carried out. On the contrary, when the measurement on mode *in* has been carried out indeed, the quadratures of the output mode becomes

$$\left[\frac{1}{\sqrt{R}}\hat{x}_{in} - \sqrt{\frac{1-R}{R}}s_{in}\right] + i\left[\sqrt{R}\hat{p}_{in} + \sqrt{1-R}\hat{p}_1\right], \tag{5.112}$$

where s_{in} represents the measurement result of the measurement variable $\hat{s}_{in}$. By performing the feed-forward operation (displacement operation) based on the measurement result:

$$\hat{X}_1\left(\sqrt{\frac{1-R}{R}}s_{in}\right), \tag{5.113}$$

we get

$$\frac{1}{\sqrt{R}}\hat{x}_{in} + i\left[\sqrt{R}\hat{p}_{in} + \sqrt{1-R}\hat{p}_1\right]. \tag{5.114}$$

This is the general input-output relationship of the circuit shown in Fig. 5.13.

5.3.4.2 Case with *p*-Squeezed State in Mode 1

Although the discussion above is independent of the initial state in mode 1, we consider a special case where mode 1 is in a p-squeezed state. In the following, we use quadrature operators which specify particular quantum states (Sect. 3.2.8).

We describe the p-squeezed-state mode as

$$\hat{x}_1 + i\hat{p}_1 = e^{r_1}\hat{x}_1^{(0)} + ie^{-r_1}\hat{p}_1^{(0)}. \tag{5.115}$$

The output-mode quadratures in Fig. 5.13 are given by

$$\frac{1}{\sqrt{R}}\hat{x}_{in} + i\left[\sqrt{R}\hat{p}_{in} + \sqrt{1-R}e^{-r_1}\hat{p}_1^{(0)}\right], \tag{5.116}$$

where $e^{-r_1}\hat{p}_1^{(0)}$ represents the squeezing component of the resource mode 1. In the limit of infinite squeezing $r_1 \to \infty$, they become

$$\frac{1}{\sqrt{R}}\hat{x}_{in} + i\sqrt{R}\hat{p}_{in}. \tag{5.117}$$

It shows a squeezing operation with the squeezing operator $\hat{S}(r)$, where $r = -\log\frac{1}{\sqrt{R}}$.

5.3.4.3 Case with Different Measurement Angle

Although we have assumed above that the measurement observable is $\hat{x}$, we can extend the discussion to a general case. When we measure $\hat{x}\cos\theta + \hat{p}\sin\theta$ at the homodyne detection, the input-output relationship becomes

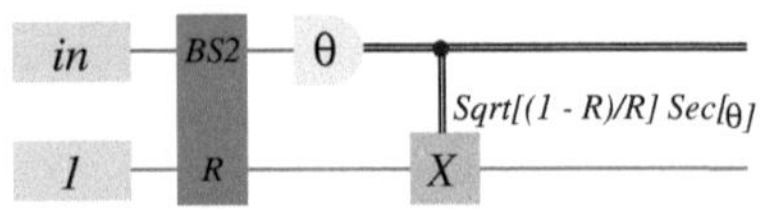

Fig. 5.14 Squeezer-based input-coupling scheme with different measurement angle

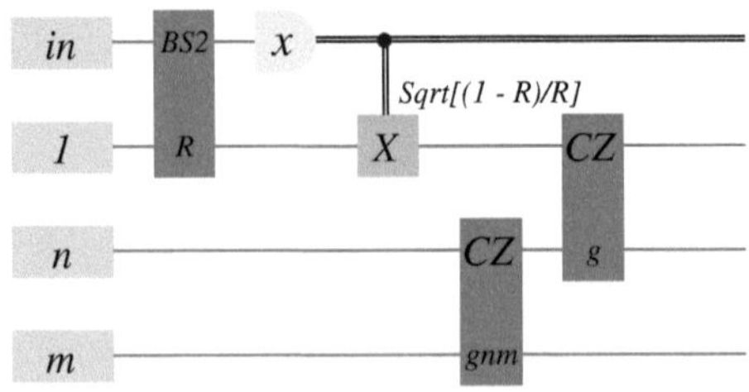

Fig. 5.15 One-way quantum computation starting with squeezer-based input-coupling scheme 1

$$\begin{pmatrix} \hat{x}_{out} \\ \hat{p}_{out} \end{pmatrix} = \begin{pmatrix} \frac{1}{\sqrt{R}} & \frac{1-R}{\sqrt{R}} \tan\theta \\ 0 & \sqrt{R} \end{pmatrix} \begin{pmatrix} \hat{x}_{in} \\ \hat{p}_{in} \end{pmatrix} + \begin{pmatrix} -\sqrt{1-R}\tan\theta\,\hat{p}_1 \\ \sqrt{1-R}\,\hat{p}_1 \end{pmatrix}. \tag{5.118}$$

Note that the gain of the feed-forward operation should be changed to $\sqrt{\frac{1-R}{R}}\sec\theta$ (Fig. 5.14).

5.3.4.4 One-Way Quantum Computation Starting with Squeezer-Based Input-Coupling Scheme

We then consider the squeezer-based input-coupling scheme for one-way quantum computation. Similar to the teleportation-based case, it is given by replacing the C_Z gate for the input coupling in Fig. 5.5 with the squeezer. The fundamental circuit is described in Fig. 5.15.

The definition of mode labels is the same as that in Sect. 5.3.2. The initial states of modes $1, n, m$ are zero eigenstates of the momentum operators, or p-squeezed states (which are infinitely squeezed in the ideal case). Note again that mode *in* can be one mode of a multi-mode input state. As one can see, Fig. 5.15 shows the beginning of a one-way quantum computation, where the information in mode *in* is first transmitted to mode 1 with a squeezing transformation applied on it. It is then coupled with a cluster state consisting of modes n and m (and the other modes) by using a C_Z gate with gain g.

Transformations of Fig. 5.15 give us the interpretation of the squeezer-based scheme. By using

$$\hat{C}_{Z1n}(g)\hat{X}_1(s) = \hat{X}_1(s)\hat{Z}_n(gs)\hat{C}_{Z1n}(g), \tag{5.119}$$

it is transformed into Fig. 5.16.

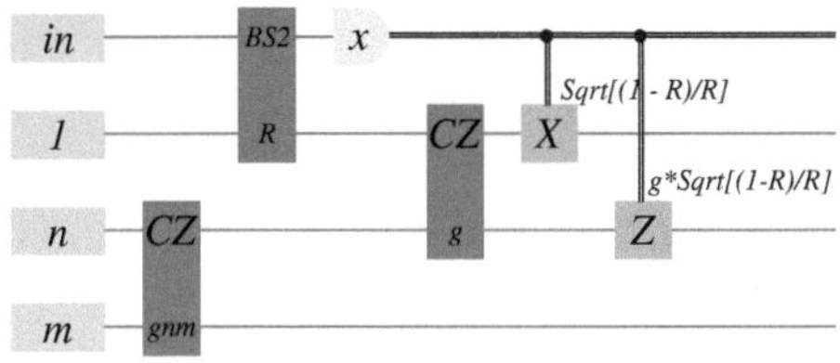

Fig. 5.16 One-way quantum computation starting with squeezer-based input-coupling scheme 2

Next, we consider an exchange of the C_Z gate on modes 1 and n, and the input-coupling interaction on modes in and 1. In the case of the C_Z-based scheme, these two gates commute because both are C_Z gates. In the case of the teleportation-based scheme, they commute as well because they do not have common modes which they work on. However, in the case of the squeezer-based scheme, they do not commute. Here, we use

$$\hat{C}_{Z1n}(g)\hat{B}^{(2)}_{in1}(\sqrt{R}) = \hat{C}_{Zin\,n}\Big(g\sqrt{\frac{R}{1-R}}\Big)\hat{B}^{(2)}_{in1}(\sqrt{R})\hat{C}_{Z1n}\Big(\frac{g}{\sqrt{1-R}}\Big), \tag{5.120}$$

leading to the revised version of the circuit shown in Fig. 5.17. Note that an additional C_Z gate on modes in and n is inserted after the exchanged two gates.

Although we have so far considered exchanges of quantum gates where unitary operations themselves are preserved, we finally transform the quantum circuit so that the transformation preserves its input-output relation when we take account of the measurement on mode in. Consider the C_Z gate on mode in and mode n:

$$\begin{pmatrix}\hat{x}'_{in}\\ \hat{p}'_{in}\\ \hat{x}'_n\\ \hat{p}'_n\end{pmatrix} = C_Z(G)\begin{pmatrix}\hat{x}_{in}\\ \hat{p}_{in}\\ \hat{x}_n\\ \hat{p}_n\end{pmatrix}, \quad C_Z(G) = \begin{pmatrix}1&0&0&0\\ 0&1&G&0\\ 0&0&1&0\\ G&0&0&1\end{pmatrix}. \tag{5.121}$$

It mixes two modes in a way that $\hat{x}_n$ and $\hat{x}_{in}$ are added to $\hat{p}_{in}$ and $\hat{p}_n$ with gain $G = g\sqrt{R/(1-R)}$, respectively. Note that the addition of $\hat{x}_n$ to $\hat{p}_{in}$ does not affect the succeeding measurement, since its measurement variable is $\hat{x}$. On the other hand,

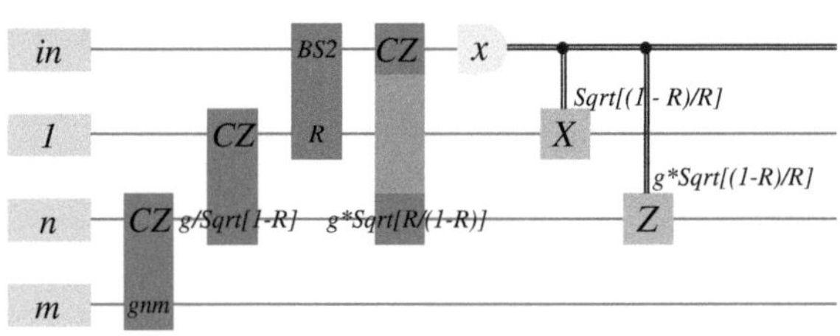

Fig. 5.17 One-way quantum computation starting with squeezer-based input-coupling scheme 3

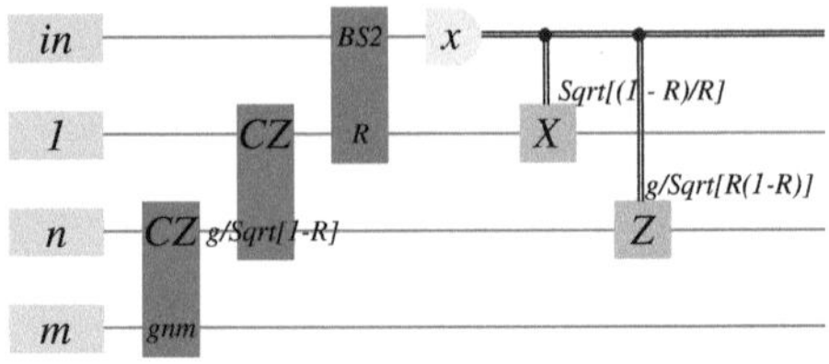

Fig. 5.18 One-way quantum computation starting with squeezer-based input-coupling scheme 4

the identical transmission of $\hat{x}_{in}$ to $\hat{p}_n$ is achieved by changing the feed-forward gain from mode *in* to mode n since $\hat{x}_{in}$ is measured right after the C_Z gate. Therefore, Fig. 5.17 can be transformed into Fig. 5.18 by removing the C_Z gate and changing the feed-forward gain.

It shows the beginning of a one-way quantum computation, where a cluster state consisting of modes 1, n, and m (and the other modes) is used as a resource, and the input mode is coupled with the cluster state using a single-mode squeezer.

The equivalence between Figs. 5.15 and 5.18 gives us the following statement on the squeezer-based input coupling scheme.

5.3.4.5 Summary of Squeezer-Based Input-Coupling Scheme

Consider a weighted cluster state. We assume that mode 1 of it is the destination of the input coupling of mode *in* (Fig. 5.19). We describe the cluster state as

$$\prod_{a\in N(1)} \hat{C}_{Z1a}\Big(\frac{g}{\sqrt{1-R}}\Big)|p=0\rangle_1|\psi\rangle_C, \tag{5.122}$$

where $|\psi\rangle_C$ represents the cluster state without mode 1. We define $|\phi\rangle_{in,L}$ to be a k-mode input state, where the label *in* represents the mode to be coupled, while L represents the other modes. By performing the input coupling with the squeezer-based scheme (Fig. 5.18), the output state becomes

$$\prod_{a\in N(1)} \hat{C}_{Z\,in,a}(g)\hat{S}_{in}(r)|\phi\rangle_{in,L}|\psi\rangle_C, \quad r=-\log\frac{1}{\sqrt{R}}, \tag{5.123}$$

or equivalently,

$$\hat{S}_{in}(r)\prod_{a\in N(1)} \hat{C}_{Z\,in,a}(ge^{-r})|\phi\rangle_{in,L}|\psi\rangle_C, \quad r=-\log\frac{1}{\sqrt{R}}, \tag{5.124}$$

which is acquired from the equivalence between Figs. 5.15 and 5.18.

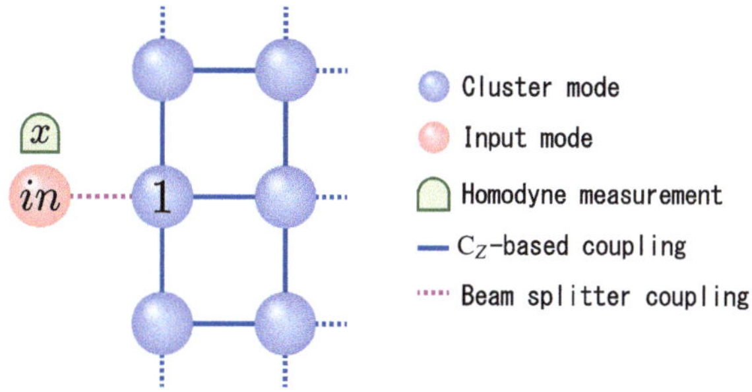

Fig. 5.19 Squeezer-based input-coupling scheme

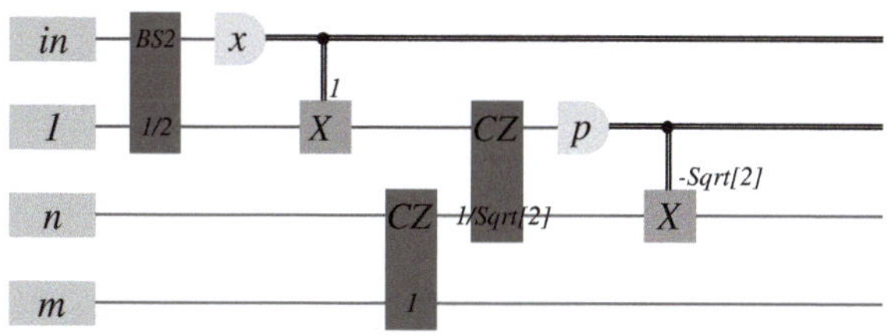

Fig. 5.20 Original circuit

In the case of $R = 1$, the squeezer-based input-coupling scheme fails, since the gain of the C_Z gate on modes 1 and $a \in N(1)$ in Eq. (5.122) becomes infinity. In the case of $R = 0$, it fails as well, since the squeezing level r in Eq. (5.123) and Eq. (5.124) becomes infinity. In the other cases of $0 < R < 1$, the input coupling is achieved successfully

An additional squeezing operation on mode *in* is applied through the input coupling. It derives from the fact that it is based on the single-mode squeezer. No matter how we choose the reflectivity R, the level of its squeezing r cannot be set to zero. In many cases, we usually choose $R = \frac{1}{2}$, for simplicity.

5.3.5 Relation Between Squeezer-Based Input-Coupling Scheme and Teleportation-Based Input-Coupling Scheme

Before we move onto the next section, we mention that the teleportation-based input-coupling scheme can be decomposed into the squeezer-based input-coupling, followed by an elementary one-way gate. It can be understood by considering a special case of Fig. 5.15, as shown in Fig. 5.20, where the reflectivity R is $R = 1/2$, the gain of the C_Z gate is $g = \sqrt{1 - R}$, and the number of modes in $N(1)$ is equal to one.

An elementary one-way gate is then performed with the operator $\hat{D} = \hat{I}$, leading to the measurement variable $\hat{p}$ in mode 1. The measurement outcome is utilized in a feed-forward process to its nearest neighbor mode n. By following the same

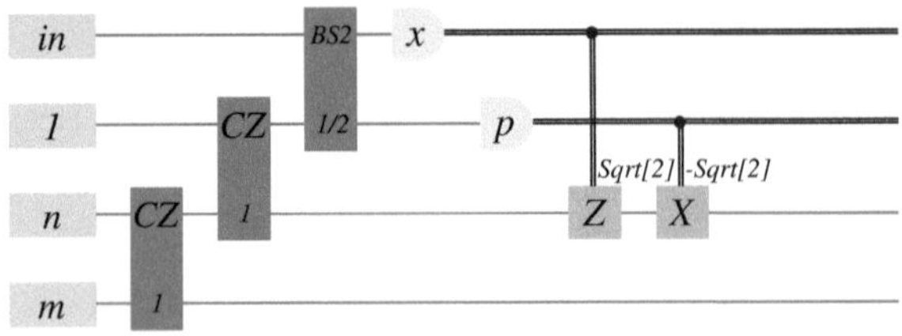

Fig. 5.21 Equivalent circuit

discussion above, we find that it can be transformed into Fig. 5.21, which is nothing but the schematic of the teleportation-based input coupling scheme.

Although the schematic with the observable $\hat{p}$ at the second measurement shows the equivalence between the squeezer-based and the teleportation-based coupling, it can be extended to $\hat{D}\hat{p}\hat{D}^{\dagger}$ with an arbitrary diagonal unitary operator $\hat{D}$. Especially, nonlinear measurements other than homodyne detections are also acceptable. On the other hand, the second measurement in the teleportation-based scheme is restricted to homodyne detections, because it is based on Bell measurements. Therefore, it can be considered as a feature of the squeezer-based input coupling scheme, which cannot be acquired by the teleportation-based input-coupling scheme.

5.4 Reshaping of Cluster States

A sufficiently large cluster state is the essential resource for one-way quantum computation. Once a desired operation is determined, the cluster state is transformed so that the operation can be carried out through the succeeding measurements. Reshaping of cluster states, which we address in this section, is a tool for transforming elementary cluster states into another shape of cluster states [5, 6].

In general, reshaping of cluster states can be categorized into two types: combining two cluster states, and removing unwanted modes in the cluster states. In the former case, a larger-scale cluster state is generated from smaller-scale cluster states. An example of its application is a case where a cluster state of appropriate-size for a computation cannot be prepared directly, although one can generate parts of it. The total cluster state can be constructed by combining them. It is known that a special-shaped cluster state of arbitrary size can be generated by combining two-mode cluster states repeatedly (Chap. 10, [7]). On the other hand of the latter case, a smaller cluster state specialized for a particular one-way quantum computation is generated from a larger-scale general cluster state. In this process, cluster modes are removed with or without preserving mode connections to their nearest neighbors.

In general, the most natural scheme to reshape cluster states is to use C_Z gates for connections, and inversed C_Z gates for disconnections (Sect. 5.4.1). To the contrary, we will discuss other schemes where only beam splitters, homodyne detections, and feed-forwards (displacements) are involved.

5.4.1 Reshaping of Cluster States with Controlled-Z Gates

An n-mode cluster state is defined to be the state which is acquired by combining n copies of $|p = 0\rangle$ states using weighted C_Z operators $\hat{C}_{Zij}(g_{ij}) = e^{\frac{i}{\hbar} g_{ij}\hat{x}_i\hat{x}_j}$. Therefore, the most natural scheme to reshape cluster states is to use C_Z operators for connections, and inversed C_Z operators $\hat{C}^{\dagger}_{Zij}(g_{ij}) = e^{-\frac{i}{\hbar} g_{ij}\hat{x}_i\hat{x}_j}$ for disconnections.

5.4.2 Erasing (Removal of Unwanted Modes)

The first tool for reshaping of cluster states is erasing of an unwanted mode with eliminating connections to its nearest neighbors.

Consider an n-mode cluster state. We define $\hat{x}_j$ and $\hat{p}_j$ to be the position and momentum operators of the cluster-state mode j, respectively, where $j = 1, \ldots, n$ represent the labels of modes. By assuming that each gain of the C_Z gate on mode j and k for generation of the cluster state is given by $g_{jk}(=g_{kj})$, the nullifiers of the cluster state becomes

$$\hat{p}_a - \sum_{a' \in N(a)} g_{aa'}\hat{x}_{a'} = 0, \quad a = 1, \ldots, n, \tag{5.125}$$

where $N(a)$ is the set of nearest neighbor modes of mode a. In the following, we assume that $N(*)$ represents the set of nearest neighbors *before* erasing.

We consider that mode k is to be erased. The erasing of mode k is achieved in the following procedure:

- Measure mode k with the observable $\hat{x}$, leading to a result s.
- Perform the feed-forward operation on each nearest neighbor mode j of mode k, which is defined as the momentum displacement operator $\hat{Z}_j(-g_{kj}s)$.

We define $\hat{x}'_m$ and $\hat{p}'_m$ to be the position and momentum operators of mode m after reshaping. When $m \notin N(k)$, they are not changed from $\hat{x}_m$ and $\hat{p}_m$. When $m \in N(k)$, the momentum operator becomes $\hat{p}'_m = \hat{p}_m - g_{km}s$, while $\hat{x}'_m$ is not changed. As a result, we acquire the new set of nullifiers:

$$\begin{aligned}
\hat{p}'_m - \sum_{a \in N(m)} g_{ma}\hat{x}'_a &= 0, \quad (m \notin N(k)), \\
\hat{p}'_m - \sum_{a \in N(m)\backslash k} g_{ma}\hat{x}'_a &= 0, \quad (m \in N(k)),
\end{aligned} \tag{5.126}$$

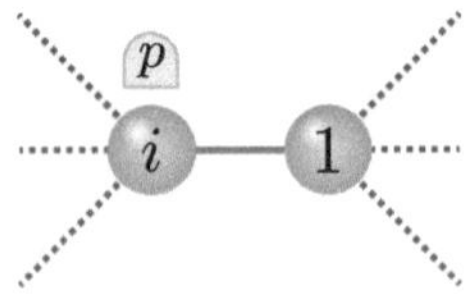

Fig. 5.22 Wire shortening

which shows mode k is successfully erased.

5.4.3 Wire Shortening (Shortening of Connections)

In the mode erasing described in the previous section, connections to nearest neighbor modes of the erased mode are eliminated. Compared with the erasing of modes, we address the scheme of wire shortening, where an unwanted mode is removed with preserving connections to its nearest neighbors. It is achieved by transferring connections between the mode to be removed and all but one nearest neighbors, to the remaining nearest neighbor mode.

Wire shortening is considered as an example of the C_Z-based input-coupling scheme, where the input state is also a part of the cluster state. Consider a cluster state described by

$$\prod_{a \in N(1)} \hat{C}_{Z\,1,a}(g_{1a})|\phi\rangle_{iL}|p=0\rangle_1|\psi\rangle_R. \tag{5.127}$$

Its graph is shown in Fig. 5.22. Here, $|\phi\rangle_{iL}$ represents the cluster state on the left side of mode i, where mode i is included. $|\psi\rangle_R$ represents the cluster state on the right side of mode 1, where mode 1 is excluded. In the following, $N(1)$ is the set of nearest neighbor modes of mode 1 *before* wire shortening. Note that mode i is a member of $N(1)$. We assume that mode i and mode 1 correspond to mode *in* and mode 1 in Sect. 5.3.2. By performing transmission of mode i to mode 1, the resulting state is given by

$$\prod_{a \in N(1)\setminus i} \hat{C}_{Z\,i,a}(g_{1a})\hat{S}_i(\log g_{1i})\hat{F}_i|\phi\rangle_{iL}|\psi\rangle_R. \tag{5.128}$$

In the case of the unity gain $g_{1i} = 1$, the additional squeezing operator vanishes. Note that an inevitable Fourier operator is also applied to mode i. It works as a phase rotation of the cluster state. However, by performing the wire shortenings twice, the phase rotation becomes a sign flip. By performing additional wire shortenings twice (four times in total), the effect of phase rotation is canceled.

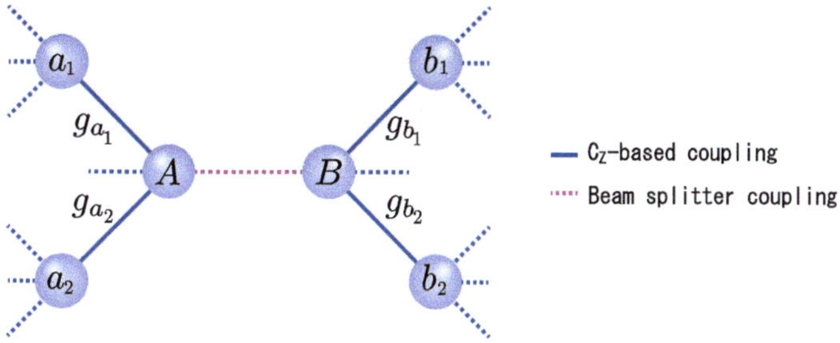

Fig. 5.23 Connection without measurement (before connection)

5.4.4 Connection Without Measurement

From this section, we move on to the second type of cluster reshaping—connections of cluster states. Here, we show a scheme to connect two cluster states without measurements.

Consider that there exist two cluster states as shown in Fig. 5.23. We label two modes to be coupled in each cluster state as A and B. We assume that the nearest neighbor modes of mode A are a_i, while those of mode B are b_i. In general, the number of nearest neighbor modes of mode A or B may be more than one. These modes are distinguished by the subscript i. We assume that the two cluster states do not have triangle structures which include mode A or mode B. A subset of nullifiers of two cluster states is given by

$$\hat{p}_A - \sum_{a_i \in N(A)} g_{a_i} \hat{x}_{a_i}, \quad \hat{p}_{a_i} - g_{a_i} \hat{x}_A - \sum_{k \in N(a_i) \setminus A} g_k \hat{x}_k, \tag{5.129}$$

$$\hat{p}_B - \sum_{b_i \in N(B)} g_{b_i} \hat{x}_{b_i}, \quad \hat{p}_{b_i} - g_{b_i} \hat{x}_B - \sum_{k \in N(b_i) \setminus B} g_k \hat{x}_k, \tag{5.130}$$

where each g with a subscript represents the corresponding edge weight of the cluster state. Note that the other nullifiers do not have position nor momentum operators of modes A and B.

The only procedure to connect these two cluster states is to combine two modes A and B by using a beam splitter. We choose $B^{(2)}_{AB}(\frac{1}{\sqrt{2}})$ as the linear transformation of the beam splitter. As a result, the above subset of nullifiers becomes

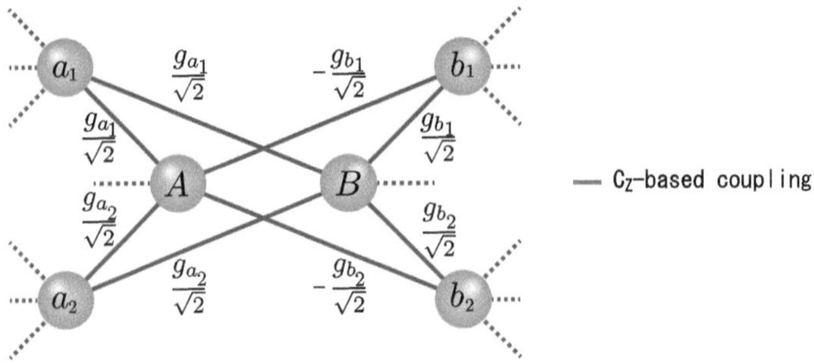

Fig. 5.24 Connection without measurement (after connection)

$$\frac{1}{\sqrt{2}}\hat{p}_A + \frac{1}{\sqrt{2}}\hat{p}_B - \sum_{a_i \in N(A)} g_{a_i}\hat{x}_{a_i}, \tag{5.131}$$

$$\hat{p}_{a_i} - \frac{g_{a_i}}{\sqrt{2}}\hat{x}_A - \frac{g_{a_i}}{\sqrt{2}}\hat{x}_B - \sum_{k \in N(a_i)\setminus A} g_k\hat{x}_k, \tag{5.132}$$

$$-\frac{1}{\sqrt{2}}\hat{p}_A + \frac{1}{\sqrt{2}}\hat{p}_B - \sum_{b_i \in N(B)} g_{b_i}\hat{x}_{b_i}, \tag{5.133}$$

$$\hat{p}_{b_i} + \frac{g_{b_i}}{\sqrt{2}}\hat{x}_A - \frac{g_{b_i}}{\sqrt{2}}\hat{x}_B - \sum_{k \in N(b_i)\setminus B} g_k\hat{x}_k, . \tag{5.134}$$

Since linear combinations of nullifiers are also nullifiers, we get the following new nullifiers from Eqs. (5.131) and (5.133):

$$\hat{p}_A - \frac{1}{\sqrt{2}}\sum_{a_i \in N(A)} g_{a_i}\hat{x}_{a_i} + \frac{1}{\sqrt{2}}\sum_{b_i \in N(B)} g_{b_i}\hat{x}_{b_i}, \tag{5.135}$$

$$\hat{p}_B - \frac{1}{\sqrt{2}}\sum_{a_i \in N(A)} g_{a_i}\hat{x}_{a_i} - \frac{1}{\sqrt{2}}\sum_{b_i \in N(B)} g_{b_i}\hat{x}_{b_i}. \tag{5.136}$$

The set of Eqs. (5.132), (5.134), (5.135), and (5.136) is the nullifier subset of the new cluster state. Note that the other nullifiers are not changed by the beam splitter transformation. Therefore, its graph is given by Fig. 5.24.

Compared to the original graph shown in Fig. 5.23, the weight g_{a_i} between mode A and its nearest neighbor mode a_i is changed to $\frac{1}{\sqrt{2}}g_{a_i}$. At the same time, mode a_i is also connected to mode B with the same weight $\frac{1}{\sqrt{2}}g_{a_i}$. In a similar manner, the weight g_{b_i} between mode B and its nearest neighbor mode b_i is changed to $\frac{1}{\sqrt{2}}g_{b_i}$, while mode b_i is connected to mode A with the sign-flipped weight $-\frac{1}{\sqrt{2}}g_{b_i}$. The sign flip originates in the law of conservation of energy at the beam splitter, thus it is unavoidable. The transformation of edges can be considered that the edges

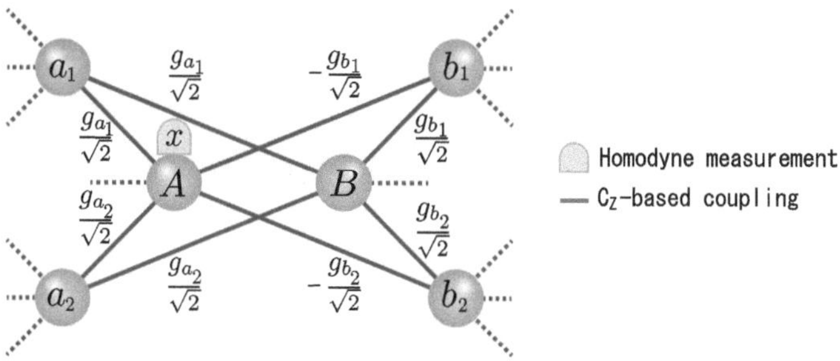

Fig. 5.25 Connection with a measurement (before connection)

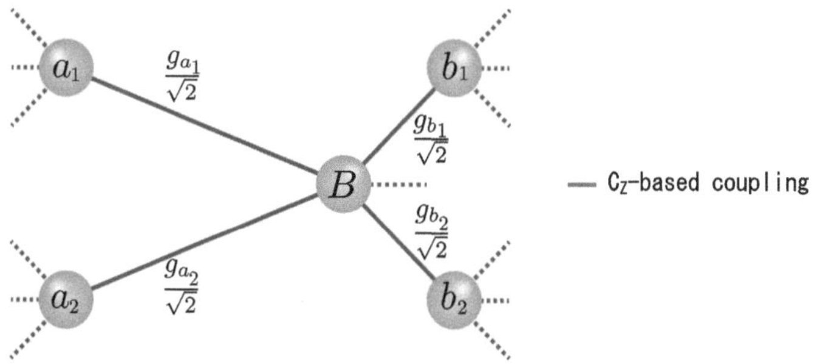

Fig. 5.26 Connection with a measurement (after connection)

between mode A (or B) and its nearest neighbor modes become shared with the nearest neighbor modes of mode B (or A).

In this connection scheme, the interaction gains are changed through the connection of cluster states. In addition, the gains between mode A and nearest neighbor modes of mode B are sign flipped. As a consequence of the latter property, unweighted cluster states cannot be created.

5.4.5 Connection with a Measurement

By adding a single-mode measurement to the cluster connection scheme without measurement, we get another type of cluster connection schemes (Fig. 5.25).

The additional procedure can be considered as the erasing of mode A. For that purpose, mode A is measured with the measurement observable $\hat{x}$. To complete the process, the feed-forward operation on each nearest neighbor mode of mode A is performed. As a result, the graph of the cluster state is transformed to Fig. 5.26.

Although we have so far described the procedure by decomposing it into two steps: connection without measurement and erasing, it can be directly understood as an example of the squeezer-based input coupling scheme, where the input state is another cluster state.

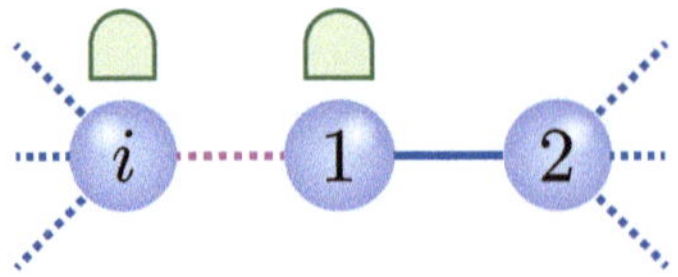

Fig. 5.27 Connection with two measurements (before connection)

We define two cluster states before connection without measurement, which is shown in Fig. 5.23, to be

$$\prod_{a_i \in N(A)} \hat{C}_{Z\,A,a_i}(g_{a_i})|p=0\rangle_A|\psi\rangle_L, \tag{5.137}$$

$$\prod_{b_i \in N(B)} \hat{C}_{Z\,B,b_i}(g_{b_i})|p=0\rangle_B|\phi\rangle_R. \tag{5.138}$$

Here, $|\psi\rangle_L$ and $|\phi\rangle_R$ represent the cluster states on the left and right sides of mode A and B, where mode A and B are excluded, respectively. The left side cluster state including mode A (Eq. (5.137)) corresponds to the cluster state in the squeezer-based coupling, while mode B in the right side cluster state (Eq. (5.138)) corresponds to the input mode to be coupled with. By performing the transmission of mode B to mode A through the squeezer-based coupling, the state becomes

$$\prod_{a_i \in N(A)} \hat{C}_{Z\,B,a_i}(\tfrac{g_{a_i}}{\sqrt{2}}) \prod_{b_i \in N(B)} \hat{C}_{Z\,B,b_i}(\tfrac{g_{b_i}}{\sqrt{2}})|p=0\rangle_B|\phi\rangle_R|\psi\rangle_L, \tag{5.139}$$

which is equivalent to Fig. 5.26. Here, we have used $\hat{S}_B(r)\hat{C}_{Z\,B,b_i}(g) = \hat{C}_{Z\,B,b_i}(ge^r)$ $\hat{S}_B(r)$ and $\hat{S}_B(r)|p=0\rangle_B = |p=0\rangle_B$.

Although the interaction gains are changed through the connection of the cluster states, the signs of gains are not flipped. Therefore, by preparing two cluster states so that $g_{a_i} = g_{b_i} = \sqrt{2}$ and the other gains are equal to 1, an unweighted cluster state can be created. This is a feature of the connection scheme with a measurement, which does not hold in the scheme without measurement.

5.4.6 Connection with Two Measurements

The last connection scheme involves two measurements on two cluster states. It can be considered as an example of the teleportation-based input coupling scheme, where the input state is another cluster state (Fig. 5.27).

Consider two cluster states, one contains mode i and the other contains mode 1 and mode 2. We assume that mode 1 does not have nearest neighbor modes other than mode 2. We define these two cluster states as

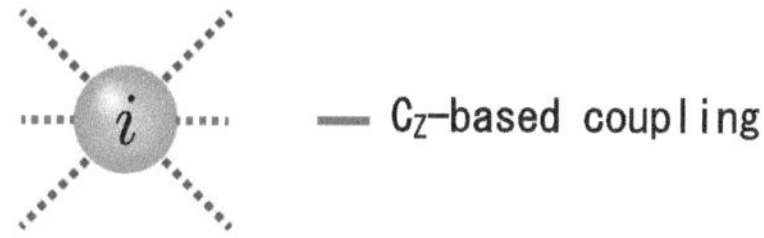

Fig. 5.28 Connection with two measurements (after connection)

$$|\phi\rangle_{iL}, \hat{C}_{Z12}(g_{12}) \prod_{a\in N(2)\backslash 1} \hat{C}_{Z2a}(g_{2a})|p=0\rangle_1|p=0\rangle_2|\psi\rangle_R. \tag{5.140}$$

Here, $|\phi\rangle_{iL}$ represents the cluster state on the left side of mode i, where mode i is included. $|\psi\rangle_R$ represents the cluster state on the right side of mode 2, where mode 1 and 2 are excluded. We assume that modes i, 1, and 2 correspond to modes *in*, 1, and 2 in the teleportation-based coupling (Sect. 5.3.3), respectively. By performing the transmission of mode i to mode 2 through the teleportation-based coupling, the state becomes

$$\prod_{a\in N(2)\backslash 1} \hat{C}_{Z\,i,a}(g_{2a})\hat{M}_{tele\,i}|\phi\rangle_{iL}|\psi\rangle_R. \tag{5.141}$$

By considering a special case of $g_{12}=1$, $R=\frac{1}{2}$, $\theta_-=\frac{\pi}{2}$, $\theta_+=0$, the operator $\hat{M}_{tele\,i}$ on mode i becomes an identity operator. As a result, the resulting cluster state is

$$\prod_{a\in N(2)\backslash 1} \hat{C}_{Z\,i,a}(g_{2a})|\phi\rangle_{iL}|\psi\rangle_R. \tag{5.142}$$

The illustration of the cluster state is shown in Fig. 5.28.

In this connection scheme, the interaction gains are not changed. Therefore, unweighted cluster states can be created by using two unweighted cluster states.

5.5 Universality of One-Way Quantum Computation

5.5.1 Universality

We have described universality of continuous-variable (CV) quantum computation in Sect. 3.6.1. An arbitrary quantum computation can be achieved by combining the following three components:

1. arbitrary one-mode Gaussian operation,
2. a two-mode Gaussian operation,
3. a one-mode non-Gaussian operation.

In this section, we briefly summarize how universality can be achieved in one-way quantum computation.

5.5.2 Arbitrary One-Mode Gaussian Operations

5.5.2.1 Displacement Operations

We take a momentum displacement operator $\hat{Z}(s) = e^{\frac{i}{\hbar}s\hat{x}}$ for the operator $\hat{D}(\hat{x})$ in the elementary circuit for one-way quantum computation (one-mode teleportation circuit, Sect. 4.2.3). By using Eq. (4.56), the observable is given by

$$\hat{p}' = \hat{p} + f'(\hat{x}) = \hat{p} + s. \tag{5.143}$$

It can be achieved by a homodyne detection with the observable $\hat{p}$, followed by the addition of s. By using Eq. (4.58), the output-mode quadratures become

$$\hat{x}_{out} + i\hat{p}_{out} = -(\hat{p}^{(in)} + s) + i(\hat{x}^{(in)}). \tag{5.144}$$

Note that we have assumed that the resource state is an ideal state $|p = 0\rangle$.

5.5.2.2 One-Mode Symplectic Operations with C_Z-Based Input-Coupling Scheme

We consider implementation of one-mode symplectic operations:

$$\begin{pmatrix} \hat{x}_{out} \\ \hat{p}_{out} \end{pmatrix} = \begin{pmatrix} a & b \\ c & d \end{pmatrix} \begin{pmatrix} \hat{x}^{(in)} \\ \hat{p}^{(in)} \end{pmatrix}, \quad ad - bc = 1, \tag{5.145}$$

where $\hat{x}^{(in)} + i\hat{p}^{(in)}$ shows the annihilation operator of an input mode. The condition $ad - bc = 1$ derives from the conservation of commutation relation. Therefore, the degrees of freedom of one-mode symplectic operations is three.

5.5.2.3 One-Step Operation

We use a quadratic phase operator:

$$\hat{D}(\hat{x}) = e^{\frac{i}{2\hbar}\kappa\hat{x}^2}, \tag{5.146}$$

for the diagonal operator $\hat{D}(\hat{x})$. By using Eq. (4.56), the observable is given by

$$\hat{p}' = \hat{p} + f'(\hat{x}) = \hat{p} + \kappa\hat{x}. \tag{5.147}$$

It can be achieved by a homodyne detection. By using Eq. (4.58), the quadrature operators of the output mode become

$$\hat{x}_{out} + i\,\hat{p}_{out} = -(\hat{p}^{(in)} + \kappa\hat{x}^{(in)}) + i\hat{x}^{(in)}. \tag{5.148}$$

Note that we have assumed that the resource state is an ideal state $|p = 0\rangle$. By using matrices, it can be formulated as

$$\begin{pmatrix}\hat{x}_{out}\\ \hat{p}_{out}\end{pmatrix} = M(\kappa)\begin{pmatrix}\hat{x}^{(in)}\\ \hat{p}^{(in)}\end{pmatrix} = \begin{pmatrix}-\kappa & -1\\ 1 & 0\end{pmatrix}\begin{pmatrix}\hat{x}^{(in)}\\ \hat{p}^{(in)}\end{pmatrix} = \begin{pmatrix}0 & -1\\ 1 & 0\end{pmatrix}\begin{pmatrix}1 & 0\\ \kappa & 1\end{pmatrix}\begin{pmatrix}\hat{x}^{(in)}\\ \hat{p}^{(in)}\end{pmatrix}. \tag{5.149}$$

It shows that $\hat{F}\hat{D}(\hat{x})$ is implemented, which corresponds to Eq. (4.53).

Note that the discussion above is based on the κ representation of the homodyne detection. Thus, a measurement with the observable $\hat{x}$ cannot be formulated. However, since the measurement with the observable $\hat{x}$ in the C_Z-based input-coupling scheme corresponds to the erasing of the mode (Sect. 5.4.2), operations cannot be implemented. Therefore, we do not have to take account of this case.

5.5.2.4 Multi-Step Operations

By concatenating the operators $\hat{F}\hat{D}_k(\hat{x})$, we can acquire an arbitrary one-mode symplectic operators.

One Step and Two Steps

The degrees of freedom of the matrices $M(\kappa_1)$ and $M(\kappa_2)M(\kappa_1)$ for one-step and two-step operations are one and two, respectively. On the other hand, the degrees of freedom of one-mode symplectic operations is three. Therefore, universality for one-mode symplectic operations cannot be achieved by using one-step and two-step operations.

Three Steps

The degrees of freedom of the matrix $M(\kappa_3)M(\kappa_2)M(\kappa_1)$ for a three-step operation is three. On the other hand, the degrees of freedom of one-mode symplectic operations is three. Therefore, there is a possibility that universality for one-mode symplectic operations can be achieved by using three-step operations.

From the equation

$$M(\kappa_3)M(\kappa_2)M(\kappa_1) = \begin{pmatrix} -\kappa_3(\kappa_2\kappa_1 - 1) + \kappa_1 & -\kappa_3\kappa_2 + 1 \\ \kappa_2\kappa_1 - 1 & \kappa_2 \end{pmatrix} = \begin{pmatrix} a & b \\ c & d \end{pmatrix}, \quad (5.150)$$

we find that we can implement an arbitrary one-mode symplectic operation with $d \neq 0$ by choosing

$$\kappa_1 = \frac{1+c}{d}, \quad \kappa_2 = d, \quad \kappa_3 = \frac{1-b}{d}. \quad (5.151)$$

When $d = 0$, we have to choose $\kappa_2 = 0$, leading to the operations:

$$\begin{pmatrix} a & b \\ c & d \end{pmatrix} = \begin{pmatrix} \kappa_3 + \kappa_1 & 1 \\ -1 & 0 \end{pmatrix}. \quad (5.152)$$

Therefore, although we can achieve an arbitrary $a \in \mathbb{R}$ by choosing κ_1 or κ_3 appropriately, we cannot set b, c other than $b = 1, c = -1$.

In summary, we cannot implement the following operations:

$$\left\{ \begin{pmatrix} a & b \\ c & d \end{pmatrix} \middle| ad - bc = 1, d = 0, b \neq 1 \right\}. \quad (5.153)$$

Therefore, universality is not achieved.

Four Steps

We solve the equation

$$M(\kappa_4)M(\kappa_3)M(\kappa_2)M(\kappa_1) \quad (5.154)$$

$$= \begin{pmatrix} \kappa_4(\kappa_3(\kappa_2\kappa_1 - 1) - \kappa_1) - (\kappa_2\kappa_1 - 1) & -\kappa_4(-\kappa_3\kappa_2 + 1) - \kappa_2 \\ -\kappa_3(\kappa_2\kappa_1 - 1) + \kappa_1 & -\kappa_3\kappa_2 + 1 \end{pmatrix} = \begin{pmatrix} a & b \\ c & d \end{pmatrix}. \quad (5.155)$$

When $\kappa_3 \neq 0$, we get

$$\kappa_2 = \frac{1-d}{\kappa_3}, \quad \kappa_3 = c - d\kappa_1, \quad \kappa_4 = \frac{1 - a + b\kappa_1}{\kappa_3}. \quad (5.156)$$

Here, we can choose κ_1 so that $\kappa_3 \neq 0$ for all $a, b, c, d, ad - bc = 1$. Therefore, an arbitrary one-mode symplectic operation can be achieved.

In summary, we can implement an arbitrary one-mode symplectic operation by using a four-step one-mode teleportation circuit with a four-mode linear cluster state as a resource.

5.5.2.5 One-Mode Symplectic Operations with Teleportation-Based Input-Coupling Scheme

In experiment, we usually use the teleportation-based input-coupling scheme for one-way quantum computation. Here, we discuss universality for one-mode symplectic operations with the teleportation-based input-coupling scheme.

5.5.2.6 Two-Step Operation

In the following, we utilize the κ representation for homodyne measurements. By using Eq. (5.90), we find that the operation is given by

$$M_{tele}(\kappa_1, \kappa_2) = \frac{1}{\kappa_1 - \kappa_2} \begin{pmatrix} -2\kappa_1\kappa_2 & -\kappa_1 - \kappa_2 \\ \kappa_1 + \kappa_2 & 2 \end{pmatrix}, \tag{5.157}$$

where we have assumed that $R = \frac{1}{2}$. Since a quantum teleportation involves two homodyne detections (Bell measurement), the degrees of freedom of the operation is two.

5.5.2.7 Three-Step Operation

As a next step, we perform an operation achieved by a one-mode teleportation circuit with $\hat{D} = e^{\frac{i}{2\hbar}\kappa\hat{x}^2}$ after the operation by the teleportation-based input-coupling circuit. The total operation is given by

$$M(\kappa_3)M_{tele}(\kappa_1, \kappa_2) = \frac{1}{\kappa_1 - \kappa_2} \begin{pmatrix} 2\kappa_3\kappa_2\kappa_1 - \kappa_1 - \kappa_2 & \kappa_3\kappa_2 + \kappa_3\kappa_1 - 2 \\ -2\kappa_2\kappa_1 & -\kappa_2 - \kappa_1 \end{pmatrix}, \tag{5.158}$$

where we have utilized the κ representation for all homodyne measurements.

In the following, we acquire κ_1, κ_2, and κ_3 for $\begin{pmatrix} a & b \\ c & d \end{pmatrix}$, $ad - bc = 1$.

Case with c ≠ 0 and d ≠ ±1

$$\kappa_1 = \frac{c}{1 + d}, \quad \kappa_2 = \frac{-c}{1 - d}, \quad \kappa_3 = \frac{d - a}{c}. \tag{5.159}$$

Case with c = 0 (ad = 1, b ∈ ℝ)

- $\kappa_1\kappa_2 = 0$ should be satisfied.
- When $\kappa_1 = 0$, we get $M(\kappa_3)M_{tele}(\kappa_1 = 0, \kappa_2) = \begin{pmatrix} 1 & -(\kappa_3\kappa_2 - 2)/\kappa_2 \\ 0 & 1 \end{pmatrix}$, thus only $d = 1$ can be achieved.
- When $\kappa_2 = 0$, we get $M(\kappa_3)M_{tele}(\kappa_1, \kappa_2 = 0) = \begin{pmatrix} -1 & (\kappa_3\kappa_1 - 2)/\kappa_1 \\ 0 & -1 \end{pmatrix}$, thus only $d = -1$ can be achieved.
- Therefore, operations with $d \in \mathbb{R} \setminus 0$ other than $d = \pm 1$ cannot be achieved.

Case with d = 1

- Since $-\dfrac{\kappa_1 + \kappa_2}{\kappa_1 - \kappa_2} = 1$, we have to choose $\kappa_1 = 0$.
- Since $M(\kappa_3)M_{tele}(\kappa_1 = 0, \kappa_2) = \begin{pmatrix} 1 & -(\kappa_3\kappa_2 - 2)/\kappa_2 \\ 0 & 1 \end{pmatrix}$, only $c = 0$ can be achieved.
- Therefore, operations with $c \in \mathbb{R}$ other than $c = 0$ cannot be achieved.

Case with d = −1

- Since $-\dfrac{\kappa_1 + \kappa_2}{\kappa_1 - \kappa_2} = -1$, we have to choose $\kappa_2 = 0$.
- Since $M(\kappa_3)M_{tele}(\kappa_1, \kappa_2 = 0) = \begin{pmatrix} -1 & (\kappa_3\kappa_1 - 2)/\kappa_1 \\ 0 & -1 \end{pmatrix}$, only $c = 0$ can be achieved.
- Therefore, operations with $c \in \mathbb{R}$ other than $c = 0$ cannot be achieved.

Therefore, by using the κ representation, we find that the following operation cannot be implemented:

$$\left\{ A = \begin{pmatrix} a & b \\ c & d \end{pmatrix} \in Sp(2, \mathbb{R}) \middle| (c = 0 \text{ and } d \neq \pm 1, 0) \text{ or } (d = \pm 1 \text{ and } c \neq 0) \right\}. \tag{5.160}$$

Note that homodyne measurements with the observables $\hat{x}$ cannot be formulated in the κ representation. Thus, we have to consider this case separately.

Case with Measurement of $\hat{x}$ Instead of Using κ_3

Similar to the case with the C_Z-based input-coupling, the measurement with the observable $\hat{x}$ in a one-mode teleportation circuit is prohibited since it corresponds to erasing of a mode.

Case with Measurement of $\hat{x}$ Instead of Using κ_1

In the teleportation-based input coupling, we can choose $\hat{x}$ as a measurement variable. When we measure $\hat{x}$ at the first measurement which corresponds to κ_1, we cannot choose the observable $\hat{x}$ at the second measurement which corresponds to κ_2. This is because $\sin\theta_- \neq 0$ should be satisfied in the teleportation-based input-coupling scheme. By considering that we cannot choose the observable $\hat{x}$ at the third measurement which corresponds to κ_3, we find that we can describe all cases of the second and the third measurements by using the κ representation. The total operation is given by

$$M(\kappa_3)M(\text{measurement of } \hat{x}, \kappa_2) = \begin{pmatrix} 2\kappa_2\kappa_3 - 1 & \kappa_3 \\ -2\kappa_2 & -1 \end{pmatrix}. \tag{5.161}$$

Thus, we find that we can implement operations with $d = -1$ by choosing κ_2 and κ_3 appropriately.

Case with Measurement of $\hat{x}$ Instead of Using κ_2

Lastly, when we measure $\hat{x}$ at the second measurement which corresponds to κ_2, we cannot choose the observable $\hat{x}$ at the first measurement which corresponds to κ_1. We find that we can describe all cases of the first and the third measurements by using the κ representation. The total operation is given by

$$M(\kappa_3)M(\kappa_1, \text{measurement of } \hat{x}) = \begin{pmatrix} 1 - 2\kappa_1\kappa_3 & -\kappa_3 \\ 2\kappa_1 & 1 \end{pmatrix}. \tag{5.162}$$

Thus, we find that we can implement operations with $d = 1$ by choosing κ_1 and κ_3 appropriately.

In summary, we find that the following operation cannot be implemented:

$$\left\{ A = \begin{pmatrix} a & b \\ c & d \end{pmatrix} \in Sp(2, \mathbb{R}) \middle| c = 0 \text{ and } d \neq \pm 1, 0 \right\}. \tag{5.163}$$

Therefore, universality is not achieved.

5.5.2.8 Four-Step Operation

We consider that we add another one-mode teleportation circuit. The operation is given by

$$M(\kappa_4)M(\kappa_3)M_{tele}(\kappa_1,\kappa_2) = \frac{1}{\kappa_1-\kappa_2} \begin{pmatrix} -\kappa_4(2\kappa_3\kappa_2\kappa_1-\kappa_1-\kappa_2)+2\kappa_2\kappa_1 & -\kappa_4(\kappa_3\kappa_2+\kappa_3\kappa_1-2)+\kappa_2+\kappa_1 \\ 2\kappa_3\kappa_2\kappa_1-\kappa_1-\kappa_2 & \kappa_3\kappa_2+\kappa_3\kappa_1-2 \end{pmatrix}. \tag{5.164}$$

In the following, we acquire κ_1, κ_2, κ_3, and κ_4 for $\begin{pmatrix} a & b \\ c & d \end{pmatrix}$, $ad-bc=1$.[1]

Case with d $\neq$ 0

- By choosing κ_1 appropriately, we can set $1-c+2d\kappa_1 \neq 0$, $\kappa_1 \neq 0$, $-c+d\kappa_1 \neq 0$.
- We choose $\kappa_2 = \dfrac{(1+c)\kappa_1}{1-c+2d\kappa_1}$, $\kappa_3 = \dfrac{1-c+d\kappa_1}{\kappa_1}$, $\kappa_4 = \dfrac{a+(1-b)\kappa_1}{-c+d\kappa_1}$.

Case with d = 0 and c $\neq$ 1

- $c=d=0$ is not satisfied since $ad-bc=1$.
- We choose $\kappa_1 \neq 0$.
- We choose $\kappa_2 = \dfrac{(1+c)\kappa_1}{1-c}$, $\kappa_3 = \dfrac{1-c}{\kappa_1}$, $\kappa_4 = \dfrac{a+(1-b)\kappa_1}{-c}$.

Case with d = 0 and c = 1

- We choose $\kappa_1 = 0$, $\kappa_2 \neq 0$, $\kappa_3 = \dfrac{2}{\kappa_2}$, $\kappa_4 = -a$.

In summary, we can implement an arbitrary one-mode symplectic operation by using a teleportation-based input-coupling circuit, followed by two-step one-mode teleportation circuit. The resource state for this operation is a four-mode linear cluster state.

[1] Although we can use observables $\hat{x}$ at the first and the second measurements which correspond to κ_1 and κ_2, we will omit these cases because universality will be achieved without using these observables.

5.5.3 Arbitrary Multi-Mode Gaussian Operations

We consider implementation of a two-mode Gaussian operation. The elementary circuit is given in Fig. 5.29.

We define

$$\hat{x}_j + i\hat{p}_j, \quad j = 1, 2, a, b, \tag{5.165}$$

as the annihilation operator of each mode j. Here, mode a and mode b represent two input modes, while mode 1 and mode 2 represent two resource modes. The input-output relationship in the Heisenberg picture is given by

$$\begin{pmatrix} \hat{x}'_1 \\ \hat{p}'_1 \\ \hat{x}'_2 \\ \hat{p}'_2 \end{pmatrix} = \begin{pmatrix} 0 & -1 & 0 & 0 \\ 1 & 0 & 0 & -1 \\ 0 & 0 & 0 & -1 \\ 0 & -1 & 1 & 0 \end{pmatrix} \begin{pmatrix} \hat{x}_a \\ \hat{p}_a \\ \hat{x}_b \\ \hat{p}_b \end{pmatrix} + \begin{pmatrix} 0 \\ \hat{p}_1 \\ 0 \\ \hat{p}_2 \end{pmatrix}. \tag{5.166}$$

Note that Eq. (5.166) holds independently of initial states.

We here assume that modes 1 and 2 are initially in $|p = 0\rangle$. Since $\hat{p}_1 = 0$ and $\hat{p}_2 = 0$, Eq. (5.166) represents the ideal operation:

$$\begin{pmatrix} \hat{x}'_1 \\ \hat{p}'_1 \\ \hat{x}'_2 \\ \hat{p}'_2 \end{pmatrix} = \begin{pmatrix} 0 & -1 & 0 & 0 \\ 1 & 0 & 0 & -1 \\ 0 & 0 & 0 & -1 \\ 0 & -1 & 1 & 0 \end{pmatrix} \begin{pmatrix} \hat{x}_a \\ \hat{p}_a \\ \hat{x}_b \\ \hat{p}_b \end{pmatrix}. \tag{5.167}$$

The matrix which shows the operation can be decomposed into

$$\begin{pmatrix} 0 & -1 & 0 & 0 \\ 1 & 0 & 0 & -1 \\ 0 & 0 & 0 & -1 \\ 0 & -1 & 1 & 0 \end{pmatrix} = \begin{pmatrix} 1 & 0 & 0 & 0 \\ 0 & 1 & 1 & 0 \\ 0 & 0 & 1 & 0 \\ 1 & 0 & 0 & 1 \end{pmatrix} \begin{pmatrix} 0 & -1 & 0 & 0 \\ 1 & 0 & 0 & 0 \\ 0 & 0 & 1 & 0 \\ 0 & 0 & 0 & 1 \end{pmatrix} \begin{pmatrix} 1 & 0 & 0 & 0 \\ 0 & 1 & 0 & 0 \\ 0 & 0 & 0 & -1 \\ 0 & 0 & 1 & 0 \end{pmatrix}. \tag{5.168}$$

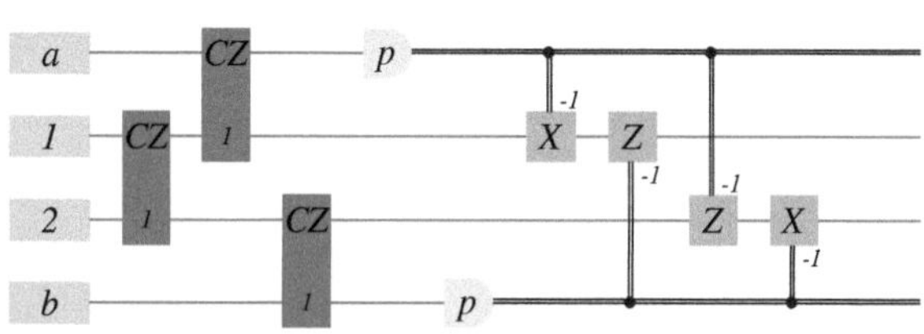

Fig. 5.29 Elementary circuit for two-mode Gaussian operation

Thus, the corresponding operator is $\hat{C}_{Z\,ab}\hat{F}_a\hat{F}_b$. By combining with implementation of an arbitrary one-mode Gaussian operations, we can achieve an arbitrary multi-mode Gaussian operations.

Figure 5.29 can be transformed into Fig. 5.30 by moving the controlled-Z gate on modes 1 and 2 to the end of the circuit.

Figure 5.30 shows a circuit where a controlled-Z gate is applied to the output modes of two C_Z-based input-coupling circuits. Since a Fourier transformation is applied to the input mode of a C_Z-based input-coupling circuit, the operator $\hat{C}_{Z\,ab}\hat{F}_a\hat{F}_b$ achieved by Fig. 5.29 is consistent with that by Fig. 5.30.

5.5.4 One-Mode Non-Gaussian Operations

5.5.4.1 Operators in One-Mode Teleportation Circuit

Since the order of the operator $\hat{D}$ in Sect. 4.2.3 can be larger than or equal to three, it is obvious that one-mode non-Gaussian operation can be implemented in one-way quantum computation. However, it is not obvious how such measurements can be implemented in experiments. Recently, it is reported that a cubic phase gate can be achieved by using a photon counting [5, 8, 9].

5.5.4.2 Photon Counting and Cubic Phase Gate

Consider a circuit for generation of a cubic phase state, which is shown in Fig. 5.31.

The initial states in modes 1 and 2 are $|p = 0\rangle$, or p-squeezed states. By combining these modes with a controlled-Z gate, a two-mode cluster state is generated. A displacement operator is applied to mode 1, where the amount of displacement is sufficiently larger than the squeezing parameter of the squeezed states. A photon counting is performed on mode 1, leading to a measurement result n. The output state in mode 2 is an approximation of a cubic phase state:

$$|\psi\rangle_2 \simeq e^{i\eta(n)\hat{x}^3}|p = 0\rangle. \tag{5.169}$$

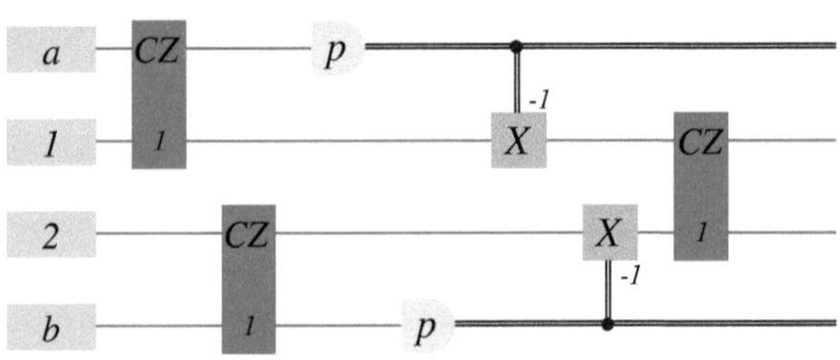

Fig. 5.30 Equivalent circuit of elementary circuit for two-mode Gaussian operation

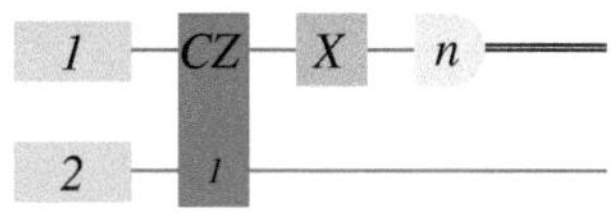

Fig. 5.31 Generation of cubic phase state

Here, $\eta(n)$ is a function of the measurement result n. In order to generate a cubic phase state $e^{i\gamma\hat{x}^3}|p=0\rangle$ with γ which is independent of the measurement result, the random term should be eliminated by using

$$\hat{S}^\dagger(r(n))e^{i\eta(n)\hat{x}^3}\hat{S}(r(n)) = e^{i\gamma\hat{x}^3}, \quad r(n) = -\frac{1}{3}\log\frac{\gamma}{\eta(n)}. \tag{5.170}$$

Thus, consider the circuit shown in Fig. 5.32.

Mode *in* shows an input mode, while mode 2 corresponds to the output mode 2 in Fig. 5.31. Thus, Fig. 5.32 shows the following procedure. First, a squeezing operator $\hat{S}(r(n))$ is applied to the input state. Next, a cubic phase operator is applied to the output state of the previous step via offline scheme (Sect. 4.2.2, Fig. 4.10) by using the cubic phase state as a resource, which is generated by the circuit shown in Fig. 5.31. Finally, another squeezing operator $\hat{S}^\dagger(r(n))$ is applied to the output state of the previous step. Up to Fourier transformations, the output state becomes

$$\hat{S}^\dagger(r(n))e^{i\eta(n)\hat{x}^3}\hat{S}(r(n))|\psi_{in}\rangle = e^{i\gamma\hat{x}^3}|\psi_{in}\rangle, \tag{5.171}$$

where $|\psi_{in}\rangle$ represents the input state in mode *in*.

Since the operator in the offline scheme is a cubic phase operator, the required feed-forward operations are members of one-mode Gaussian operations. In addition, squeezing operations are also one-mode Gaussian operations. In general, an arbitrary one-mode Gaussian operation can be achieved by using a four-mode linear cluster state as a resource state. Therefore, the circuit shown in Fig. 5.32 can be achieved by Fig. 5.33a, whose graph representation is given by Fig. 5.33b.

5.6 One-Way Quantum Computation with $\hat{\delta}$ Representation

As the final issue of this chapter, we discuss imperfection in one-way quantum computation. Although one can consider several origins of errors, we focus only

Fig. 5.32 Cubic phase gate

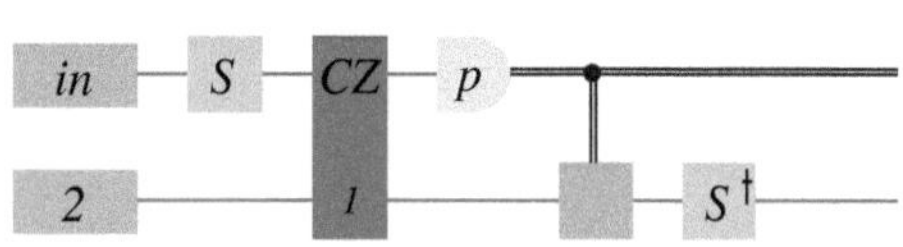

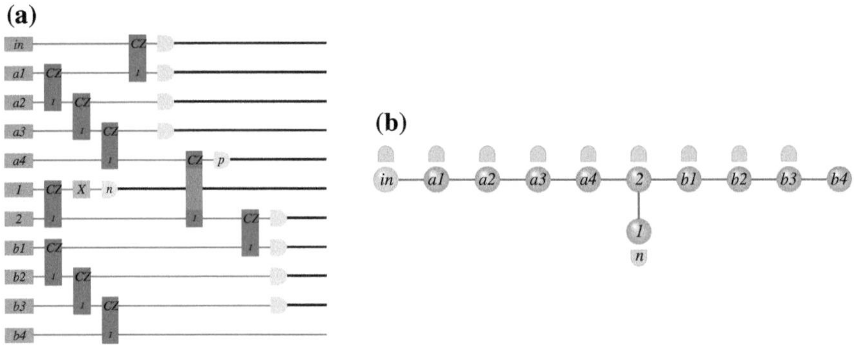

Fig. 5.33 Cubic phase gate with cluster state. **a** Quantum circuit, **b** graph

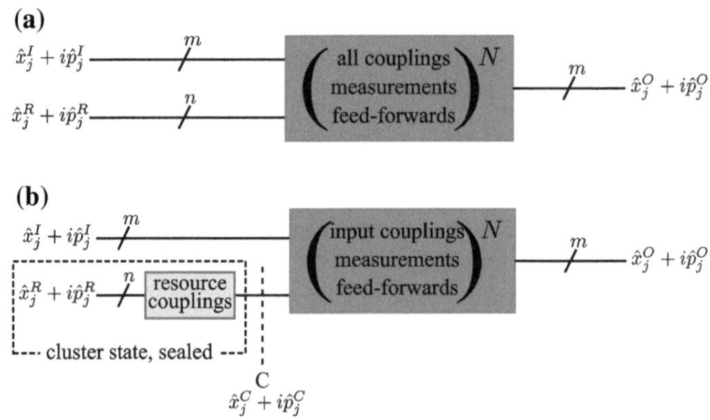

Fig. 5.34 Abstract diagram of cascaded elementary gates. **a** Cascaded elementary gates, **b** converted version of cascaded elementary gates

on limited levels of resource squeezing, which is dominant in recent experimental demonstrations of one-way quantum computations.

In all building blocks for one-way quantum computation we have so far discussed, including the elementary one-mode and two-mode one-way gates, and three types of input coupling schemes, errors are described by additional momentum operators of resource modes in the annihilation operators of the output modes. Since an arbitrary quantum computation can be constructed by a concatenation of these blocks, it is obvious that imperfections of the total computation can also be described by additional momentum operators of the resource modes.

In the discussion above, we have assumed that each resource state is coupled with the previous output state using a C_Z gate after all previous operations are finished. It is discrepant with one-way quantum computation where the resource is prepared as a multi-party entangled cluster state. Its discrepancy can be removed by reordering of the C_Z gates.

Consider one-way quantum computation on an m-mode input state with n-mode resource states. Although we usually assume that these resources are $|p = 0\rangle$ or p-squeezed vacuum states implicitly, the following discussion does not depend on their actual states. We define the position and momentum operators of these resource modes to be $\hat{x}_j^R$ and $\hat{p}_j^R$, respectively, where $j = 1, \ldots, n$. At the first stage, we assume that the elementary building blocks are just cascaded (Fig. 5.34a). From the discussion above, the position and momentum operators of the output modes ($\hat{x}_j^O$ and $\hat{p}_j^O$ for $j = 1, \ldots, m$) can be described by the position and momentum operators of the input mode operators ($\hat{x}_j^I$ and $\hat{p}_j^I$ for $j = 1, \ldots, m$) and the momentum operators of the resource modes ($\hat{p}_j^R$):

$$\hat{x}_j^O = F_j(\hat{x}_1^I, \ldots, \hat{x}_m^I, \hat{p}_1^I, \ldots, \hat{p}_m^I, \hat{p}_1^R, \ldots, \hat{p}_n^R), \tag{5.172}$$

$$\hat{p}_j^O = G_j(\hat{x}_1^I, \ldots, \hat{x}_m^I, \hat{p}_1^I, \ldots, \hat{p}_m^I, \hat{p}_1^R, \ldots, \hat{p}_n^R). \tag{5.173}$$

We consider changes to the orders of the feed-forwards and the C_Z gates. First, all C_Z operators commute with each other. Next, each original feed-forward is a displacement in phase space, which is described by displacement operators $\hat{X}_k(s)$ and $\hat{Z}_k(s)$. Since $\hat{X}_k(s)$ and $\hat{Z}_k(s)$ are members of the Pauli group, and $\hat{C}_{Zij}$ is a member of the Clifford group, there exists a Pauli operator $\hat{P}_{ijk}$ which satisfies

$$\hat{C}_{Zij}\hat{X}_k(s) = \hat{P}_{ijk}\hat{C}_{Zij}, \tag{5.174}$$

where $\hat{P}_{ijk}$ is given by

$$\hat{P}_{ijk} = \begin{cases} \hat{X}_i(s)\hat{Z}_j(s) & (i = k) \\ \hat{X}_j(s)\hat{Z}_i(s) & (j = k) \\ \hat{X}_k(s) & (\text{else}) \end{cases}. \tag{5.175}$$

Therefore, we can convert the original quantum circuit (Fig. 5.34a) to a newer one (Fig. 5.34b) where all n-mode resource states are initially entangled with a network of C_Z gates, leading to a cluster state. After coupling of an input state with the cluster state, the measurements and the feed-forwards are carried out. In this conversion processes, we can preserve all feed-forwards in members of the Pauli group.

We define position and momentum operators after the C_Z-based couplings of the resource modes to be $\hat{x}_j^C$ and $\hat{p}_j^C$, respectively. They are given by

$$\hat{x}_j^C + i\hat{p}_j^C = \hat{x}_j^R + i\left(\hat{p}_j^R + \sum_{k \in N(j)} \hat{x}_k^R\right), \tag{5.176}$$

where $N(j)$ represents the set of nearest neighbor modes of mode j. Thus, we can describe the resource mode operators $\hat{p}_j^R$ by means of the cluster mode operators $\hat{x}_j^C$

and $\hat{p}_j^C$:

$$\hat{p}_j^R = \hat{p}_j^C - \sum_{k\in N(j)} \hat{x}_k^C. \tag{5.177}$$

We define $\hat{\delta}_j^C \equiv \hat{p}_j^C - \sum_{k\in N(j)} \hat{x}_k^C$, which corresponds to the nullifiers of the cluster state. We can replace $\hat{p}_j^R$ with $\hat{\delta}_j^C$ in Eqs. (5.172) and (5.173), leading to

$$\hat{x}_j^O = F_j(\hat{x}_1^I, \ldots, \hat{x}_m^I, \hat{p}_1^I, \ldots, \hat{p}_m^I, \hat{\delta}_1^C, \ldots, \hat{\delta}_n^C), \tag{5.178}$$

$$\hat{p}_j^O = G_j(\hat{x}_1^I, \ldots, \hat{x}_m^I, \hat{p}_1^I, \ldots, \hat{p}_m^I, \hat{\delta}_1^C, \ldots, \hat{\delta}_n^C). \tag{5.179}$$

The important thing is that the additional terms in Eqs. (5.178) and (5.179) are described not by the original resource modes ($\hat{p}_j^R$) but by the cluster modes ($\hat{x}_j^C$ and $\hat{p}_j^C$). Since the discussion above does not depend on the initial states at (R), the state at the cluster modes (C) can be in any state. However, in order to achieve an ideal one-way quantum computation, the cluster state should be chosen so that it satisfies $\hat{\delta}_j^C = 0$, or $\hat{\delta}_j^C \to 0$ in the limit of infinite squeezing. Note that it does not necessarily mean that the state at (C) should be prepared as an ideal cluster state or a canonical cluster state. For example, it can be prepared as a Gaussian cluster state, which is generated by combining squeezed states on beam splitters. In this case, the state at (R) is considered to be that acquired by applying the inversed C_Z network operations on the Gaussian cluster state. Since Eqs. (5.178) and (5.179) hold in all cases, and the Gaussian cluster state satisfies $\hat{\delta}_j^C \to 0$, an ideal operation is achieved in the limit of infinite squeezing.

By describing the quadrature operators of the output modes (O) by using those of the input modes (I) and the cluster modes (C), we can discuss one-way quantum computation without referring to the state of the resource mode (R). The common property of the cluster state is $\hat{\delta}_j^C = 0$ or $\hat{\delta}_j^C \to 0$. Therefore, any state which satisfies this nullifier condition can be used as a resource for one-way quantum computation without any discriminations. No matter how a cluster state is generated—for example, as a canonical cluster state, or as a Gaussian cluster state—the same input output relations (Eqs. (5.178) and (5.179)) are satisfied. In addition, measurement bases and feed-forwards for a specific operation do not depend on the schemes of cluster generations.

The difference of the states at the cluster mode (C) is represented by $\hat{\delta}_j^C$. For example, a canonical cluster state and a Gaussian cluster state have different nullifier values. The difference of errors in one-way quantum computation is explained by the difference of $\hat{\delta}_j^C$.

References

1. Zhang, J., Braunstein, S.L.: Continuous-variable Gaussian analog of cluster states. Phys. Rev. A **73**, 032318 (2006)
2. van Loock, Peter, Weedbrook, Christian, Mile, Gu: Building Gaussian cluster states by linear optics. Phys. Rev. A **76**, 032321 (2007)
3. Reck, M., Zeilinger, A., Bernstein, H.J., Bertani, P.: Experimental realization of any discrete unitary operator. Phys. Rev. Lett **73**, 58 (1994)
4. Braunstein, S.L.: Squeezing as an irreducible resource. Phys. Rev. A **71**, 055801 (2005)
5. Gu, M., Weedbrook, C., Menicucci, N.C., Ralph, T.C., van Loock, P.: Quantum computing with continuous-variable clusters. Phys. Rev. A **79**, 062318 (2009)
6. Miwa, Y., Ukai, R., Yoshikawa, J., Filip, R., van Loock, P., Furusawa, A.: Demonstration of cluster-state shaping and quantum erasure for continuous variables. Phys. Rev. A **82**, 032305 (2010)
7. Menicucci, N.C.: Temporal-mode continuous-variable cluster states using linear optics. Phys. Rev. A **83**, 062314 (2011)
8. Gottesman, D., Kitaev, A., Preskill, J.: Encoding a qubit in an oscillator. Phys. Rev. A **64**, 012310 (2001)
9. Furusawa, A., van Loock, P.: Quantum Teleportation and Entanglement. Wiley, Berlin (2011)

Chapter 6
Experimental Generation of Optical Continuous-Variable Cluster States

6.1 Theory on Cluster State Generation

6.1.1 Two-Mode Cluster State

6.1.1.1 Experimental Setup

Figure 6.1 shows a schematic of out experimental setup for generation of a two-mode cluster state.

In the following, we describe theories on generation of the two-mode cluster state in both cases with and without losses.

6.1.1.2 Without Losses

An approximation of a two-mode cluster state can be generated by combining two p-squeezed states on a beam splitter where its reflectivity and relative phase are chosen appropriately (Gaussian cluster state, Sect. 5.2.2). We employ this scheme for our experimental generation of the two-mode cluster state. In the following, we formulate its procedure by using quadrature operators which specify particular quantum states (Sect. 3.2.8).

We assume that the initial states in modes 1 and 2 are vacuum states. We describe these modes as

$$\hat{x}_k^{(0)} + i\hat{p}_k^{(0)}, \quad k = 1, 2, \tag{6.1}$$

where the subscript k denotes the mode number, while the superscripts (0) represent that these modes are in vacuum states.

As a next step, we apply p-squeezing operators on both modes. We can describe these two squeezed states as

$$\hat{x}_k^{(r)} + i\hat{p}_k^{(r)} = e^{r_k}\hat{x}_k^{(0)} + ie^{-r_k}\hat{p}_k^{(0)}, \quad k = 1, 2, \tag{6.2}$$

R. Ukai, *Multi-Step Multi-Input One-Way Quantum Information Processing with Spatial and Temporal Modes of Light*, Springer Theses,
DOI 10.1007/978-4-431-55019-8_6

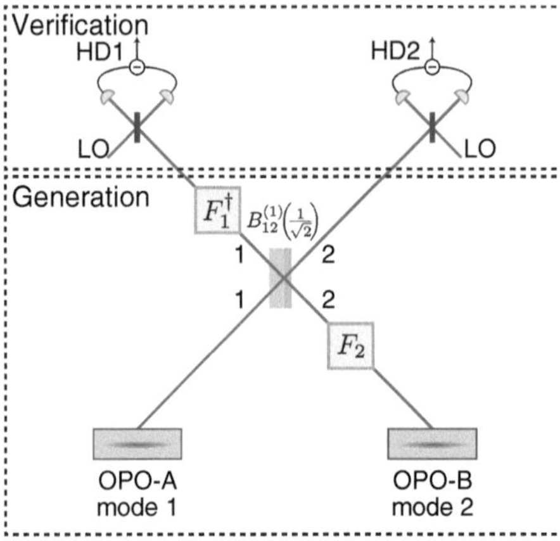

Fig. 6.1 Experimental setup for generation of two-mode cluster state, *OPO* Optical Parametric Oscillator, *HD* Homodyne Detector, *LO* Local Oscillator

where r_k represents the squeezing parameter for mode k, while the superscripts (r) represent that these modes are in squeezed states.

In general, an approximation of an N-mode cluster state can be generated by combining N p-squeezed states via an appropriate network of beam splitters (Sect. 5.2.2). Since an operation defined by a beam splitter network is a member of photon-number-preserving Gaussian operations, it can be formulated by a linear transformation of a column vector of the annihilation operators, where the corresponding matrix is a unitary matrix (Sect. 3.5.8). Note that it is not determined uniquely. In our case of the two-mode cluster state generation, it can be generated by using a unitary matrix U:

$$U = \begin{pmatrix} \frac{1}{\sqrt{2}}i & \frac{1}{\sqrt{2}} \\ \frac{1}{\sqrt{2}} & \frac{1}{\sqrt{2}}i \end{pmatrix}, \tag{6.3}$$

leading to a linear transformation of an annihilation operator vector:

$$\begin{pmatrix} \hat{x}_1^{(c)} + i\hat{p}_1^{(c)} \\ \hat{x}_2^{(c)} + i\hat{p}_2^{(c)} \end{pmatrix} = U \begin{pmatrix} \hat{x}_1^{(r)} + i\hat{p}_1^{(r)} \\ \hat{x}_2^{(r)} + i\hat{p}_2^{(r)} \end{pmatrix}, \tag{6.4}$$

where the superscripts (c) represent that these modes are in a two-mode cluster state. We then get

$$\begin{pmatrix} \hat{x}_1^{(c)} + i\hat{p}_1^{(c)} \\ \hat{x}_2^{(c)} + i\hat{p}_2^{(c)} \end{pmatrix} = \begin{pmatrix} -\frac{1}{\sqrt{2}}e^{-r_1}\hat{p}_1^{(0)} + \frac{1}{\sqrt{2}}e^{r_2}\hat{x}_2^{(0)} \\ \frac{1}{\sqrt{2}}e^{r_1}\hat{x}_1^{(0)} - \frac{1}{\sqrt{2}}e^{-r_2}\hat{p}_2^{(0)} \end{pmatrix} + i \begin{pmatrix} \frac{1}{\sqrt{2}}e^{r_1}\hat{x}_1^{(0)} + \frac{1}{\sqrt{2}}e^{-r_2}\hat{p}_2^{(0)} \\ \frac{1}{\sqrt{2}}e^{-r_1}\hat{p}_1^{(0)} + \frac{1}{\sqrt{2}}e^{r_2}\hat{x}_2^{(0)} \end{pmatrix}, \tag{6.5}$$

which leads to the values of the nullifiers:

$$\begin{cases} \hat{\delta}_1 \equiv \hat{p}_1^{(c)} - \hat{x}_2^{(c)} = \sqrt{2}e^{-r_2}\hat{p}_2^{(0)} & \text{(i)} \\ \hat{\delta}_2 \equiv \hat{p}_2^{(c)} - \hat{x}_1^{(c)} = \sqrt{2}e^{-r_1}\hat{p}_1^{(0)} & \text{(ii)} \end{cases}. \tag{6.6}$$

Importantly, $\hat{\delta}_k$ becomes zero for each k in the limit of infinite squeezing $r_l \to \infty$ for both l. Therefore, we find that mode 1 and mode 2 are in a two-mode cluster state.

6.1.1.3 With Losses

As a next step, we will describe generation of a two-mode cluster state when an experimental setup has non-zero losses.

There are several sources of losses, including intra-cavity losses of OPOs, propagation losses, and losses at homodyne detections. However, for simplicity, we use a model of losses that there are no losses except for homodyne detections.[1] It can be formulated by placing virtual beam splitters just before homodyne detectors.

We assume that the energy loss in mode k ($k = 1, 2$) is $1-\eta_k$, where η_k represents the energy efficiency. We define mode kL as the loss mode, which is initially in a vacuum state, for mode k. The energy loss in mode k is then formulated by combining mode k and mode kL via a beam splitter $B^{(4)}_{k,kL}(\sqrt{\eta_k})$. By extracting mode 1 and mode 2 after these beam splitter couplings with the loss modes, we get cluster modes with losses as

$$\begin{aligned} \begin{pmatrix} \hat{x}_1^{(cl)} + i\hat{p}_1^{(cl)} \\ \hat{x}_2^{(cl)} + i\hat{p}_2^{(cl)} \end{pmatrix} &= \begin{pmatrix} \sqrt{\eta_1} & 0 \\ 0 & \sqrt{\eta_2} \end{pmatrix} \begin{pmatrix} \hat{x}_1^{(c)} + i\hat{p}_1^{(c)} \\ \hat{x}_2^{(c)} + i\hat{p}_2^{(c)} \end{pmatrix} + \begin{pmatrix} \sqrt{1-\eta_1} & 0 \\ 0 & \sqrt{1-\eta_2} \end{pmatrix} \begin{pmatrix} \hat{x}_{1L}^{(0)} + i\hat{p}_{1L}^{(0)} \\ \hat{x}_{2L}^{(0)} + i\hat{p}_{2L}^{(0)} \end{pmatrix} \\ &= \begin{pmatrix} -\sqrt{\frac{\eta_1}{2}}e^{-r_1}\hat{p}_1^{(0)} + \sqrt{\frac{\eta_1}{2}}e^{r_2}\hat{x}_2^{(0)} + \sqrt{1-\eta_1}\hat{x}_{1L}^{(0)} \\ \sqrt{\frac{\eta_2}{2}}e^{r_1}\hat{x}_1^{(0)} - \sqrt{\frac{\eta_2}{2}}e^{-r_2}\hat{p}_2^{(0)} + \sqrt{1-\eta_2}\hat{x}_{2L}^{(0)} \end{pmatrix} \\ &\quad + i \begin{pmatrix} \sqrt{\frac{\eta_1}{2}}e^{r_1}\hat{x}_1^{(0)} + \sqrt{\frac{\eta_1}{2}}e^{-r_2}\hat{p}_2^{(0)} + \sqrt{1-\eta_1}\hat{p}_{1L}^{(0)} \\ \sqrt{\frac{\eta_2}{2}}e^{-r_1}\hat{p}_1^{(0)} + \sqrt{\frac{\eta_2}{2}}e^{r_2}\hat{x}_2^{(0)} + \sqrt{1-\eta_2}\hat{p}_{2L}^{(0)} \end{pmatrix}, \end{aligned} \tag{6.7}$$

which leads to the values of the cluster nullifiers:

$$\begin{cases} \hat{\delta}_1^{(l)} \equiv \hat{p}_1^{(cl)} - \hat{x}_2^{(cl)} = (\sqrt{\frac{\eta_1}{2}} - \sqrt{\frac{\eta_2}{2}})e^{r_1}\hat{x}_1^{(0)} + (\sqrt{\frac{\eta_1}{2}} + \sqrt{\frac{\eta_2}{2}})e^{-r_2}\hat{p}_2^{(0)} \\ \qquad + \sqrt{1-\eta_1}\hat{p}_{1L}^{(0)} - \sqrt{1-\eta_2}\hat{x}_{2L}^{(0)} \quad \text{(i)} \\ \hat{\delta}_2^{(l)} \equiv \hat{p}_2^{(cl)} - \hat{x}_1^{(cl)} = (\sqrt{\frac{\eta_2}{2}} - \sqrt{\frac{\eta_1}{2}})e^{r_2}\hat{x}_2^{(0)} + (\sqrt{\frac{\eta_1}{2}} + \sqrt{\frac{\eta_2}{2}})e^{-r_1}\hat{p}_1^{(0)} \\ \qquad + \sqrt{1-\eta_2}\hat{p}_{2L}^{(0)} - \sqrt{1-\eta_1}\hat{x}_{1L}^{(0)} \quad \text{(ii)} \end{cases}. \tag{6.8}$$

When losses are asymmetrical ($\eta_1 \neq \eta_2$), the nullifier values have the anti-squeezing components $\hat{x}_k^{(0)}$ of the squeezed states. We also find that the nullifier values do not

[1] Another useful model is that there are no losses except for generation of squeezed states. It can be formulated by placing virtual beam splitters just after ideal squeezed-state resources.

become zero in the limit of infinite squeezing $r \to \infty$ when there are finite losses ($\eta_k < 1$).

On the other hand, when losses are symmetrical ($\eta_k = \eta$), we get

$$\begin{cases} \hat{\delta}_1^{(l)} \equiv \hat{p}_1^{(cl)} - \hat{x}_2^{(cl)} = \sqrt{2\eta}e^{-r_2}\hat{p}_2^{(0)} + \sqrt{1-\eta}\hat{p}_{1L}^{(0)} - \sqrt{1-\eta}\hat{x}_{2L}^{(0)} & \text{(i)} \\ \hat{\delta}_2^{(l)} \equiv \hat{p}_2^{(cl)} - \hat{x}_1^{(cl)} = \sqrt{2\eta}e^{-r_1}\hat{p}_1^{(0)} + \sqrt{1-\eta}\hat{p}_{2L}^{(0)} - \sqrt{1-\eta}\hat{x}_{1L}^{(0)} & \text{(ii)} \end{cases}. \tag{6.9}$$

Therefore, the values of the nullifiers do not have the anti-squeezing components. However, they do not become zero in the limit of infinite squeezing $r \to \infty$ when there are finite losses ($\eta_k < 1$).

6.1.1.4 Variances of Nullifiers and Effective Squeezing Level

For simplicity, we here assume that the squeezing parameters and losses are symmetrical ($r_k = r, \eta_k = \eta$). In this case, the variances of the nullifiers are given by

$$\langle \Delta^2 \hat{\delta}_k^{(l)} \rangle = \frac{\hbar}{2}\left[2\eta e^{-2r} + 2(1-\eta)\right]. \tag{6.10}$$

In a special case where there is no losses (efficiency $\eta = 1$), it becomes

$$\langle \Delta^2 \hat{\delta}_k \rangle = \frac{\hbar}{2} \times 2e^{-2r}. \tag{6.11}$$

We define the *effective squeezing parameter* $r^{(e)}$ and the *effective squeezing level* $x^{(e)}$ as the squeezing parameter and the squeezing level which satisfy

$$\langle \Delta^2 \hat{\delta}_k^{(l)} \rangle = \frac{\hbar}{2} \times 2e^{-2r^{(e)}} = \frac{\hbar}{2} \times 2 \times 10^{-\frac{x^{(e)}}{10}}. \tag{6.12}$$

That is, they are the squeezing parameter and the squeezing level of a pure (loss-less) squeezed state which gives us the same variance as the lossy squeezed states.

Although we have defined the effective squeezing level by using the setup of the two-mode cluster state generation, and by assuming that the squeezing parameters and losses are symmetrical, we can extend the discussion above to a general case. That is, we define the *effective squeezing parameter* $r^{(e)}$ and the *effective squeezing level* $x^{(e)}$ as the squeezing parameter and the squeezing level of a pure (loss-less) squeezed state which gives us the same value as the lossy squeezed states. For example, it can be evaluated by a variance of a nullifier, or a logarithmic negativity.

We next consider two operators which are acquired by flipping the signs of the nullifiers:

$$\hat{\delta}_{1+} \equiv \hat{p}_1^{(c)} + \hat{x}_2^{(c)}, \quad \hat{\delta}_{2+} \equiv \hat{p}_2^{(c)} + \hat{x}_1^{(c)}. \tag{6.13}$$

When the squeezing levels and losses are symmetrical ($\eta_k = \eta, r_k = r$), the variance of each operator $\hat{\delta}_{k+}$ becomes

$$\langle \Delta^2 \hat{\delta}^{(l)}_{k+} \rangle = \frac{\hbar}{2} \left[2\eta e^{2r} + 2(1-\eta) \right]. \tag{6.14}$$

In a special case where there is no losses (efficiency $\eta = 1$), it becomes

$$\langle \Delta^2 \hat{\delta}_{k+} \rangle = \frac{\hbar}{2} \times 2e^{2r}. \tag{6.15}$$

We define the *effective squeezing parameter* $r^{(ae)}$ and the *effective squeezing level* $x^{(ae)}$ *of anti-squeezing* as the squeezing parameter and the squeezing level which satisfy

$$\langle \Delta^2 \hat{\delta}^{(l)}_{k+} \rangle = \frac{\hbar}{2} \times 2e^{2r^{(ae)}} = \frac{\hbar}{2} \times 2 \times 10^{\frac{x^{(ae)}}{10}}. \tag{6.16}$$

In analysis of experiment, we mainly utilize the effective squeezing parameter $r^{(e)}$ and effective squeezing level $x^{(e)}$, not $r^{(ae)}$ nor $x^{(ae)}$. This is because input-output relationships in one-way quantum computation are usually constructed so that anti-squeezing components of resource modes are perfectly canceled out in output modes. Therefore, errors in one-way quantum computations originate in squeezing components of resource modes at the outputs (Sect. 5.6). $r^{(ae)}$ and $x^{(ae)}$ will be used only in analysis of losses in this chapter.

6.1.1.5 Covariance Matrix

The covariance matrix (Sect. 3.2.6) of the two-mode cluster state with finite losses can be calculated by using Eq. (6.7). We utilize the *xpxp* notation for the covariance matrix (Sect. 3.2.5). We get the covariance matrix V:

$$V = \frac{\hbar}{2} \begin{pmatrix} v_{11} & 0 & 0 & v_{14} \\ 0 & v_{22} & v_{23} & 0 \\ 0 & v_{23} & v_{33} & 0 \\ v_{14} & 0 & 0 & v_{44} \end{pmatrix}, \tag{6.17}$$

where

$$v_{11} = \frac{1}{2}[\eta_1(e^{-2r_1} + e^{2r_2} - 2) + 2], \tag{6.18}$$

$$v_{14} = \frac{1}{2}\sqrt{\eta_1\eta_2}[e^{2r_2} - e^{-2r_1}], \tag{6.19}$$

$$v_{22} = \frac{1}{2}[\eta_1(e^{2r_1} + e^{-2r_2} - 2) + 2], \tag{6.20}$$

$$v_{23} = \frac{1}{2}\sqrt{\eta_1\eta_2}[e^{2r_1} - e^{-2r_2}], \tag{6.21}$$

$$v_{33} = \frac{1}{2}[\eta_2(e^{2r_1} + e^{-2r_2} - 2) + 2], \tag{6.22}$$

$$v_{44} = \frac{1}{2}[\eta_2(e^{-2r_1} + e^{2r_2} - 2) + 2]. \tag{6.23}$$

When the efficiencies are symmetrical ($\eta_k = \eta$), it becomes

$$V = \frac{\hbar}{2}\begin{pmatrix} v_{11} & 0 & 0 & v_{14} \\ 0 & v_{22} & v_{23} & 0 \\ 0 & v_{23} & v_{22} & 0 \\ v_{14} & 0 & 0 & v_{11} \end{pmatrix}, \tag{6.24}$$

where

$$v_{11} = \frac{1}{2}[\eta(e^{-2r_1} + e^{2r_2} - 2) + 2], \tag{6.25}$$

$$v_{14} = \frac{1}{2}\eta[e^{2r_2} - e^{-2r_1}], \tag{6.26}$$

$$v_{22} = \frac{1}{2}[\eta(e^{2r_1} + e^{-2r_2} - 2) + 2], \tag{6.27}$$

$$v_{23} = \frac{1}{2}\eta[e^{2r_1} - e^{-2r_2}]. \tag{6.28}$$

When the squeezing parameters are also symmetrical ($r_k = r$), it becomes

$$V = \frac{\hbar}{2}\begin{pmatrix} v_{11} & 0 & 0 & v_{14} \\ 0 & v_{11} & v_{14} & 0 \\ 0 & v_{14} & v_{11} & 0 \\ v_{14} & 0 & 0 & v_{11} \end{pmatrix}, \tag{6.29}$$

where

$$v_{11} = \eta(\cosh 2r - 1) + 1, \quad v_{14} = \eta \sinh 2r. \tag{6.30}$$

In the special case of $\eta = 1$, it becomes

$$V = \frac{\hbar}{2}\begin{pmatrix} \cosh 2r & 0 & 0 & \sinh 2r \\ 0 & \cosh 2r & \sinh 2r & 0 \\ 0 & \sinh 2r & \cosh 2r & 0 \\ \sinh 2r & 0 & 0 & \cosh 2r \end{pmatrix}. \tag{6.31}$$

Table 6.1 Mode correspondence between theory and experiment (generation of two-mode cluster state)

OPO	Mode
OPO-A	Mode 1
OPO-B	Mode 2

6.1.1.6 Decomposition into Beam Splitter

We then decompose the transformation defined by Eq. (6.3) into a beam splitter and rotations in phase space. Its decomposition does not determined uniquely. One solution for this is given by

$$U = F_1^\dagger B_{12}^{(1)}\left(\frac{1}{\sqrt{2}}\right) F_2, \tag{6.32}$$

where F_k represents the matrix of the Fourier transformation on mode k. It is equivalent to an identity matrix except that

$$F_k = \left(i\right), \quad F_k^\dagger = \left(-i\right) \tag{6.33}$$

for the (k, k) entry of the matrix F_k. $B_{ij}^{(n)}(\sqrt{R})$ represents the matrix of the beam splitter with the energy reflectivity R on mode i and mode j. By extracting the (i, i), (i, j), (j, i), and (j, j) entries of the beam splitter matrices, they are given by

$$B_{ij}^{(1)}(\sqrt{R}) = \begin{pmatrix} -\sqrt{R} & \sqrt{T} \\ \sqrt{T} & \sqrt{R} \end{pmatrix}, \quad B_{ij}^{(2)}(\sqrt{R}) = \begin{pmatrix} \sqrt{T} & -\sqrt{R} \\ \sqrt{R} & \sqrt{T} \end{pmatrix}, \tag{6.34}$$

$$B_{ij}^{(3)}(\sqrt{R}) = \begin{pmatrix} \sqrt{T} & \sqrt{R} \\ -\sqrt{R} & \sqrt{T} \end{pmatrix}, \quad B_{ij}^{(4)}(\sqrt{R}) = \begin{pmatrix} \sqrt{R} & \sqrt{T} \\ \sqrt{T} & -\sqrt{R} \end{pmatrix}, \tag{6.35}$$

where T represents the energy transmissivity of the beam splitter, which satisfies $T + R = 1$. In this case,

$$F_1 = \begin{pmatrix} i & \\ & 1 \end{pmatrix}, \quad F_2 = \begin{pmatrix} 1 & \\ & i \end{pmatrix}, \quad B_{12}^{(1)}\left(\tfrac{1}{\sqrt{2}}\right) = \begin{pmatrix} -\frac{1}{\sqrt{2}} & \frac{1}{\sqrt{2}} \\ \frac{1}{\sqrt{2}} & \frac{1}{\sqrt{2}} \end{pmatrix}. \tag{6.36}$$

6.1.1.7 Mode Correspondence Between Theory and Experiment

Squeezed states are experimentally generated by using OPOs. Correspondences between the modes in theory and the OPOs in experiment are given by Table 6.1.

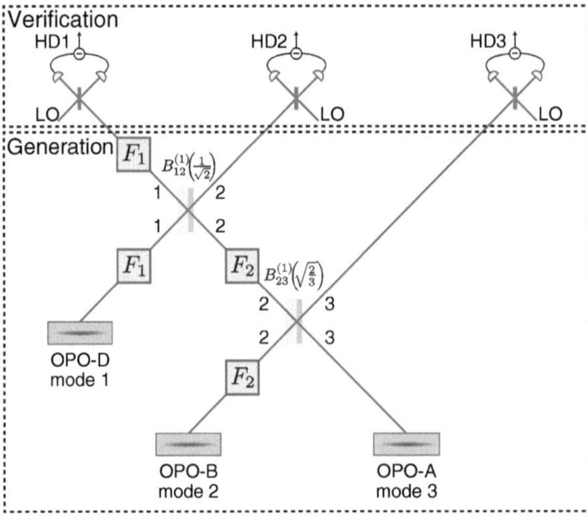

Fig. 6.2 Experimental setup for generation of three-mode linear cluster state

6.1.2 Three-Mode Linear Cluster State

6.1.2.1 Experimental Setup

Figure 6.2 shows a schematic of our experimental setup for generation of a three-mode linear cluster state.

In the following, we describe theories on generation of the three-mode linear cluster state. For simplicity, we assume that there are no propagation losses.

6.1.2.2 Without Losses

An approximation of a three-mode linear cluster state can be generated by combining three p-squeezed states on two beam splitters where their reflectivities and relative phases are chosen appropriately (Gaussian cluster state, Sect. 5.2.2). We employ this scheme for our experimental generation of the three-mode linear cluster state. In the following, we formulate its procedure by using quadrature operators which specify particular quantum states (Sect. 3.2.8).

We assume that the initial states in modes 1, 2 and 3 are vacuum states. We describe these modes as

$$\hat{x}_k^{(0)} + i\hat{p}_k^{(0)}, \quad k = 1, 2, 3, \tag{6.37}$$

where the subscript k denotes the mode number, while the superscripts (0) represent that these modes are in vacuum states.

As a next step, we apply p-squeezing operators on these modes. We can describe the squeezed states as

$$\hat{x}_k^{(r)} + i\hat{p}_k^{(r)} = e^{r_k}\hat{x}_k^{(0)} + ie^{-r_k}\hat{p}_k^{(0)}, \quad k = 1, 2, 3, \tag{6.38}$$

where r_k represents the squeezing parameter for mode k, while the superscripts (r) represent that these modes are in squeezed states.

An approximation of a three-mode linear cluster state can be generated by combining three p-squeezed states on an appropriate network of beam splitters. For example, we can generate it by using a unitary matrix U:

$$U = \begin{pmatrix} \frac{1}{\sqrt{2}} & \frac{1}{\sqrt{3}}i & -\frac{1}{\sqrt{6}} \\ \frac{1}{\sqrt{2}}i & \frac{1}{\sqrt{3}} & \frac{1}{\sqrt{6}}i \\ 0 & \frac{1}{\sqrt{3}}i & \frac{2}{\sqrt{6}} \end{pmatrix}, \tag{6.39}$$

leading to a linear transformation of an annihilation operator vector:

$$\begin{pmatrix} \hat{x}_1^{(c)} + i\hat{p}_1^{(c)} \\ \hat{x}_2^{(c)} + i\hat{p}_2^{(c)} \\ \hat{x}_3^{(c)} + i\hat{p}_3^{(c)} \end{pmatrix} = U \begin{pmatrix} \hat{x}_1^{(r)} + i\hat{p}_1^{(r)} \\ \hat{x}_2^{(r)} + i\hat{p}_2^{(r)} \\ \hat{x}_3^{(r)} + i\hat{p}_3^{(r)} \end{pmatrix}, \tag{6.40}$$

where the superscripts (c) represent that these modes are in a three-mode linear cluster state. We then get

$$\begin{pmatrix} \hat{x}_1^{(c)} + i\hat{p}_1^{(c)} \\ \hat{x}_2^{(c)} + i\hat{p}_2^{(c)} \\ \hat{x}_3^{(c)} + i\hat{p}_3^{(c)} \end{pmatrix} = \begin{pmatrix} \frac{1}{\sqrt{2}}e^{r_1}\hat{x}_1^{(0)} - \frac{1}{\sqrt{3}}e^{-r_2}\hat{p}_2^{(0)} - \frac{1}{\sqrt{6}}e^{r_3}\hat{x}_3^{(0)} \\ -\frac{1}{\sqrt{2}}e^{-r_1}\hat{p}_1^{(0)} + \frac{1}{\sqrt{3}}e^{r_2}\hat{x}_2^{(0)} - \frac{1}{\sqrt{6}}e^{-r_3}\hat{p}_3^{(0)} \\ - \frac{1}{\sqrt{3}}e^{-r_2}\hat{p}_2^{(0)} + \frac{2}{\sqrt{6}}e^{r_3}\hat{x}_3^{(0)} \end{pmatrix} + i\begin{pmatrix} \frac{1}{\sqrt{2}}e^{-r_1}\hat{p}_1^{(0)} + \frac{1}{\sqrt{3}}e^{r_2}\hat{x}_2^{(0)} - \frac{1}{\sqrt{6}}e^{-r_3}\hat{p}_3^{(0)} \\ \frac{1}{\sqrt{2}}e^{r_1}\hat{x}_1^{(0)} + \frac{1}{\sqrt{3}}e^{-r_2}\hat{p}_2^{(0)} + \frac{1}{\sqrt{6}}e^{r_3}\hat{x}_3^{(0)} \\ \frac{1}{\sqrt{3}}e^{r_2}\hat{x}_2^{(0)} + \frac{2}{\sqrt{6}}e^{-r_3}\hat{p}_3^{(0)} \end{pmatrix}, \tag{6.41}$$

which leads to the values of the nullifiers:

$$\begin{cases} \hat{\delta}_1 \equiv \hat{p}_1^{(c)} - \hat{x}_2^{(c)} = \sqrt{2}e^{-r_1}\hat{p}_1^{(0)} & \text{(i)} \\ \hat{\delta}_2 \equiv \hat{p}_2^{(c)} - \hat{x}_1^{(c)} - \hat{x}_3^{(c)} = \sqrt{3}e^{-r_2}\hat{p}_2^{(0)} & \text{(ii)} \\ \hat{\delta}_3 \equiv \hat{p}_3^{(c)} - \hat{x}_2^{(c)} = \frac{1}{\sqrt{2}}e^{-r_1}\hat{p}_1^{(0)} + \frac{3}{\sqrt{6}}e^{-r_3}\hat{p}_3^{(0)} & \text{(iii)} \end{cases}. \tag{6.42}$$

Table 6.2 Mode correspondence between theory and experiment (generation of three-mode linear cluster state)

OPO	Mode
OPO-A	Mode 3
OPO-B	Mode 2
OPO-D	Mode 1

Importantly, $\hat{\delta}_k$ becomes zero for each k in the limit of infinite squeezing $r_l \to \infty$ for all l. Therefore, we find that modes 1, 2, and 3 are in a three-mode linear cluster state.

6.1.2.3 Decomposition into Beam Splitters

In general, an arbitrary unitary transformation formulated by a linear transformation of N-mode annihilation operators can be implemented by at most $N(N-1)/2$ beam splitters [1] (Sect. 3.5.8). Thus, the operation formulated as Eq. (6.39) is necessarily achieved by using at most three beam splitters. However, by decomposing the unitary matrix U into

$$U = F_1 B_{12}^{(1)}\left(\frac{1}{\sqrt{2}}\right) F_1 F_2 B_{23}^{(1)}\left(\sqrt{\frac{2}{3}}\right) F_2, \tag{6.43}$$

we can generate the three-mode linear cluster state by using only two beam splitters.

6.1.2.4 Mode Correspondence Between Theory and Experiment

Squeezed states are experimentally generated by using OPOs. Correspondences between the modes in theory and the OPOs in experiment are given by Table 6.2.

6.1.3 Four-Mode Linear Cluster State

6.1.3.1 Experimental Setup

Figure 6.3 shows a schematic of our experimental setup for generation of a four-mode linear cluster state.

In the following, we describe theories on generation of the four-mode linear cluster state. For simplicity, we assume that there are no propagation losses.

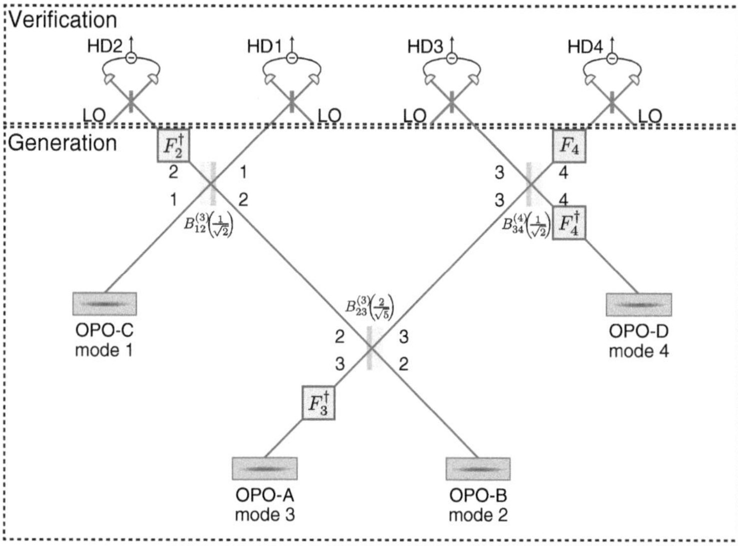

Fig. 6.3 Experimental setup for generation of four-mode linear cluster state

6.1.3.2 Without Losses

An approximation of a four-mode linear cluster state can be generated by combining four p-squeezed states on three beam splitters where their reflectivities and relative phases are chosen appropriately (Gaussian cluster state, Sect. 5.2.2). We employ this scheme for our experimental generation of the four-mode linear cluster state. In the following, we formulate its procedure by using quadrature operators which specify particular quantum states (Sect. 3.2.8).

We assume that the initial states in modes 1, 2, 3, and 4 are vacuum states. We describe these modes as

$$\hat{x}_k^{(0)} + i\hat{p}_k^{(0)}, \quad k = 1, 2, 3, 4, \tag{6.44}$$

where the subscript k denotes the mode number, while the superscripts (0) represent that these modes are in vacuum states.

As a next step, we apply p-squeezing operators on these modes. We can describe the squeezed states as

$$\hat{x}_k^{(r)} + i\hat{p}_k^{(r)} = e^{r_k}\hat{x}_k^{(0)} + ie^{-r_k}\hat{p}_k^{(0)}, \quad k = 1, 2, 3, 4, \tag{6.45}$$

where r_k represents the squeezing parameter for mode k, while the superscripts (r) represent that these modes are in squeezed states.

An approximation of a four-mode linear cluster state can be generated by combining four p-squeezed states on an appropriate network of beam splitters. For example, we can generate it by using a unitary matrix U:

$$U = \begin{pmatrix} \frac{1}{\sqrt{2}} & \frac{1}{\sqrt{10}} & -\frac{2}{\sqrt{10}}i & 0 \\ \frac{1}{\sqrt{2}}i & -\frac{1}{\sqrt{10}}i & -\frac{2}{\sqrt{10}} & 0 \\ 0 & -\frac{2}{\sqrt{10}} & -\frac{1}{\sqrt{10}}i & -\frac{1}{\sqrt{2}}i \\ 0 & -\frac{2}{\sqrt{10}}i & \frac{1}{\sqrt{10}} & -\frac{1}{\sqrt{2}} \end{pmatrix}, \tag{6.46}$$

leading to a linear transformation of an annihilation operator vector:

$$\begin{pmatrix} \hat{x}_1^{(c)} + i\hat{p}_1^{(c)} \\ \hat{x}_2^{(c)} + i\hat{p}_2^{(c)} \\ \hat{x}_3^{(c)} + i\hat{p}_3^{(c)} \\ \hat{x}_4^{(c)} + i\hat{p}_4^{(c)} \end{pmatrix} = U \begin{pmatrix} \hat{x}_1^{(r)} + i\hat{p}_1^{(r)} \\ \hat{x}_2^{(r)} + i\hat{p}_2^{(r)} \\ \hat{x}_3^{(r)} + i\hat{p}_3^{(r)} \\ \hat{x}_4^{(r)} + i\hat{p}_4^{(r)} \end{pmatrix}, \tag{6.47}$$

where the superscripts (c) represent that these modes are in a four-mode linear cluster state. We then get

$$\begin{pmatrix} \hat{x}_1^{(c)} + i\hat{p}_1^{(c)} \\ \hat{x}_2^{(c)} + i\hat{p}_2^{(c)} \\ \hat{x}_3^{(c)} + i\hat{p}_3^{(c)} \\ \hat{x}_4^{(c)} + i\hat{p}_4^{(c)} \end{pmatrix} = \begin{pmatrix} \frac{1}{\sqrt{2}}e^{r_1}\hat{x}_1^{(0)} + \frac{1}{\sqrt{10}}e^{r_2}\hat{x}_2^{(0)} + \frac{2}{\sqrt{10}}e^{-r_3}\hat{p}_3^{(0)} \\ -\frac{1}{\sqrt{2}}e^{-r_1}\hat{p}_1^{(0)} + \frac{1}{\sqrt{10}}e^{-r_2}\hat{p}_2^{(0)} - \frac{2}{\sqrt{10}}e^{r_3}\hat{x}_3^{(0)} \\ -\frac{2}{\sqrt{10}}e^{r_2}\hat{x}_2^{(0)} + \frac{1}{\sqrt{10}}e^{-r_3}\hat{p}_3^{(0)} + \frac{1}{\sqrt{2}}e^{-r_4}\hat{p}_4^{(0)} \\ \frac{2}{\sqrt{10}}e^{-r_2}\hat{p}_2^{(0)} + \frac{1}{\sqrt{10}}e^{r_3}\hat{x}_3^{(0)} - \frac{1}{\sqrt{2}}e^{r_4}\hat{x}_4^{(0)} \end{pmatrix} + i\begin{pmatrix} \frac{1}{\sqrt{2}}e^{-r_1}\hat{p}_1^{(0)} + \frac{1}{\sqrt{10}}e^{-r_2}\hat{p}_2^{(0)} - \frac{2}{\sqrt{10}}e^{r_3}\hat{x}_3^{(0)} \\ \frac{1}{\sqrt{2}}e^{r_1}\hat{x}_1^{(0)} - \frac{1}{\sqrt{10}}e^{r_2}\hat{x}_2^{(0)} - \frac{2}{\sqrt{10}}e^{-r_3}\hat{p}_3^{(0)} \\ -\frac{2}{\sqrt{10}}e^{-r_2}\hat{p}_2^{(0)} - \frac{1}{\sqrt{10}}e^{r_3}\hat{x}_3^{(0)} - \frac{1}{\sqrt{2}}e^{r_4}\hat{x}_4^{(0)} \\ -\frac{2}{\sqrt{10}}e^{r_2}\hat{x}_2^{(0)} + \frac{1}{\sqrt{10}}e^{-r_3}\hat{p}_3^{(0)} - \frac{1}{\sqrt{2}}e^{-r_4}\hat{p}_4^{(0)} \end{pmatrix}, \tag{6.48}$$

which leads to the values of nullifiers:

$$\begin{cases} \hat{\delta}_1 \equiv \hat{p}_1^{(c)} - \hat{x}_2^{(c)} = \sqrt{2}e^{-r_1}\hat{p}_1^{(0)} & \text{(i)} \\ \hat{\delta}_2 \equiv \hat{p}_2^{(c)} - \hat{x}_1^{(c)} - \hat{x}_3^{(c)} = -\frac{5}{\sqrt{10}}e^{-r_3}\hat{p}_3^{(0)} - \frac{1}{\sqrt{2}}e^{-r_4}\hat{p}_4^{(0)} & \text{(ii)} \\ \hat{\delta}_3 \equiv \hat{p}_3^{(c)} - \hat{x}_2^{(c)} - \hat{x}_4^{(c)} = \frac{1}{\sqrt{2}}e^{-r_1}\hat{p}_1^{(0)} - \frac{5}{\sqrt{10}}e^{-r_2}\hat{p}_2^{(0)} & \text{(iii)} \\ \hat{\delta}_4 \equiv \hat{p}_4^{(c)} - \hat{x}_3^{(c)} = -\sqrt{2}e^{-r_4}\hat{p}_4^{(0)} & \text{(iv)} \end{cases}. \tag{6.49}$$

Importantly, $\hat{\delta}_k$ becomes zero for each k in the limit of infinite squeezing $r_l \to \infty$ for all l. Therefore, we find that modes 1, 2, 3, and 4 are in a four-mode linear cluster state.

6.1.3.3 Decomposition into Beam Splitters

In general, an arbitrary unitary transformation formulated by a linear transformation of N-mode annihilation operators can be implemented by at most $N(N-1)/2$ beam splitters [1] (Sect. 3.5.8). Thus, the operation formulated as Eq. (6.46) is necessarily achieved by using at most six beam splitters. However, by decomposing the unitary matrix U into

$$U = F_4 F_2^\dagger B_{34}^{(4)}\left(\frac{1}{\sqrt{2}}\right) B_{12}^{(3)}\left(\frac{1}{\sqrt{2}}\right) B_{23}^{(3)}\left(\frac{2}{\sqrt{5}}\right) F_3^\dagger F_4^\dagger, \tag{6.50}$$

we can generate the four-mode linear cluster state by using only three beam splitters.

6.1.3.4 Mode Correspondence Between Theory and Experiment

Squeezed states are experimentally generated by using OPOs. Correspondences between the modes in theory and the OPOs in experiment are given by Table 6.3.

6.2 Theory on Measurement of Covariance Matrix

We consider a measurement of the covariance matrix of an N-mode quantum state. We assume that the expectation values $\langle \hat{x}_i \rangle$, $\langle \hat{p}_i \rangle$ of quadrature operators $\hat{x}_i$, $\hat{p}_i$ are zero.

Based on mode indices i, j $(i \neq j)$, the elements of the covariance matrix can be categorized into the following three cases:

- $\langle \hat{x}_i^2 \rangle$, $\langle \hat{p}_i^2 \rangle$: variances of quadrature operators.
- $\langle \hat{x}_i \hat{p}_i + \hat{p}_i \hat{x}_i \rangle /2$: correlation between quadrature operators of single mode.
- $\langle \hat{x}_i \hat{x}_j \rangle$, $\langle \hat{x}_i \hat{p}_j \rangle$, $\langle \hat{p}_i \hat{x}_j \rangle$, $\langle \hat{p}_i \hat{p}_j \rangle$: correlations between quadrature operators of two modes.

They are acquired experimentally in the following procedure.

Table 6.3 Mode correspondence between theory and experiment (generation of four-mode linear cluster state)

OPO	Mode
OPO-A	Mode 3
OPO-B	Mode 2
OPO-C	Mode 1
OPO-D	Mode 4

6.2.1 Measurement of $\langle \hat{x}_i^2 \rangle$, $\langle \hat{p}_i^2 \rangle$

We can detect $\hat{x}_i$ or $\hat{p}_i$ via a homodyne measurement on mode i by adjusting the relative phase between the signal beam and the local oscillator beam to be 0 or 90°. By measuring the power of the signal from the homodyne detector using a spectrum analyzer, we can get $\langle \hat{x}_i^2 \rangle$ or $\langle \hat{p}_i^2 \rangle$.

6.2.2 Measurement of $\langle \hat{x}_i \hat{p}_i + \hat{p}_i \hat{x}_i \rangle / 2$

We can detect $(\hat{x}_i + \hat{p}_i)/\sqrt{2}$ or $(\hat{x}_i - \hat{p}_i)/\sqrt{2}$ via a homodyne measurement on mode i by adjusting the relative phase between the signal beam and the local oscillator beam to be 45° or −45°. By measuring the power of the signal from the homodyne detector using a spectrum analyzer, we can get

$$\left\langle \left(\frac{\hat{x}_i + \hat{p}_i}{\sqrt{2}} \right)^2 \right\rangle = \frac{1}{2} \left[\langle \hat{x}_i^2 \rangle + \langle \hat{x}_i \hat{p}_i + \hat{p}_i \hat{x}_i \rangle + \langle \hat{p}_i^2 \rangle \right], \tag{6.51}$$

$$\left\langle \left(\frac{\hat{x}_i - \hat{p}_i}{\sqrt{2}} \right)^2 \right\rangle = \frac{1}{2} \left[\langle \hat{x}_i^2 \rangle - \langle \hat{x}_i \hat{p}_i + \hat{p}_i \hat{x}_i \rangle + \langle \hat{p}_i^2 \rangle \right]. \tag{6.52}$$

We can get $\langle \hat{x}_i \hat{p}_i + \hat{p}_i \hat{x}_i \rangle / 2$ from the difference of these two variables:

$$\frac{\langle \hat{x}_i \hat{p}_i + \hat{p}_i \hat{x}_i \rangle}{2} = \frac{1}{2} \left[\left\langle \left(\frac{\hat{x}_i + \hat{p}_i}{\sqrt{2}} \right)^2 \right\rangle - \left\langle \left(\frac{\hat{x}_i - \hat{p}_i}{\sqrt{2}} \right)^2 \right\rangle \right]. \tag{6.53}$$

6.2.3 Measurement of $\langle \hat{x}_i \hat{x}_j \rangle$, $\langle \hat{x}_i \hat{p}_j \rangle$, $\langle \hat{p}_i \hat{x}_j \rangle$, $\langle \hat{p}_i \hat{p}_j \rangle$

We consider the measurement of $\langle \hat{x}_i \hat{x}_j \rangle$, for example. We can detect $\hat{x}_i$ and $\hat{x}_j$ via homodyne measurements on mode i and mode j by adjusting both relative phases between the signal beams and the local oscillator beams to be 0°. Addition of these two signals from the homodyne detectors gives us the measurement result of $\hat{x}_i + \hat{x}_j$. By measuring the power of the signal using a spectrum analyzer, we can get

$$\langle (\hat{x}_i + \hat{x}_j)^2 \rangle = \langle \hat{x}_i^2 \rangle + 2 \langle \hat{x}_i \hat{x}_j \rangle + \langle \hat{x}_j^2 \rangle. \tag{6.54}$$

In a similar manner, we can detect $\hat{x}_i$ and $-\hat{x}_j$ via homodyne measurements on mode i and mode j by adjusting each relative phase between the signal beam and the local oscillator beam to be 0° or 180°. Addition of these two signals from the homodyne detectors gives us the measurement result of $\hat{x}_i - \hat{x}_j$. By measuring the power of the signal using a spectrum analyzer, we can get

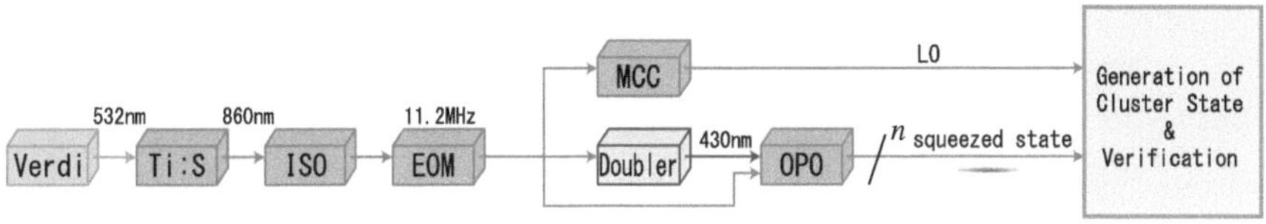

Fig. 6.4 From laser to oscillators

$$\langle(\hat{x}_i - \hat{x}_j)^2\rangle = \langle\hat{x}_i^2\rangle - 2\langle\hat{x}_i\hat{x}_j\rangle + \langle\hat{x}_j^2\rangle. \tag{6.55}$$

We can get $\langle\hat{x}_i\hat{x}_j\rangle$ from the difference of these two variables:

$$\langle\hat{x}_i\hat{x}_j\rangle = \frac{1}{4}\Big[\langle(\hat{x}_i + \hat{x}_j)^2\rangle - \langle(\hat{x}_i - \hat{x}_j)^2\rangle\Big]. \tag{6.56}$$

Similar to the measurement of $\langle\hat{x}_i\hat{x}_j\rangle$ which is achieved by homodyne measurements on modes (i, j) with relative phases of (0, 0), (0, 180) degrees, $\langle\hat{x}_i\hat{p}_j\rangle$, $\langle\hat{p}_i\hat{x}_j\rangle$, and $\langle\hat{p}_i\hat{p}_j\rangle$ are acquired by homodyne measurements with relative phases of (0, 90), (0, −90) degrees, (90, 0), (90, 180) degrees, and (90, 90), (90, −90) degrees, respectively. In experiment, we acquire only one of $\langle\hat{x}_i\hat{p}_j\rangle$ and $\langle\hat{p}_j\hat{x}_i\rangle$, since $\langle\hat{x}_i\hat{p}_j\rangle = \langle\hat{p}_j\hat{x}_i\rangle$.

6.3 Experiment Components

6.3.1 From Laser to Oscillators

We use an optical table RS4000 (Newport, 1,500 mm × 4,200 mm) with vibration isolators I-2000 (Newport). The table is covered with a windshield in order to obstruct the flow of air. It has an effect as well that it keeps out the dust from the optical setup.

Figure 6.4 shows the experimental setup from the laser to oscillators. The laser resource is a Ti:Sapphire (Ti:S) laser MBR110 (COHERENT) which is pumped by an Nd:YVO_4 laser Verdi-V10 (COHERENT). The wavelength of the Ti:S laser is set to be 860 nm, while the output power is about 1.8 W. An optical isolator (ISO, FI850-5SV, Linos) is placed at the output of the laser in order to prevent retroreflection beams going back to the laser. The beam then passes through an electro-optical modulator (EOM, PM25, Linos), where phase modulation of 11.2 MHz is applied. It is utilized to lock all the optical cavities via the Pound-Drever-Hall locking (FM sideband locking) technique (Sect. 6.4.2).

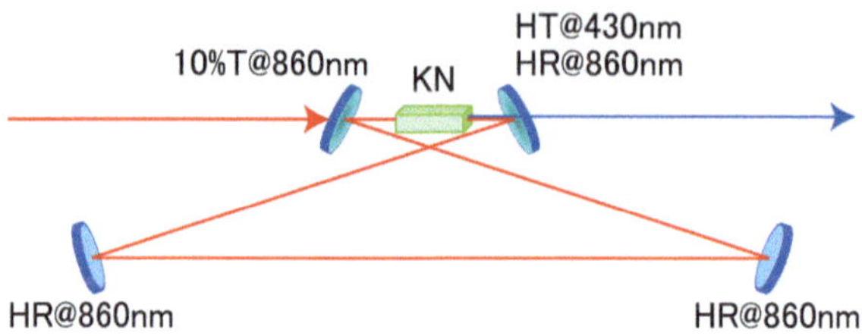

Fig. 6.5 Doubler

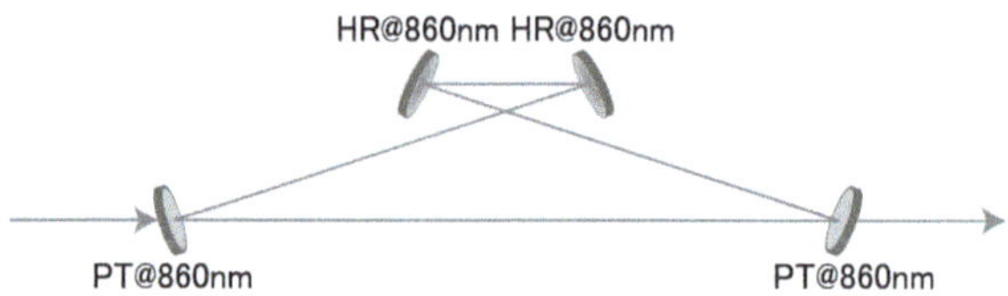

Fig. 6.6 Mode cleaning cavity

6.3.2 Doubler

A doubler is an optical cavity which generates an optical beam with the wavelength of 430 nm. It is utilized as a pump beam for generation of a squeezed state by an OPO.

The schematic of our doubler is shown in Fig. 6.5. It is a bow-tie cavity with the total length of 500 mm. It consists of two flat mirrors and two concave mirrors with the radius of curvature of 50 mm. For the fundamental beam with the wavelength of 860 nm, two flat mirrors and one of the concave mirrors have high reflectivities (HR mirrors), while the energy transmissivity of the other concave mirror is 10 %. The fundamental beam is injected into the cavity through the 10 %T coupler. The concave mirror which is HR for 860 nm is specially coated so that it has high transmissivity for 430 nm beam. Thus, it works as the output coupler of the doubler.

A 3 mm × 3 mm × 10 mm $KNbO_3$ crystal (VLOC) is located between the two concave mirrors. It works as a nonlinear crystal for second harmonic generation (SHG), leading to the frequency-doubled beam with the wavelength of 430 nm. The phase matching condition is achieved by adjusting temperature of the crystal. For this purpose, it is placed in a copper crystal holder with a Peltier element. In our experiment, the phase matching temperature is about 22 °C, while the refractive index is $n = 2.278$. We acquired about 450 mW frequency-doubled beam from about 850 mW fundamental beam.

6.3.3 Mode Cleaning Cavity

A mode cleaning cavity (MCC) is utilized to shaping the spatial mode of the fundamental beam from the laser. Since the laser beam potentially has higher Gaussian modes of Gaussian beams, we extract the 00-mode ingredient via the MCC.

The schematic of our MCC is shown in Fig. 6.6. It is a bow-tie cavity with the total length of 500 mm. In order for mechanical stability, four mirror mounts are fixed on an L-shaped angle made of aluminum. The output of the MCC serves as the following beams:

- Alignment beams for OPOs.
- Local oscillator beams for homodyne measurements.

In experiments in Chap. 7 and later, it also serves as the following beams:

- Input states for two-mode operations.
- Displacement beams.

6.3.4 Optical Parametric Oscillator

An optical parametric oscillator (OPO) is a cavity with which we can generate a squeezed light.

6.3.4.1 Configuration

Figure 6.7 shows the schematic of our OPO.

It is a bow-tie cavity with the total length of 500 mm. It consists of two flat mirrors and two concave mirrors with the radius of curvature of 50 mm. For the fundamental beam with the wavelength of 860 nm, one of the flat mirrors and two concave mirrors are HR mirrors, while the energy transmissivity of the other flat mirror is 12 %. The partial transmissivity mirror serves as the output coupler of the squeezed state. In order to inject the probe beam and the lock beam, the HR flat mirror has a small amount of transmissivity $T = 150\,\text{ppm} = 0.015\,\%$. On the other hand, the reflectivity of each concave mirror is set to be $R > 99.99\,\%$ for the fundamental beam, while it has high transmissivity for the frequency-doubled beam.

A 1 mm × 1 mm × 10 mm periodically poled $KTiOPO_4$ (PPKTP) nonlinear crystal (Raicol Crystals) is located between the two concave mirrors. It is known that the blue light induced infrared absorption (BLIIRA) of PPKTP crystal is small, thus it is the standard crystal in our laboratory. By using the same setup as the doubler, phase matching condition is achieved by adjusting temperature of the crystal. Phase matching temperature is about 40 °C.

6.3.4.2 Beams

The following four beams are injected into an OPO.

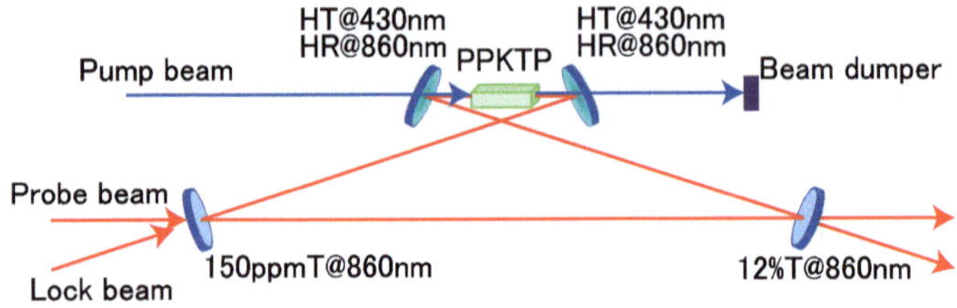

Fig. 6.7 Optical parametric oscillator

6.3.4.3 Pump Beam

The first beam is the *pump beam* with the wavelength of 430 nm. It is injected into the OPO via one of the concave mirrors. The direction of this pump beam becomes the reference to circling of the other beams.

6.3.4.4 Probe Beam

The second beam is the *probe beam* with the wavelength of 860 nm. It is injected into the OPO via the HR flat mirror in the same direction to the pump beam. The output of this probe beam from the output coupler is utilized to lock relative phases, such as the phase of the pump beam, and the phases of local oscillator beams in homodyne detections.

6.3.4.5 Lock Beam

The third beam is the *lock beam* with the wavelength of 860 nm. It is injected into the OPO via the HR flat mirror in the opposite direction to the pump beam. It is utilized to lock the total length of the OPO. In order to avoid interference between the probe beam and the lock beam, its spatial mode is converted from 00 mode to approximate 01 mode by using a glass plate. However the total length of the OPO with which the 00 mode beam resonates is different from the length with which the 01 mode beam resonates. Thus its frequency is shifted in advance by using an acousto-optic modulator (AOM) so that both the 00 mode beam without frequency shift and the 01 mode beam with frequency shift resonate simultaneously with the same total length of the OPO.

6.3.4.6 Alignment Beam

The last beam is the *alignment beam* with the wavelength of 860 nm. It is injected into the OPO via the 12 %T flat mirror in the opposite direction to the pump beam.

Different from the other three beams above, which are always utilized during measurements, the alignment beam is used only when we adjust the OPO.

Since the lock beam and the probe beam are injected into the OPO via the HR flat mirror, it is difficult to build the OPO using these two beams. Thus, we first use the alignment beam injected via the 12 %T flat mirror in order to build the OPO. After we have adjusted the OPO so that the alignment beam resonates, we lock the length of the OPO using the reflected beam of the alignment beam. Since the transmitted beam propagates in the opposite direction to the probe beam, we can adjust the optical path of the probe beam. After the adjustment of the probe beam, we can adjust the optical path of the lock beam because the reflected beam of the probe beam propagates in the opposite direction to the lock beam. Finally, we inject the alignment beam into the OPO with the phase matching condition satisfied. By the process of SHG, a beam with the wavelength of 430 nm is outputted from one of the concave mirrors. Since it propagates in the opposite direction to the pump beam, we can adjust the optical path of the pump beam.

6.3.5 Mode of Quantum State

The center of the frequency band of the squeezed state generated by our OPO is equivalent to that of the laser.

In general, a squeezed state is *generated* with the frequency band whose center frequency is equivalent to that of the laser,[2] while the bandwidth is equivalent to that of the OPO. In our experiment, the half width at half maximum (HWHM) of the bandwidth is about 6 MHz. The frequency band of a cluster state we *generate* is equivalent to that of the squeezed state.

To the contrary, we *utilize* a *sideband* squeezed state with the frequency band whose center frequency is ± 1 MHz away from the center frequency of the laser, while the HWHM of the sidebands is 15 kHz. The wave packet of a quantum state is a 1 MHz sine wave with an envelope of about 70 μs time duration. It derives from the setting of the spectrum analyzer (SA) we use in our experiment: center frequency is 1 MHz, resolution bandwidth is 30 kHz (full width at half maximum, FWHM), and we use the zero span mode.

The advantage of adopting the sideband as the frequency band is that the experimental setup for one-way quantum computation becomes simpler (Sect. 7.3.3).

---

[2] To be precise, squeezed states are generated at several frequency bands with different center frequencies corresponding to the frequency comb of the cavity as far as the phase matching condition (Footnote 2 continued)
is satisfied. However, they will be filtered out by homodyne detections because they detect interference between the signal beams and the local oscillator beams with the laser frequency, and because the detectors are not broadband enough to detect them.

6.3.6 *Homodyne Measurement*

In this thesis, quantum states are measured by homodyne measurements (Sect. 2.4). the schematic of a homodyne measurement is shown in Fig. 6.8.

A homodyne measurement is achieved in the following procedure. The signal beam ($\hat{a}$) and a local oscillator (LO) beam (α), which is an intense coherent beam, are combined via a 50:50 beam splitter (half beam splitter, HBS). The outputs ($\hat{a}_1$, $\hat{a}_2$) are converted to photocurrents via photodiodes. The measurement result is acquired by calculating the difference of these photocurrents.

In experiment, we have to adjust the reflectivity of the beam splitter precisely. For that purpose, we in advance apply 1 MHz intensity modulation onto the local oscillator beam, while we shut off the signal beam. The reflectivity is then adjusted by rotating the beam splitter so that the 1 MHz signal is not detected at the output of the homodyne detector. The level of this cancellation is evaluated by the power ratio of the 1 MHz signals between the case where one of the photodiodes is disabled by shutting off the beam in front of the photodiode, and the case where both photodiodes are enabled. The typical level is about 40 dB, which is sufficient to measure quantum states.

6.4 Locking the Cavities and Phases

In this section, we mention experimental techniques to lock total lengths of optical cavities and relative phases between two beams.

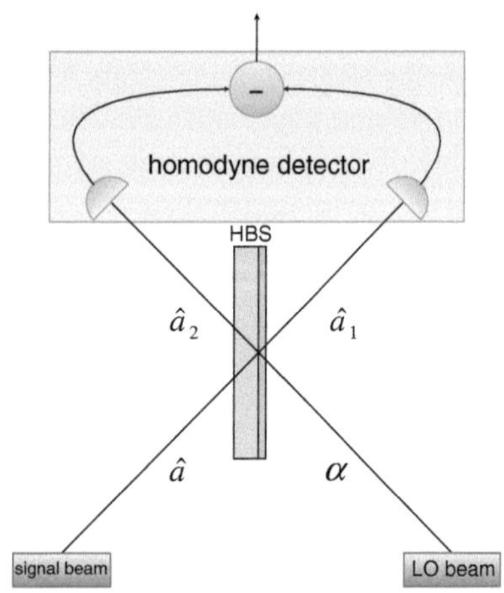

Fig. 6.8 Homodyne measurement

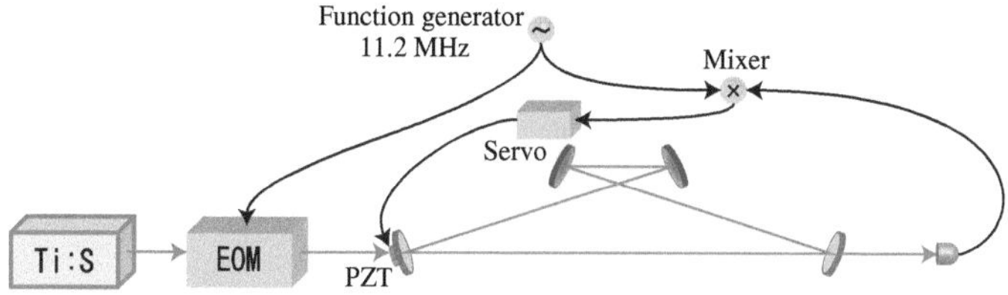

Fig. 6.9 FM sideband locking (locking of MCC)

6.4.1 Phase Modulation

The FM sideband locking and the AC locking are achieved by using a phase-modulated beam. In this subsection, we briefly describe the phase modulation.

The phase modulation with the frequency f is experimentally realized by oscillating the optical path length of a beam with the same frequency of f. In our laboratory, we utilize two types of optical setups. One is that the optical path length is oscillated by using a mirror with a piezo electric transducer (PZT), on which an electrical signal of a sine wave with the frequency f is applied. Since the length of the PZT is changed according to the electrical signal, phase modulation can be applied to the beam. Note that we can neglect distortion of the optical axis caused by the change of the PZT length because displacement of the PZT is considerably shorter than the radius of the beam. In this thesis, we utilize the PZT-based phase modulation for the AC locking.

The other is that the optical path length is oscillated by using an electro-optical modulator (EOM), on which an electrical signal of a sine wave with the frequency f is applied. The optical path length is modulated via the process of the Pockels effect: the electric field produces birefringence of an optical medium, where its refractive index is changed in proportional to the electric field [2]. In this thesis, we utilize the EOM-based phase modulation for the FM sideband locking.

In both cases of the PZT-based and the EOM-based phase modulation, it is formulated as

$$E_{in} = E_0 e^{i\omega t} \rightarrow E_{mod} = E_0 e^{i(\omega t + \beta \sin \Omega t)}, \tag{6.57}$$

where β is the modulation depth, while $\Omega = 2\pi f$ is the angular frequency of the phase modulation. When β is small enough, it can be expanded in the following way:

$$E_{mod} \simeq E_0 e^{i\omega t}(1 + i\beta \sin \Omega t) = E_0 e^{i\omega t}\left(1 + \frac{\beta}{2} e^{i\Omega t} - \frac{\beta}{2} e^{-i\Omega t}\right). \tag{6.58}$$

It shows that the sideband signal with angular frequency of $\pm\Omega$ is generated.

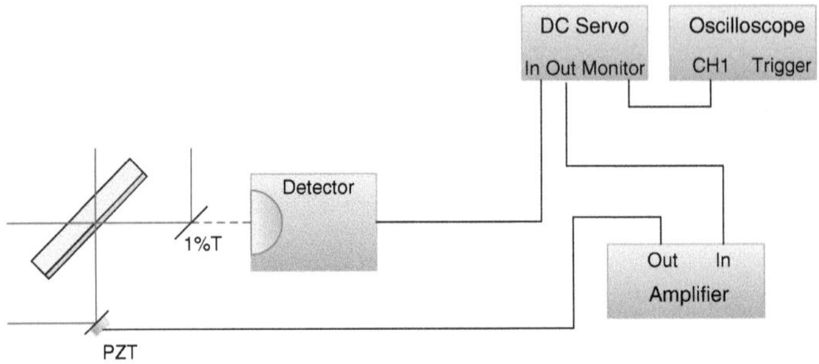

Fig. 6.10 DC locking

Fig. 6.11 Phase relationship achieved by DC locking

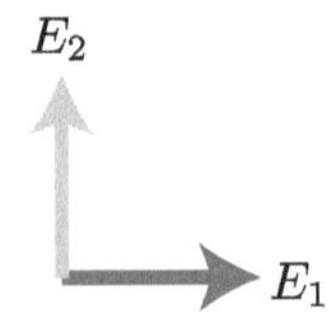

6.4.2 FM Sideband Locking

Figure 6.9 shows the basic structure of FM sideband locking system for an optical cavity. A phase-modulated beam passes through the cavity, and then is detected by a detector. The electrical signal from the detector is sent to a mixer, with which it is multiplied by the original modulation signal. The output signal of the mixer is given by

$$V_{err} = \frac{\mathrm{d}|F(L)|^2}{\mathrm{d}L}, \tag{6.59}$$

where L is the total length of the cavity, while $F(L)$ is the amplitude of the transmitted beam. Therefore, V_{err} works as the error signal for locking of the cavity length so that it is an integer multiple of the wavelength of the laser beam.

6.4.3 DC Locking

We utilize the *DC locking* system for the purpose of adjusting the relative phase between two laser beams to be 90 or 270° (Fig. 6.10).

We assume that E_1 and E_2 represent the amplitudes of two laser beams, while θ is the relative phase between them. These two beams are combined on a beam splitter. The power of its output is given by

$$P = |E_1|^2 + |E_2|^2 + 2E_1E_2 \cos\theta. \tag{6.60}$$

1 % of the interfered beam is picked up by a 1 % transmissivity beam splitter, followed by converted to an electrical signal via a photo detector. It works as the error signal for locking of the relative phase between two laser beams to be 90 or 270°. The choice of 90 or 270° is achieved via the polarity switch in the feedback servo (DC servo) controller (Fig. 6.11).

6.4.4 AC Locking

We utilize the *AC locking* system for the purpose of adjusting the relative phase between two laser beams to be 0° or 180° (Fig. 6.12).

We assume that E_1 and E_2 represent the amplitudes of two laser beams, while θ is the relative phase between them. After one of these beams is phase-modulated, they are combined on a beam splitter. The power of its output is given by

$$P = |E_1|^2 + |E_2|^2 + 2E_1E_2(\cos\theta + \beta \sin\theta \sin\Omega t), \tag{6.61}$$

where β represents the modulation depth, while Ω is the angular frequency of the phase modulation. 1 % of the interfered beam is picked up by a 1 % transmissivity beam splitter, followed by converted to an electrical signal via a photo detector. It is sent to a mixer, with which it is multiplied by the modulation signal. The output of the mixer is given by

$$V = 2E_1E_2\beta \sin\theta(1 + \sin 2\Omega t) + (\text{component of } \sin\Omega t). \tag{6.62}$$

By using a low pass filter (LPF) which filters out all components with angular frequency of Ω and higher, we get the error signal for locking of the relative phase between two laser beams to be 0° or 180°. The choice of 0° or 180° is achieved via the polarity switch in the feedback servo (AC servo) controller (Fig. 6.13).

The same technique can be utilized to lock the phase between a pump beam and a probe beam in an OPO. It can also be utilized in order to lock the relative phase between a probe beam and a local oscillator beam at a homodyne detection to be 0° or 180°.

6.4.5 Locking the Relative Phases for Generation of Two-Mode Cluster State

Figure 6.14 shows arrangement of probe beams and phase modulations for generation of a two-mode cluster state. By using Fig. 6.14, we can determine how each relative phase between two beams is locked.

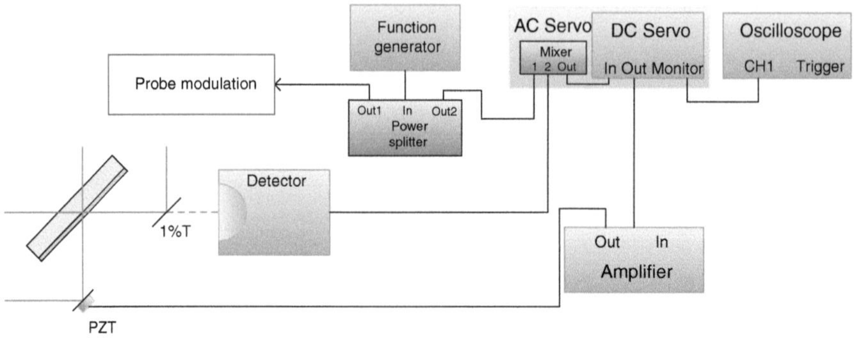

Fig. 6.12 AC locking

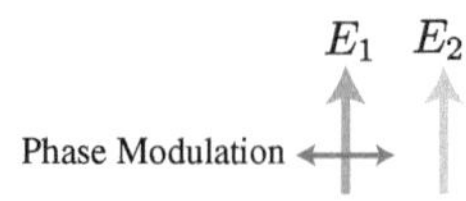

Fig. 6.13 Phase relationship achieved by AC locking

6.4.5.1 Probe Beams and Phase Modulations

Probe beams for OPO-A and OPO-B are phase modulated with the frequency of 210 and 140 kHz, respectively. These phase modulations are achieved via PZTs on their optical paths.

6.4.5.2 Locking the Parametric Gains

In this thesis, we lock the phase of the pump beam for an OPO so that the parametric gain has the minimum value. It is achieved by using the AC locking technique with an error signal which is demodulated from the transmitted probe beam with the original modulation signal (sine wave with the frequency of 210 kHz for OPO-A, while 140 kHz for OPO-B).

The relationship between the squeezed state and the phase modulation is described under each OPO in Fig. 6.14. Since the parametric gain is adjusted so that it has the minimum value, the probe beam points to the squeezing component of the squeezed beam, while the modulation signal is orthogonal to the probe beam. We intentionally choose this arrangement of phases so that the modulation signal is amplified via the optical parametric process, leading to the high signal-to-noise (S/N) ratio, and thus stability of locking system.

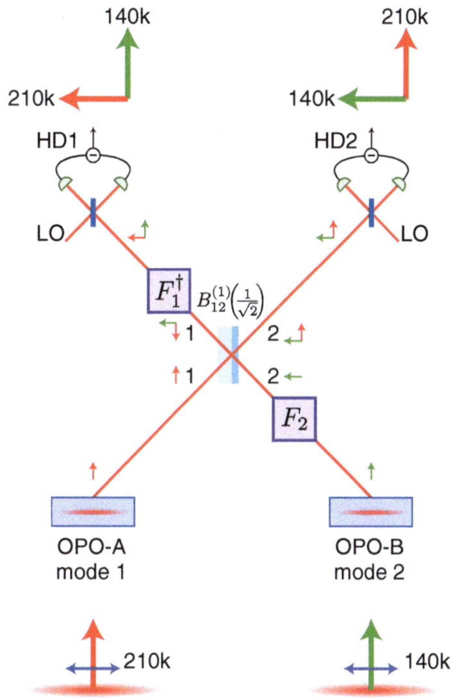

Fig. 6.14 Probe beams and phase modulations for generation of two-mode cluster state

6.4.5.3 Fourier Transformations and Beam Splitter

In Fig. 6.14, the relationship of the probe beams in each optical path is described by arrows in phase space. At the output of each OPO, the probe beam points to the p axis. Th probe beams from different OPOs are distinguished by colors of arrows.

The Fourier transformation on mode 2 rotates the arrow with 90° in a counterclockwise direction. At the beam splitter $B_{12}^{(1)}\left(\frac{1}{\sqrt{2}}\right) = \frac{1}{\sqrt{2}}\begin{pmatrix} -1 & 1 \\ 1 & 1 \end{pmatrix}$, the phase of the probe beam from mode 1 to mode 1 is rotated with 180° in phase space, which corresponds to the sign of the (1, 1) entry of the matrix. The inversed Fourier transformation is then applied to mode 1, which rotates the arrows with 90° in a clockwise direction.

6.4.5.4 Locking the Relative Phase on the Beam Splitter for Generation of Cluster State

As is shown in Fig. 6.14, we lock the relative phase between two probe beams from OPO-A and OPO-B so that it is 90°. This locking is achieved by using the DC locking technique. Since the relative phase between the probe beam and the squeezed state in each OPO is locked in advance, the relative phase between two squeezed states from these two OPOs becomes locked indirectly.

Table 6.4 Phase lockings for generation of two-mode cluster state

Relative phase	Technique, demodulation frequency
Parametric gain of OPO-A	AC locking, 210 kHz
Parametric gain of OPO-B	AC locking, 140 kHz
beam splitter $B_{12}^{(1)}(\frac{1}{\sqrt{2}})$	DC locking
Homodyne measurement in mode 1	AC locking, x: 210 kHz, p: 140 kHz
Homodyne measurement in mode 2	AC locking, x: 140 kHz, p: 210 kHz

6.4.5.5 Locking the Relative Phase on the Beam Splitter for Homodyne Detection

We here consider locking of the relative phase between the signal beam and the local oscillator beam on the half beam splitter for the homodyne detection in mode 1. We assume that we measure the position component of mode 1 ($\hat{x}_1$). For this purpose, we adjust the phase of the local oscillator beam so that it points to the x axis in phase space. Since the probe beam from OPO-A points to the x axis as well, this phase locking can be achieved by using the AC locking technique with the demodulation signal with frequency of 210 kHz, which have been utilized to modulate the probe beam of OPO-A. Whether it is locked to 0° or 180° is determined by the polarity switch in the feedback servo (AC servo) controller. In order to measure the momentum component of mode 1 ($\hat{p}_1$), we use the AC locking technique with the demodulation signal with frequency of 140 kHz, which have been utilized to modulate the probe beam of OPO-B. In a similar manner, we can lock the phase of the local oscillator beam for the homodyne detection in mode 2.

6.4.5.6 Summary of Phase Locking

Table 6.4 shows phase locking techniques for generation of a two-mode cluster state.

6.4.6 Locking the Relative Phases for Generation of Three-Mode Linear Cluster State

Figure 6.15 shows arrangement of probe beams and phase modulations for generation of a three-mode linear cluster state.

The locking system for generation of a three-mode linear cluster state is similar to that for generation of a two-mode cluster state. In the following, we mention differences between these two experiments, and additions to the two-mode cluster state experiment.

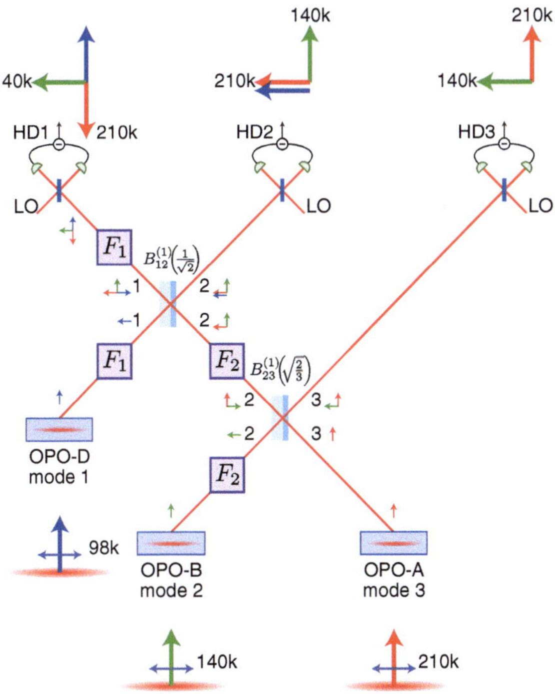

Fig. 6.15 Probe beams and phase modulations for generation of three-mode linear cluster state

6.4.6.1 Probe Beams and Phase Modulations

Probe beams for OPO-A, OPO-B and OPO-C are phase modulated with the frequency of 210, 140 and 98 kHz, respectively.

Table 6.5 Phase lockings for generation of three-mode linear cluster state

Relative phase	Technique, Demodulation frequency
Parametric gain of OPO-A	AC locking, 210 kHz
Parametric gain of OPO-B	AC locking, 140 kHz
Parametric gain of OPO-D	AC locking, 98 kHz
Beam splitter $B_{23}^{(1)}(\sqrt{2/3})$	DC locking
beam splitter $B_{12}^{(1)}(\frac{1}{\sqrt{2}})$	AC locking, 210 kHz
Homodyne measurement in mode 1	AC locking, x: 140 kHz, p: 210 kHz
Homodyne measurement in mode 2	AC locking, x: 210 kHz, p: 140 kHz
Homodyne measurement in mode 3	AC locking, x: 140 kHz, p: 210 kHz

6.4.6.2 Locking the Relative Phase on the Beam Splitter $B_{12}^{(1)}(\frac{1}{\sqrt{2}})$

One of the two input beams to the beam splitter $B_{12}^{(1)}(\frac{1}{\sqrt{2}})$ comes from the OPO-D. The other input beam is an output of the beam splitter $B_{23}^{(1)}\left(\sqrt{\frac{2}{3}}\right)$, where two beams from OPO-A and OPO-B are combined. As is shown in Fig. 6.15, two probe beams from OPO-A and OPO-D are to be combined parallelly on the beam splitter of $B_{12}^{(1)}(\frac{1}{\sqrt{2}})$. This phase locking can be achieved by using the AC locking technique with the demodulation signal with the frequency of 210 kHz, which have been utilized to modulate the probe beam of OPO-A.

6.4.6.3 Summary of Phase Locking

Table 6.5 shows phase locking techniques for generation of a three-mode linear cluster state.

6.4.7 Locking the Relative Phases for Generation of Four-Mode Linear Cluster State

Figure 6.16 shows arrangement of probe beams and phase modulations for generation of a four-mode linear cluster state.

6.4.7.1 Summary of Phase Locking

Table 6.6 shows phase locking techniques for generation of four-mode linear cluster state.

6.5 Measurement Results of Two-Mode Cluster State

6.5.1 Preparation

6.5.1.1 Adjustment of Interference Visibilities

Table 6.7 shows adjustment results of interference visibilities.

In this table, "OPO-A" and "OPO-B" show probe beams for the corresponding OPOs, while "LO-1" and "LO-2" show local oscillator beams for the corresponding homodyne measurements, respectively. Note that the interferences between "OPO-

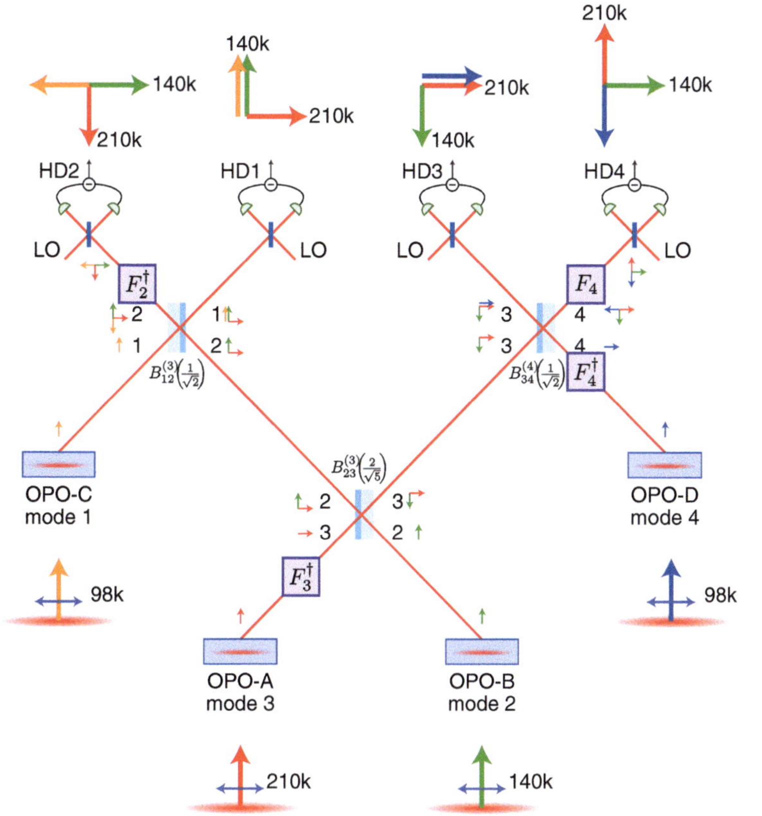

Fig. 6.16 Probe beams and phase modulations for generation of four-mode linear cluster state

Table 6.6 Phase lockings for generation of four-mode linear cluster state

Relative phase	Technique, demodulation frequency
Parametric gain of OPO-A	AC locking, 210 kHz
Parametric gain of OPO-B	AC locking, 140 kHz
Parametric gain of OPO-C	AC locking, 98 kHz
Parametric gain of OPO-D	AC locking, 98 kHz
Beam splitter $B_{23}^{(3)}(\frac{2}{\sqrt{5}})$	DC locking
Beam splitter $B_{12}^{(3)}(\frac{1}{\sqrt{2}})$	AC locking, 140 kHz
Beam splitter $B_{34}^{(4)}(\frac{1}{\sqrt{2}})$	AC locking, 210 kHz
Homodyne measurement in mode 1	AC locking, x: 210 kHz, p: 140 kHz
Homodyne measurement in mode 2	AC locking, x: 140 kHz, p: 210 kHz
Homodyne measurement in mode 3	AC locking, x: 210 kHz, p: 140 kHz
Homodyne measurement in mode 4	AC locking, x: 140 kHz, p: 210 kHz

Table 6.7 Interference visibilities for generation of two-mode cluster state

Beam splitter	Beams used for adjustment	Visibility (%)
Cluster state generation	OPO-A, OPO-B	99.1
Homodyne measurement 1	OPO-A, LO-1	97.2
Homodyne measurement 2	OPO-B, LO-2	96.1

A" and "LO-2", and between "OPO-B" and "LO-1" are adjusted indirectly via the interference at the beam splitter of cluster state generation.

6.5.1.2 Adjustment of Parametric Gains

Adjustment of the parametric gain for each OPO is achieved by alignment of the pump beam and temperature control of the nonlinear crystal. Table 6.8 shows adjustment results of the parametric gains.

The difference between two phase matching temperatures derives from individual specificity of the nonlinear crystals and temperature control systems.

6.5.1.3 Balance Between Homodyne Detections

The output amplitude of a homodyne detector (Sect. 2.4) depends on the intensity of the local oscillator and the circuit constant of the detector. In order to measure quantum correlations between two modes accurately, we have to uniform the gains of homodyne detectors. The balance between two homodyne detectors is adjusted in the following two steps.

6.5.1.4 First Step

We apply phase modulation with the frequency of 1 MHz to the probe beam for OPO-A. Since we have assumed that the probe beam points to the p axis, the phase modulation points to the x axis. Note that, OPO-A corresponds to mode 1. Thus, we can consider that the state of mode 1 is a coherent state with amplitude in $\hat{x}_1$. We rewrite Eq. (6.5) so that it shows the general input-output relationship which does

Table 6.8 Parametric gains for generation of two-mode cluster state

OPO	Parametric gain (maximum)	Phase matching temperature (o)
OPO-A	7.9	40.5
OPO-B	8.4	40.2

not depend on the specific input state:

$$\begin{pmatrix} \hat{x}_1' + i\,\hat{p}_1' \\ \hat{x}_2' + i\,\hat{p}_2' \end{pmatrix} = \begin{pmatrix} -\frac{1}{\sqrt{2}}\hat{p}_1 + \frac{1}{\sqrt{2}}\hat{x}_2 \\ \frac{1}{\sqrt{2}}\hat{x}_1 - \frac{1}{\sqrt{2}}\hat{p}_2 \end{pmatrix} + i \begin{pmatrix} \frac{1}{\sqrt{2}}\hat{x}_1 + \frac{1}{\sqrt{2}}\hat{p}_2 \\ \frac{1}{\sqrt{2}}\hat{p}_1 + \frac{1}{\sqrt{2}}\hat{x}_2 \end{pmatrix}. \tag{6.63}$$

Since $\frac{1}{\sqrt{2}}\hat{x}_1$ appears in $\hat{p}_1'$ and $\hat{x}_2'$, we obtain $\hat{p}_1'$ and $-\hat{x}_2'$ by adjusting the phases of local oscillator beams to 90° and 180°. Two outputs of the homodyne detectors are added by a power adder. We measure the output of the power adder by using a spectrum analyzer, with which we detect the power of the component with the frequency of 1 MHz. When two signals from the homodyne detectors are added with identical gains and same phases, no 1 MHz signal will be detected by the spectrum analyzer. In order to adjust the power of the signal, step attenuators are placed between one of the homodyne detectors and the power adder, while the relative phase is adjusted by changing the length of cables.

Figure 6.17 shows adjustment results. Trace "HD1" shows the measurement result when the signal from homodyne detector 1 is connected to the power adder, while that from homodyne detector 2 is not connected. We find intense 1 MHz signal is detected. Trace "HD1 - HD2" shows the measurement result when the both signals from two homodyne detectors are connected to the power adder. The 1 MHz signals are canceled out, and almost no signal is detected.

We mention that the levels at sideband frequency (1.2 MHz, for example) are different from each other. This is because we carried out the balance adjustment with pumping OPOs, thus squeezed states are generated. In the trace of "HD1", variance of $\hat{p}_1' = \frac{1}{\sqrt{2}}\hat{x}_1 + \frac{1}{\sqrt{2}}\hat{p}_2$, which is given by $\frac{1}{2}\langle \Delta^2 \hat{x}_1 \rangle + \frac{1}{2}\langle \Delta^2 \hat{p}_2 \rangle$, is detected. In the trace of "HD1 - HD2", on the other hand, variance of $\hat{p}_1' - \hat{x}_2' = \sqrt{2}\hat{p}_2$, which is

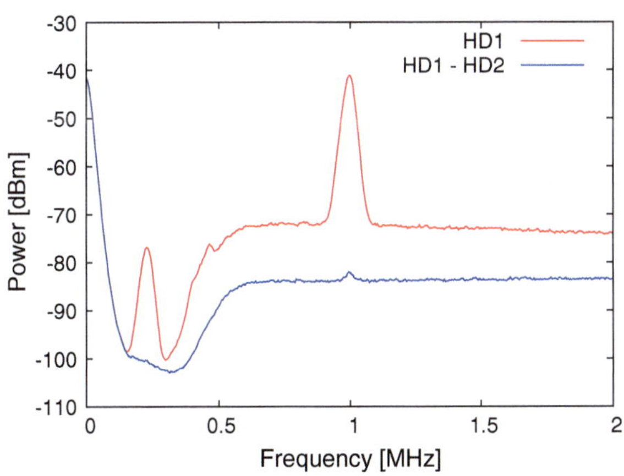

Fig. 6.17 Balance between homodyne detections

given by $2\langle \Delta^2 \hat{p}_2 \rangle$, is detected. Since $\hat{x}_1$ and $\hat{p}_2$ show anti-squeezing and squeezing components, the power in trace "HD1" is larger than that in trace "HD1 - HD2".

6.5.1.5 Second Step

Although the balance between two homodyne detectors is almost adjusted, there might exist some error derived from experimental imperfection. Since propagation efficiency (or equivalently, energy loss) from OPO-A to homodyne detector 1 is different from that to homodyne detector 2, the powers of coherent states detected by two homodyne detectors are different from each other. This error is compensated by uniforming the shot noise level (SNL), which is achieved by adjusting the step attenuators placed between one of the homodyne detectors and the power adder. In our experiment, adjustment of up to 0.3 dB is carried out.

6.5.2 *Measurement Results and Their Analysis ($PG1 = 7$, $PG2 = 7$)*

In this subsection, we show measurement results and analysis of generation of the two-mode cluster state when both parametric gains are adjusted to 7 ($PG1 = 7, PG2 = 7$).

6.5.2.1 Measurement Results of Variances

Figure 6.18 shows measurement results of variances which will be utilized in order to evaluate van Loock-Furusawa entanglement criterion as well as to acquire the covariance matrix.

Each result is listed in Table 6.9.

In Fig. 6.18 and Table 6.9, all results are normalized by the values which are acquired by using vacuum states as inputs for all homodyne detections. Therefore, 0 dB in single-mode measurements (such as measurements of $\hat{x}_1$ or $(\hat{x}_2 + \hat{p}_2)/\sqrt{2}$) corresponds to the shot noise level (SNL$= \frac{\hbar}{2}$), while 0 dB in two-mode measurements (such as $\hat{x}_1 + \hat{x}_2$) corresponds to $2 \times$ SNL.

6.5.2.2 Entanglement Detection via van Loock-Furusawa Criterion

By using parts of measurement results in Table 6.9, we can evaluate entanglement of the generated state via van Loock-Furusawa criterion. In the case of the two-mode cluster state, a sufficient condition of entanglement (Sects. 3.7.4 and 5.1.4) is given by

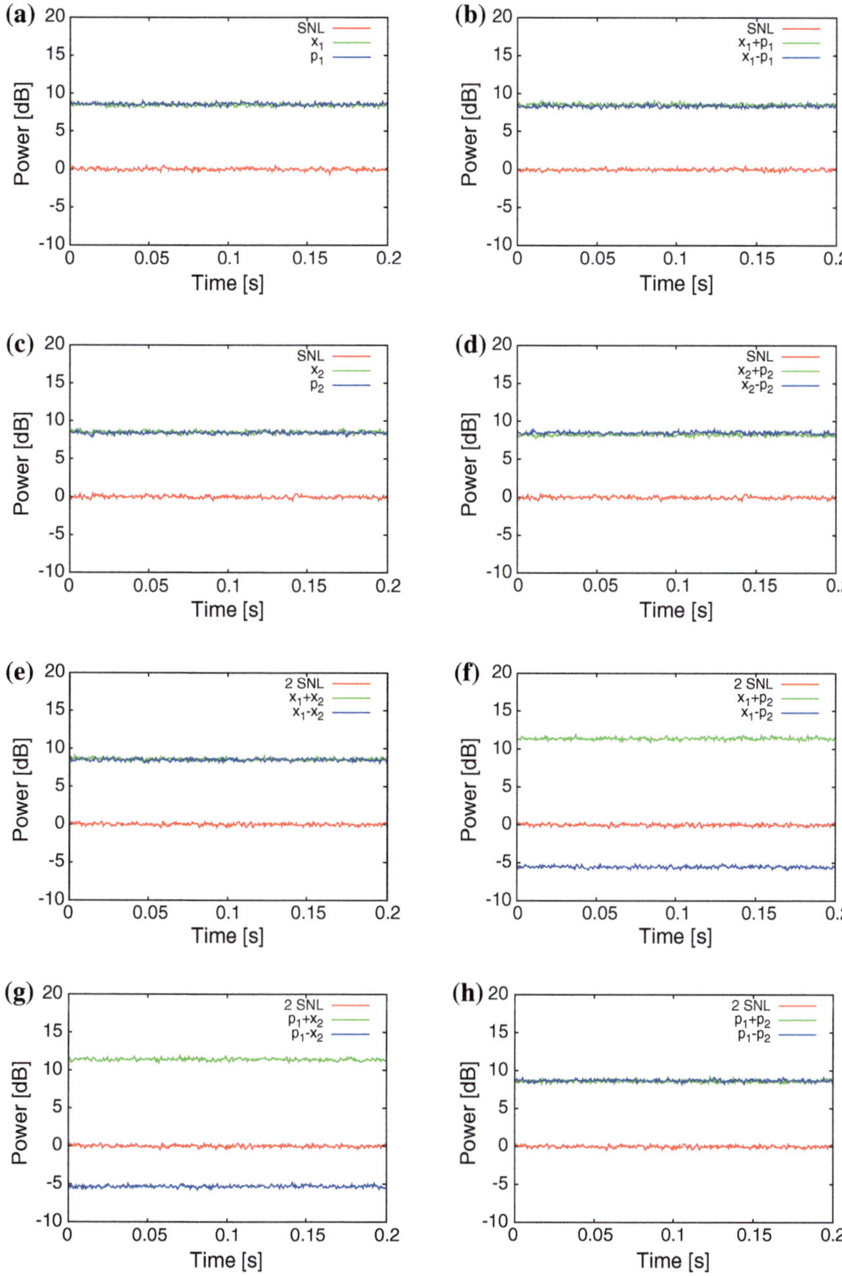

Fig. 6.18 Measurement results of variances ($PG1 = 7$, $PG2 = 7$). **a** $\hat{x}_1$, $\hat{p}_1$, **b** $\hat{x}_1 \pm \hat{p}_1$, **c** $\hat{x}_2$, $\hat{p}_2$, **d** $\hat{x}_2 \pm \hat{p}_2$, **e** $\hat{x}_1 \pm \hat{x}_2$, **f** $\hat{x}_1 \pm \hat{p}_2$, **g** $\hat{p}_1 \pm \hat{x}_2$, **h** $\hat{p}_1 \pm \hat{p}_2$

Table 6.9 Measurement results of variances ($PG1 = 7$, $PG2 = 7$)

Measurement Variable	Variance (dB) (Relative to vacuum inputs)	Measurement Variable	Variance (dB) (Relative to (vacuum inputs)
$\hat{x}_1$	8.48 ± 0.03	$\hat{x}_2$	8.53 ± 0.03
$\hat{p}_1$	8.59 ± 0.03	$\hat{p}_2$	8.44 ± 0.03
$(\hat{x}_1 + \hat{p}_1)/\sqrt{2}$	8.55 ± 0.03	$(\hat{x}_2 + \hat{p}_2)/\sqrt{2}$	8.25 ± 0.03
$(\hat{x}_1 - \hat{p}_1))/\sqrt{2}$	8.37 ± 0.03	$(\hat{x}_2 - \hat{p}_2)/\sqrt{2}$	8.49 ± 0.03
$\hat{x}_1 + \hat{x}_2$	8.61 ± 0.03	$\hat{p}_1 + \hat{x}_2$	11.48 ± 0.03
$\hat{x}_1 - \hat{x}_2$	8.55 ± 0.03	$\hat{p}_1 - \hat{x}_2$	-5.28 ± 0.03
$\hat{x}_1 + \hat{p}_2$	11.40 ± 0.03	$\hat{p}_1 + \hat{p}_2$	8.65 ± 0.03
$\hat{x}_1 - \hat{p}_2$	-5.52 ± 0.03	$\hat{p}_1 - \hat{p}_2$	8.75 ± 0.03

$$\langle \Delta^2(\hat{p}_1 - \hat{x}_2)\rangle + \langle \Delta^2(\hat{p}_2 - \hat{x}_1)\rangle < 2\hbar. \tag{6.64}$$

By using the measurement results of $\hat{p}_1 - \hat{x}_2$ and $\hat{x}_1 - \hat{p}_2$, we get

$$\langle \Delta^2(\hat{p}_1 - \hat{x}_2)\rangle = \hbar \times 10^{\frac{-5.28}{10}} = (0.297 \pm 0.002)\hbar, \tag{6.65}$$

$$\langle \Delta^2(\hat{p}_2 - \hat{x}_1)\rangle = \hbar \times 10^{\frac{-5.52}{10}} = (0.281 \pm 0.002)\hbar, \tag{6.66}$$

leading to

$$\langle \Delta^2(\hat{p}_1 - \hat{x}_2)\rangle + \langle \Delta^2(\hat{p}_2 - \hat{x}_1)\rangle = (0.577 \pm 0.003)\hbar < 2\hbar. \tag{6.67}$$

Therefore, the inseparability (entanglement) of the generated cluster state is verified.

6.5.2.3 Estimation of Squeezing Levels and Propagation Efficiency by Variances

We estimate squeezing levels and energy propagation efficiency by using measurement results of nullifiers: $\hat{p}_1 - \hat{x}_2$ and $\hat{p}_2 - \hat{x}_1$, and operators which are acquired by changing the signs of the nullifiers: $\hat{p}_1 + \hat{x}_2$, $\hat{p}_2 + \hat{x}_1$.

For simplicity, we assume that squeezing parameters and energy propagation efficiencies are symmetrical ($r_k = r$, $\eta_k = \eta$). In this case, the variances of $\hat{p}_1 - \hat{x}_2$ and $\hat{p}_1 + \hat{x}_2$ become identical to those of $\hat{p}_2 - \hat{x}_1$ and $\hat{p}_2 + \hat{x}_1$, respectively. Since their experimental results have different values because of asymmetry of experimental setup, we use the averaged values -5.40 dB and 11.44 dB in the following discussion. They correspond to the effective squeezing level $x^{(e)} = 5.40$ defined by Eq. (6.12), and the effective squeezing level of anti-squeezing $x^{(ae)} = 11.44$ defined by Eq. (6.16).

By using $x^{(e)}$ and $x^{(ae)}$, we can estimate the squeezing parameter r (note that it is not the *effective* squeezing parameter) and the energy efficiency η. Since

$$\frac{\hbar}{2}\left[2\eta e^{-2r} + 2(1-\eta)\right] = \frac{\hbar}{2} \times 2 \times 10^{-\frac{x^{(e)}}{10}}, \quad \frac{\hbar}{2}\left[2\eta e^{2r} + 2(1-\eta)\right] = \frac{\hbar}{2} \times 2 \times 10^{\frac{x^{(ae)}}{10}}, \tag{6.68}$$

we acquire

$$\eta = 0.75, \quad r = 1.45. \tag{6.69}$$

We then compare these estimated values with the other experimental parameters. First, we consider the squeezing level. We have adjusted the parametric gains of both OPO-A and OPO-B so that they are equal to 7. In general, a parametric gain G_+ and a squeezing level without experimental losses have the following relationship:

$$G_+ = \frac{1}{(1-x)^2}, \quad 10^{-\frac{a}{10}} = e^{-2r} = 1 - \frac{4x}{(1+x)^2 + f^2}, \quad f = \frac{f_m}{f_{opo}}, \tag{6.70}$$

where $f_m = 1$ MHz is the measurement frequency, while $f_{opo} = 6.0$ MHz is the bandwidth (HWHM) of an OPO. Therefore, we get the squeezing parameter $r = 1.37$ for the parametric gain $G_+ = 7$. Next, we consider the energy efficiency. In our experimental setup, the escape efficiency is about 96 % (derived from the intra-cavity loss of 0.5 %), the propagation efficiency is about 92 % on average, and the interference efficiency is about 91 % on average (derived from the imperfect visibilities). Thus, the total energy efficiency is acquired as the product of these efficiencies: 80 %. Therefore, the squeezing parameter and the energy efficiency estimated by these experimental parameters are consistent with those acquired by $x^{(e)}$ and $x^{(ae)}$.

6.5.2.4 Covariance Matrix

By using all measurement results in Table 6.9, we can acquire all elements of the covariance matrix (Sect. 6.2). We use the system of units with $\hbar = \frac{1}{2}$, for simplicity.

6.5.2.5 Element of V_{11}

By using the measurement result of $\hat{x}_1$, we get

$$V_{11} = \langle \Delta^2 \hat{x}_1 \rangle = \frac{\hbar}{2} \times 10^{\frac{8.48}{10}} = 1.76. \tag{6.71}$$

Table 6.10 Variables acquired by covariance matrix ($PG1 = 7$, $PG2 = 7$)

Variable	Result
Determinant of 1st order principal submatrix: $\det V_1 = V_{11}$	1.76 ± 0.01
Determinant of 2nd order principal submatrix: $\det V_2 = \det A$	3.19 ± 0.03
Determinant of 3rd order principal submatrix: $\det V_3$	0.47 ± 0.08
Determinant of 4th order principal submatrix: $\det V_4 = \det V$	0.06 ± 0.02
Symplectic eigenvalue: ν_-	0.46 ± 0.05
Symplectic eigenvalue: ν_+	0.51 ± 0.04
PT symplectic eigenvalue: $\tilde{\nu}_-$	0.068 ± 0.009
PT symplectic eigenvalue: $\tilde{\nu}_+$	3.481 ± 0.009
Logarithmic negativity: E_N	1.29 ± 0.12

6.5.2.6 Element of V_{12}

By using the measurement results of $(\hat{x}_1 \pm \hat{p}_1)/\sqrt{2}$, we get

$$V_{12} = \frac{1}{2}\left[\left\langle\left(\frac{\hat{x}_1 + \hat{p}_1}{\sqrt{2}}\right)^2\right\rangle - \left\langle\left(\frac{\hat{x}_1 - \hat{p}_1}{\sqrt{2}}\right)^2\right\rangle\right] = \frac{1}{2}\left[\frac{\hbar}{2} \times 10^{\frac{8.55}{10}} - \frac{\hbar}{2} \times 10^{\frac{8.37}{10}}\right] = 0.04. \tag{6.72}$$

6.5.2.7 Element of V_{14}

By using the measurement results of $\hat{x}_1 \pm \hat{p}_2$, we get

$$V_{14} = \frac{1}{4}\left[\langle(\hat{x}_1 + \hat{p}_2)^2\rangle - \langle(\hat{x}_1 - \hat{p}_2)^2\rangle\right] = \frac{1}{4}\left[\hbar \times 10^{\frac{11.40}{10}} - \hbar \times 10^{-\frac{5.52}{10}}\right] = 1.69. \tag{6.73}$$

As a result, we get the covariance matrix V:

$$V = \begin{pmatrix} 1.76 & 0.04 & 0.01 & 1.69 \\ 0.04 & 1.81 & 1.72 & -0.02 \\ 0.01 & 1.72 & 1.78 & -0.05 \\ 1.69 & -0.02 & -0.05 & 1.75 \end{pmatrix}. \tag{6.74}$$

Note that statistical error in each element is less than ± 0.01. Figure 6.19 shows its graph.

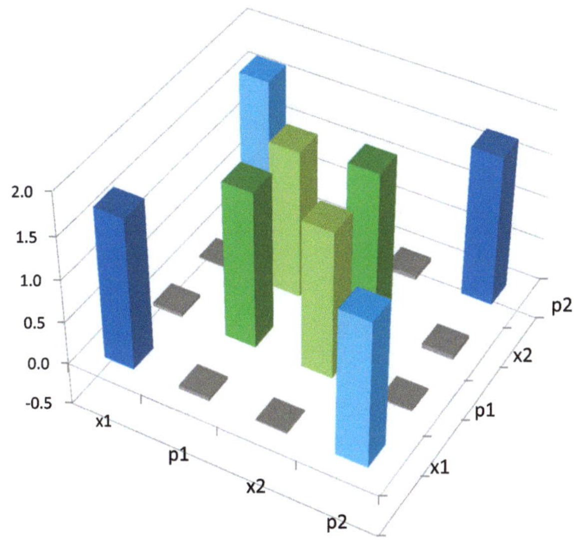

Fig. 6.19 Covariance matrix ($PG1 = 7$, $PG2 = 7$)

6.5.2.8 Variables Acquired by Covariance Matrix

Table 6.10 shows several variables (Sects. 3.2.6 and 3.7.5) which are acquired by the covariance matrix of the generated two-mode cluster state.

Since all determinants of k-th order principal submatrices are positive: $\det V_k > 0$ ($k = 1, \ldots, 4$), and the symplectic eigenvalue ν_- is $\nu_- \geq \frac{\hbar}{2}$, the symmetric matrix V certainly shows a covariance matrix of a physical state. In addition to this, since the PT symplectic eigenvalue $\tilde{\nu}_-$ satisfies $\tilde{\nu}_- < \frac{\hbar}{2}$, the physical state is an entangled state. Note that the logarithmic negativity of experiment ($E_N = 1.29 \pm 0.12$) is smaller than that of the theoretical prediction ($E_N = 1.43$). This is because the generated state has a large logarithmic negativity and large values in its covariance matrix. Thus, a tiny difference (about ± 0.01) in the covariance matrix leads to a large difference (about ± 0.1) in the logarithmic negativity.

6.5.2.9 Estimation of Squeezing Levels and Propagation Efficiency by Covariance Matrix

The theoretical prediction of the covariance matrix is given by Eq. (6.17). By performing fitting with fitting parameters of r_1, r_2, η_1, and η_2, we can estimate squeezing parameters and energy efficiencies. We here minimize

$$f = (V_{e11} - V_{t11})^2 + (V_{e14} - V_{t14})^2 + (V_{e22} - V_{t22})^2 + (V_{e23} - V_{t23})^2 + (V_{e33} - V_{t33})^2 + (V_{e44} - V_{t44})^2, \tag{6.75}$$

where V_e represents the covariance matrix acquired experimentally, while V_t represents its theoretical prediction. As a result, we get

$$\eta_1 = 0.75, \quad \eta_2 = 0.74, \quad r_1 = 1.46, \quad r_2 = 1.45, \tag{6.76}$$

which are consistent with the other estimation results. Figure 6.20 shows the theoretical covariance matrix with these estimated parameters.

6.5.3 Measurement Results and Their Analysis with Other Parametric Gains

In Sect. 6.5.2, we have shown measurement results and analysis of generation of the two-mode cluster state when both parametric gains are adjusted to 7 ($PG1 = 7$, $PG2 = 7$). In this subsection, we show experimental results with the other sets of parametric gains: $(PG1, PG2) = (7, 4), (7, 1.5), (7, 1), (4, 1), (1, 1)$.

We get the following covariance matrices:

$$V^{(7,4)} = \begin{pmatrix} 0.93 & -0.02 & 0.05 & 0.85 \\ -0.02 & 1.79 & 1.70 & -0.02 \\ 0.05 & 1.70 & 1.77 & -0.03 \\ 0.85 & -0.02 & -0.03 & 0.92 \end{pmatrix}, \quad V^{(7,1.5)} = \begin{pmatrix} 0.29 & 0.02 & 0.03 & 0.21 \\ 0.02 & 1.86 & 1.71 & -0.03 \\ 0.03 & 1.71 & 1.85 & 0.00 \\ 0.21 & -0.03 & 0.00 & 0.28 \end{pmatrix},$$

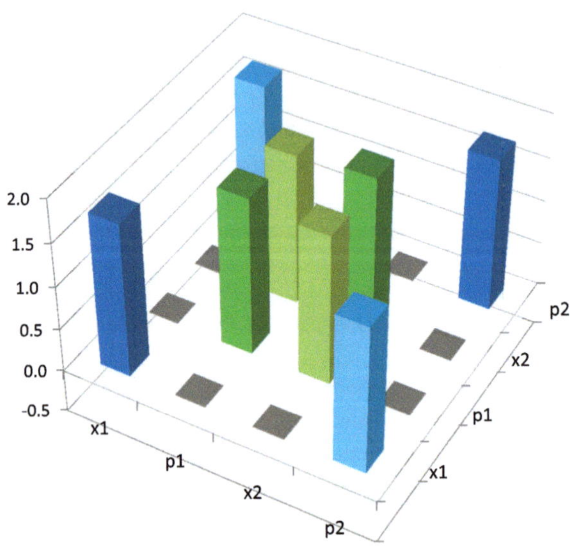

Fig. 6.20 Theoretical covariance matrix with estimated parameters

$$V^{(7,1)} = \begin{pmatrix} 0.16 & 0.05 & 0.05 & 0.09 \\ 0.05 & 1.85 & 1.59 & -0.04 \\ 0.05 & 1.59 & 1.90 & -0.05 \\ 0.09 & -0.04 & -0.05 & 0.16 \end{pmatrix}, \quad V^{(4,1)} = \begin{pmatrix} 0.17 & 0.02 & 0.02 & 0.08 \\ 0.02 & 1.03 & 0.80 & -0.02 \\ 0.02 & 0.80 & 1.07 & -0.01 \\ 0.08 & -0.02 & -0.01 & 0.17 \end{pmatrix},$$

$$V^{(1,1)} = \begin{pmatrix} 0.25 & 0.00 & 0.00 & 0.00 \\ 0.00 & 0.25 & 0.00 & 0.00 \\ 0.00 & 0.00 & 0.25 & 0.00 \\ 0.00 & 0.00 & 0.00 & 0.25 \end{pmatrix}, \tag{6.77}$$

where $V^{(i,j)}$ represents the covariance matrix of $(PG1, PG2) = (i, j)$. Statistical error in each element is less than ± 0.01.

Through Figs. 6.21, 6.22, 6.23, 6.24 and 6.25, we show experimental and theoretical covariance matrices. Note that the theoretical values are calculated by using the parametric gains which are identical to the experimental settings, as well as propagation efficiencies acquired by fitting in Sect. 6.5.2.

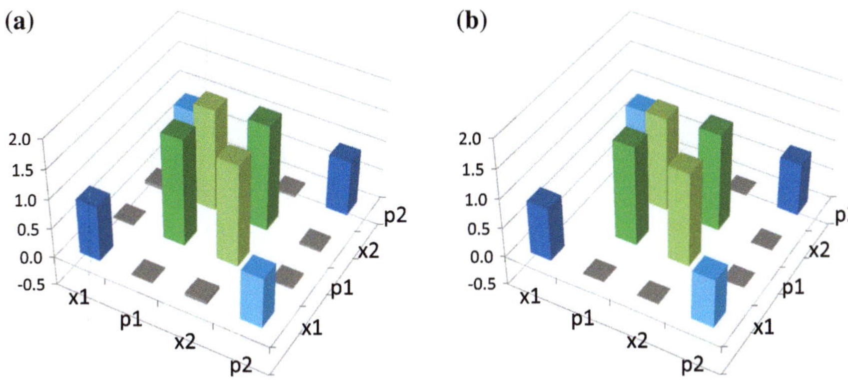

Fig. 6.21 Covariance matrices ($PG1 = 7, PG2 = 4$). **a** Measurement result, **b** Theoretical prediction

We discuss changes of covariance matrices depending on the squeezing parameters. In the following, we use Eq. (6.24) as the theoretical values where efficiencies are symmetrical, while squeezing parameters are asymmetrical, for simplicity. Note that we can use Eq. (6.24) because η_1 is almost the same to η_2.

We start from the result with ($PG1 = 7, PG2 = 7$). By decreasing $PG2$, the values of $V_{11} = V_{44}$ and $V_{14} = V_{41}$ become smaller since decrease of e^{2r_2} is dominant. On the other hand, $V_{22} = V_{33}$ and $V_{23} = V_{32}$ have little changes because their dominant term is e^{2r_1}.

By decreasing $PG1$ from ($PG1 = 7, PG2 = 1$), the values of $V_{22} = V_{33}$ and $V_{23} = V_{32}$ become smaller since decrease of e^{2r_1} is dominant.

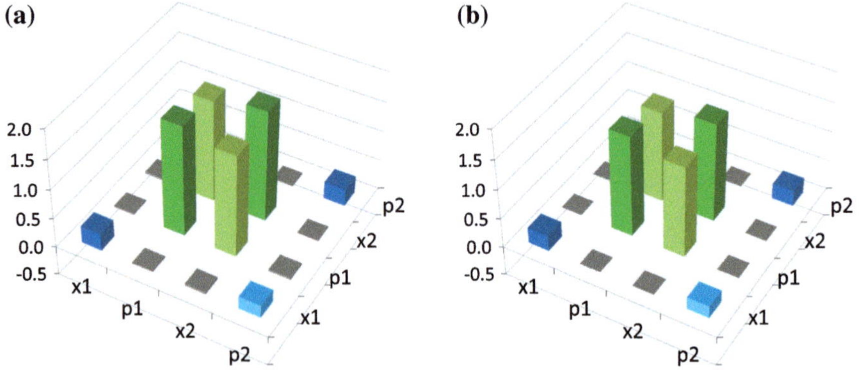

Fig. 6.22 Covariance matrices ($PG1 = 7$, $PG2 = 1.5$). **a** Measurement result, **b** Theoretical prediction

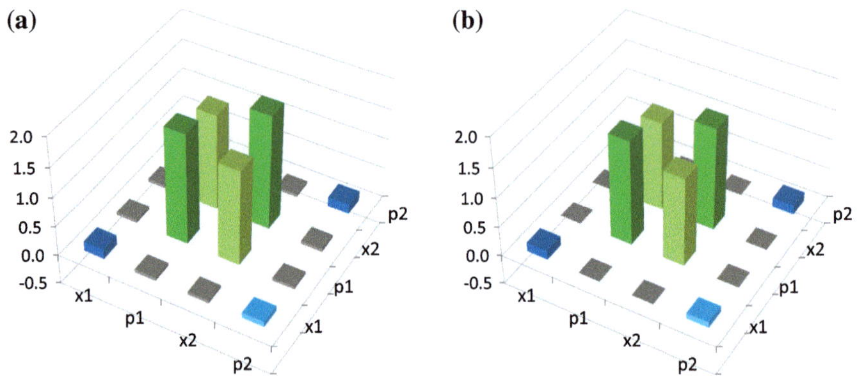

Fig. 6.23 Covariance matrices ($PG1 = 7$, $PG2 = 1$). **a** Measurement result, **b** Theoretical prediction

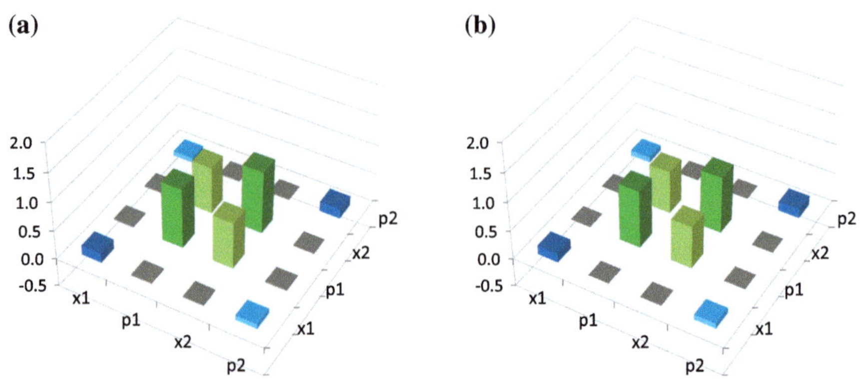

Fig. 6.24 Covariance matrices ($PG1 = 4$, $PG2 = 1$). **a** Measurement result, **b** Theoretical prediction

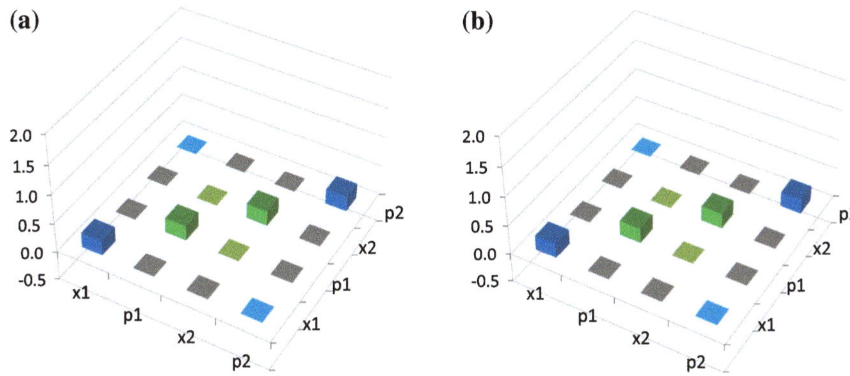

Fig. 6.25 Covariance matrices ($PG1 = 1$, $PG2 = 1$). **a** Measurement result, **b** Theoretical prediction

6.6 Measurement Results of Three-Mode Linear Cluster State

In this section, we show measurement results of our three-mode linear cluster state generation.

6.6.1 Preparation

We omit the detail of experimental procedure for generation of a three-mode linear cluster state because it is similar to that of a two-mode cluster state. Interference visibilities are shown in Table 6.11.

6.6.2 Measurement Results

Figure 6.26 shows measurement results of nullifier variances.

Each result is listed in Table 6.12.

In Fig. 6.26 and Table 6.12, all results are normalized by the values which are acquired by using vacuum states as inputs for all homodyne detections. Therefore, 0 dB in two-mode measurements corresponds to 2 × SNL, while 0 dB in three-mode measurements corresponds to 3 × SNL. Values in Table 6.12 correspond to the effective squeezing level, which we find is larger than 5 dB.

Table 6.11 Interference visibilities for generation of three-mode linear cluster state

Beam splitter	Beams used for adjustment	Visibility (%)
Cluster state generation (mode 2 and 3)	OPO-A, OPO-B	98.7
Cluster state generation (mode 1 and 2)	OPO-A, OPO-D	98.1
Homodyne measurement 1	OPO-B, LO-1	97.7
Homodyne measurement 2	OPO-B, LO-2	98.1
Homodyne measurement 3	OPO-A, LO-3	97.3

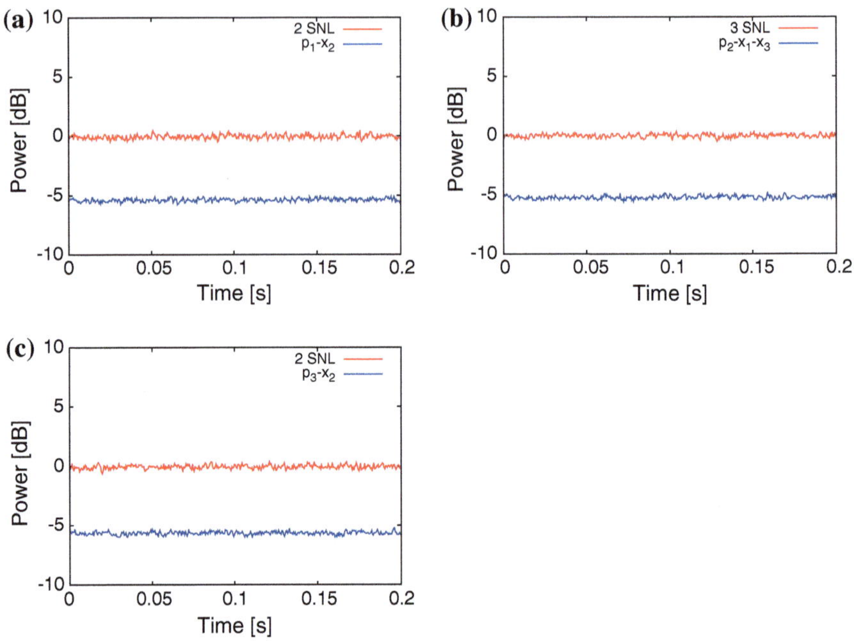

Fig. 6.26 Measurement results of variances. **a** $\hat{p}_1 - \hat{x}_2$, **b** $\hat{p}_2 - \hat{x}_1 - \hat{x}_3$, **c** $\hat{p}_3 - \hat{x}_2$

Table 6.12 Measurement results of variances

Measurement variable	Variance (dB) (relative to vacuum inputs)
$\hat{p}_1 - \hat{x}_2$	-5.33 ± 0.02
$\hat{p}_2 - \hat{x}_1 - \hat{x}_3$	-5.18 ± 0.02
$\hat{p}_3 - \hat{x}_2$	-5.61 ± 0.02

6.6.3 Entanglement Detection Via van Loock-Furusawa Criterion

By using measurement results in Table 6.12, we can evaluate entanglement of the generated state via van Loock-Furusawa criterion. In the case of the three-mode linear cluster state, a sufficient condition of entanglement is given by

$$\langle \Delta^2(\hat{p}_1 - \hat{x}_2)\rangle + \langle \Delta^2(\hat{p}_2 - \hat{x}_1 - \hat{x}_3)\rangle < 2\hbar, \tag{6.78}$$

$$\langle \Delta^2(\hat{p}_3 - \hat{x}_2)\rangle + \langle \Delta^2(\hat{p}_2 - \hat{x}_1 - \hat{x}_3)\rangle < 2\hbar. \tag{6.79}$$

When these two inequalities are satisfied simultaneously, the quantum state is fully inseparable (Sects. 3.7.4 and 5.1.4). By using Table 6.12, we get

$$\langle \Delta^2(\hat{p}_1 - \hat{x}_2)\rangle = \hbar \times 10^{\frac{-5.33}{10}} = (0.293 \pm 0.002)\hbar, \tag{6.80}$$

$$\langle \Delta^2(\hat{p}_2 - \hat{x}_1 - \hat{x}_3)\rangle = \frac{3}{2}\hbar \times 10^{\frac{-5.18}{10}} = (0.455 \pm 0.002)\hbar, \tag{6.81}$$

$$\langle \Delta^2(\hat{p}_3 - \hat{x}_2)\rangle = \hbar \times 10^{\frac{-5.61}{10}} = (0.275 \pm 0.002)\hbar, \tag{6.82}$$

leading to

$$\langle \Delta^2(\hat{p}_1 - \hat{x}_2)\rangle + \langle \Delta^2(\hat{p}_2 - \hat{x}_1 - \hat{x}_3)\rangle = (0.748 \pm 0.002)\hbar < 2\hbar, \tag{6.83}$$

$$\langle \Delta^2(\hat{p}_3 - \hat{x}_2)\rangle + \langle \Delta^2(\hat{p}_2 - \hat{x}_1 - \hat{x}_3)\rangle = (0.730 \pm 0.002)\hbar < 2\hbar. \tag{6.84}$$

Therefore, the full inseparability (entanglement) of the generated cluster state is verified.

6.7 Measurement Results of Four-Mode Linear Cluster State

In this section, we show measurement results of our four-mode linear cluster state generation.

6.7.1 Preparation

Interference visibilities and parametric gains are shown in Tables 6.13 and 6.14, respectively.

6.7.2 Measurement Results

Figure 6.27 shows measurement results of nullifier variances.

Each result is listed in Table 6.15.

Values in Table 6.15 correspond to the effective squeezing level, which we find is about 5 dB or higher.

Table 6.13 Interference visibilities for generation of four-mode linear cluster state

Beam splitter	Beams used for adjustment	Visibility (%)
Cluster state generation (mode 2 and 3)	OPO-A, OPO-B	98.3
Cluster state generation (mode 1 and 2)	OPO-A, OPO-C	97.1
Cluster state generation (mode 3 and 4)	OPO-B, OPO-D	95.3
Homodyne measurement 1	OPO-A, LO-1	95.4
Homodyne measurement 2	OPO-A, LO-2	96.7
Homodyne measurement 3	OPO-B, LO-3	95.7
Homodyne measurement 4	OPO-B, LO-4	95.3

Table 6.14 Parametric gains for generation of four-mode linear cluster state

OPO name	Mode number	Pump power (mW)	Parametric gain (G_+, G_-)	Phase matching temperature °
OPO-A	Mode 3	108	G_+=6.8, G_-=0.46	38.6
OPO-B	Mode 2	102	G_+=9.8, G_-=0.43	39.9
OPO-C	Mode 1	89	G_+=3.9, G_-=0.47	40.4
OPO-D	Mode 4	85	G_+=5.6, G_-=0.46	41.1

Table 6.15 Measurement results of variances

Measurement variable	Variance (dB) (relative to vacuum inputs)
$\hat{p}_1 - \hat{x}_2$	-5.60 ± 0.02
$\hat{p}_2 - \hat{x}_1 - \hat{x}_3$	-4.79 ± 0.02
$\hat{p}_3 - \hat{x}_2 - \hat{x}_4$	-5.25 ± 0.02
$\hat{p}_4 - \hat{x}_3$	-5.51 ± 0.02

6.7.3 Entanglement Detection Via van Loock-Furusawa Criterion

By using measurement results in Table 6.15, we can evaluate entanglement of the generated state via van Loock-Furusawa criterion. In the case of the four-mode linear cluster state, a sufficient condition of entanglement is given by

$$\langle \Delta^2(\hat{p}_1 - \hat{x}_2)\rangle + \langle \Delta^2(\hat{p}_2 - \hat{x}_1 - \hat{x}_3)\rangle < 2\hbar, \tag{6.85}$$

$$\langle \Delta^2(\hat{p}_2 - \hat{x}_1 - \hat{x}_3)\rangle + \langle \Delta^2(\hat{p}_3 - \hat{x}_2 - \hat{x}_4)\rangle < 2\hbar, \tag{6.86}$$

$$\langle \Delta^2(\hat{p}_3 - \hat{x}_2 - \hat{x}_4)\rangle + \langle \Delta^2(\hat{p}_4 - \hat{x}_3)\rangle < 2\hbar. \tag{6.87}$$

When these three inequalities are satisfied simultaneously, the quantum state is fully inseparable (Sects. 3.7.4 and 5.1.4). By using Table 6.15, we get

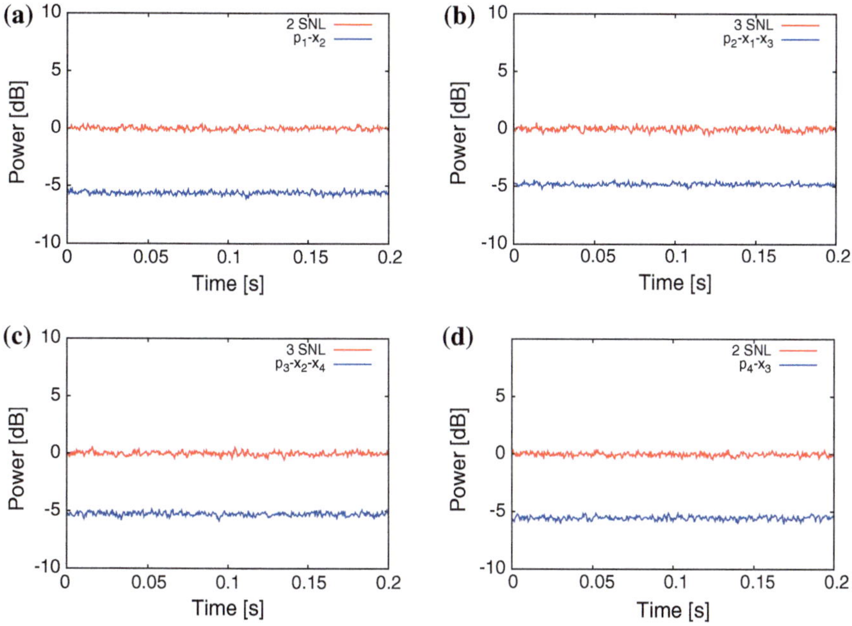

Fig. 6.27 Measurement results of variances. **a** $\hat{p}_1 - \hat{x}_2$, **b** $\hat{p}_2 - \hat{x}_1 - \hat{x}_3$, **c** $\hat{p}_3 - \hat{x}_2 - \hat{x}_4$, **d** $\hat{p}_4 - \hat{x}_3$

$$\langle \Delta^2(\hat{p}_1 - \hat{x}_2) \rangle = \hbar \times 10^{\frac{-5.60}{10}} = (0.276 \pm 0.002)\hbar, \tag{6.88}$$

$$\langle \Delta^2(\hat{p}_2 - \hat{x}_1 - \hat{x}_3) \rangle = \frac{3}{2}\hbar \times 10^{\frac{-4.79}{10}} = (0.498 \pm 0.002)\hbar, \tag{6.89}$$

$$\langle \Delta^2(\hat{p}_3 - \hat{x}_2 - \hat{x}_4) \rangle = \frac{3}{2}\hbar \times 10^{\frac{-5.25}{10}} = (0.448 \pm 0.002)\hbar, \tag{6.90}$$

$$\langle \Delta^2(\hat{p}_4 - \hat{x}_3) \rangle = \hbar \times 10^{\frac{-5.51}{10}} = (0.281 \pm 0.002)\hbar, \tag{6.91}$$

leading to

$$\langle \Delta^2(\hat{p}_1 - \hat{x}_2) \rangle + \langle \Delta^2(\hat{p}_2 - \hat{x}_1 - \hat{x}_3) \rangle = (0.774 \pm 0.003)\hbar < 2\hbar, \tag{6.92}$$

$$\langle \Delta^2(\hat{p}_2 - \hat{x}_1 - \hat{x}_3) \rangle + \langle \Delta^2(\hat{p}_3 - \hat{x}_2 - \hat{x}_4) \rangle = (0.946 \pm 0.003)\hbar < 2\hbar, \tag{6.93}$$

$$\langle \Delta^2(\hat{p}_3 - \hat{x}_2 - \hat{x}_4) \rangle + \langle \Delta^2(\hat{p}_4 - \hat{x}_3) \rangle = (0.729 \pm 0.003)\hbar < 2\hbar. \tag{6.94}$$

Therefore, the full inseparability (entanglement) of the generated cluster state is verified.

References

1. Reck, M., Zeilinger, A., Bernstein, H.J., Bertani, P.: Experimental realization of any discrete unitary operator. Phys. Rev. Lett **73**, 58 (1994)
2. Saleh, B.E.A., Teich, M.C.: Fundamentals of Photonics, 2nd edn. Wiley-Interscience, Berlin (2007)

Chapter 7
Experimental Demonstration of Controlled-Z Gate for Continuous Variables

7.1 Controlled-Z Gate Experiment

7.1.1 Operation

In the experimental demonstration of a *controlled-Z gate* for continuous-variable one-way quantum computation, we utilize a four-partite entangled state, called the *four-mode linear cluster state*, as a resource (Sect. 6.1.3). We prepare a two-mode input state independently of the cluster state. The input coupling with the cluster is achieved via the *teleportation-based input-coupling scheme* (Sect. 5.3.3). The two modes to which the input state is transmitted are equivalent to the output modes. By changing the relative phases between signal beams and local oscillator beams in homodyne detections, we can implement Gaussian operations. We choose the measurement bases so that the operation for each single mode becomes the Fourier transformation. This is the simplest operation which can be achieved by the experimental setup we use. The input–output relationship in the Heisenberg picture is given by

$$\begin{pmatrix} \hat{x}'_\alpha \\ \hat{p}'_\alpha \\ \hat{x}'_\beta \\ \hat{p}'_\beta \end{pmatrix} = C_Z F_\alpha F_\beta \begin{pmatrix} \hat{x}_\alpha \\ \hat{p}_\alpha \\ \hat{x}_\beta \\ \hat{p}_\beta \end{pmatrix}, \tag{7.1}$$

where $\hat{x}_i$ and $\hat{p}_i$ represent quadrature operators of the input mode i, while $\hat{x}'_i$ and $\hat{p}'_i$ represent those of the output mode i. The matrices C_Z, F_α, and F_β:

$$C_Z = \begin{pmatrix} 1&0&0&0 \\ 0&1&1&0 \\ 0&0&1&0 \\ 1&0&0&1 \end{pmatrix}, \quad F_\alpha = \begin{pmatrix} 0&-1&0&0 \\ 1&0&0&0 \\ 0&0&1&0 \\ 0&0&0&1 \end{pmatrix}, \quad F_\beta = \begin{pmatrix} 1&0&0&0 \\ 0&1&0&0 \\ 0&0&0&-1 \\ 0&0&1&0 \end{pmatrix} \tag{7.2}$$

R. Ukai, *Multi-Step Multi-Input One-Way Quantum Information Processing with Spatial and Temporal Modes of Light*, Springer Theses,
DOI 10.1007/978-4-431-55019-8_7

show the controlled-Z gate, the Fourier transformation on mode α, and the Fourier transformation on mode β, respectively.

7.1.2 Importance of This Experiment

We summarize importance of this experiment.

7.1.2.1 Gaussian Operation on Two-Mode Input State (Common to the Three Experiments)

This is an experimental demonstration of one-way quantum computation where Gaussian operations on two-mode input states are implemented. By combining the experimental demonstration of one-mode Gaussian operations we have reported earlier [1], we can implement an arbitrary multi-mode Gaussian operations in the framework of one-way quantum computation in principle (Sects. 3.6 and 5.5).

This is a common property to the three experiments in this thesis.

7.1.2.2 Nonlocal Gate (Common to the Three Experiments)

This is an experimental demonstration of a nonlocal gate, where target two modes are located at a distance (Sect. 8.3).

This is a common property to the three experiments in this thesis.

7.1.2.3 Minimum Setup of Two-Mode Gate with Teleportation-Based Input-Coupling Scheme (Characteristic of This Experiment)

Since two modes of a cluster state are consumed during a teleportation-based input coupling (Sect. 5.3.3), we need at least four modes of a cluster state to implement a two-mode operation where a two-mode input state is coupled with the cluster state via two teleportation-based input couplings. Therefore, this is the minimum setup of implementing a two-mode gate with teleportation-based input-coupling scheme.

7.1.2.4 Gate Teleportation (Characteristic of This Experiment)

We here describe the resource four-mode linear cluster state as $C1$–$C2$–$C3$–$C4$. Up to local phase rotations, it is equivalent to two EPR pairs ($C1$–$C2$ and $C3$–$C4$) with a controlled-Z gate interaction between them ($C2$–$C3$). In our experiment, quantum states in modes α and β are transmitted to modes $C2$ and $C3$ via quantum teleportations with EPR states $C1$–$C2$ and $C3$–$C4$, respectively. As a result, the initial controlled-Z gate between modes $C2$ and $C3$ is teleported to the two input states. Therefore, we can consider that this is an experimental demonstration of two-mode gate teleportation (Sects. 4.1.2 and 7.2.8).

7.1.2.5 Elementary Circuit for Multi-Mode Operations with Temporal Modes (Characteristic of This Experiment)

This is an experimental demonstration of the elementary circuit for multi-mode Gaussian operations with temporal modes (Sect. 10.6).

7.1.3 Comparison of Three Experiments

In order to compare three experiments in this thesis, we summarize the properties of this experiment in a common form. It is given in Table 7.1.

7.2 Theory

7.2.1 Abstract Illustration and Abstract Experimental Setup

Figures 7.1 and 7.3 show an abstract illustration and an abstract setup of the controlled-Z gate experiment.

The abstract illustration shown in Fig. 7.1 gives us a brief explanation of the experimental procedure, which we summarize in the following.

Table 7.1 Properties of controlled-Z gate in a common form

Item	Controlled-Z gate experiment
Main feature	Operation does not have excess squeezing.
Resource	Four-mode linear cluster state
Input coupling scheme	Teleportation-based input-coupling scheme
Excess 3 dB squeezing derived from input coupling	Does not exist
Operation experimentally demonstrated	$C_Z F_\alpha F_\beta$, fixed
Entanglement verification at the output	Van Loock-Furusawa criterion
Logarithmic negativity	Not measured

7.2.1.1 Cluster State and Input State

(Figure 7.2a) Blue circles show resource cluster modes, while blue solid lines show controlled-Z gates. Thus, the resource state is a four-mode linear cluster state. We define labels of two input modes to be α and β.

7.2.1.2 Input Coupling

(Figure 7.2b) Input modes α and β are coupled with cluster modes 1 and 4 by using beam splitters, respectively.

7.2.1.3 Homodyne Measurement

(Figure 7.2c) We perform homodyne measurements on modes α, β, 1, and 4, where measurement variables are x for modes α and β, while p for modes 1 and 4.

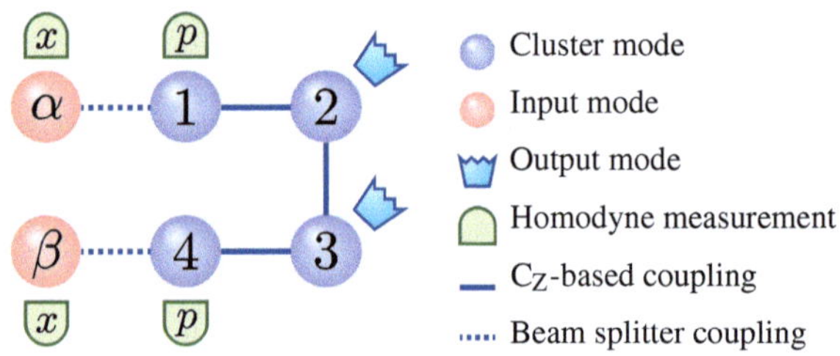

Fig. 7.1 Abstract illustration of controlled-Z gate

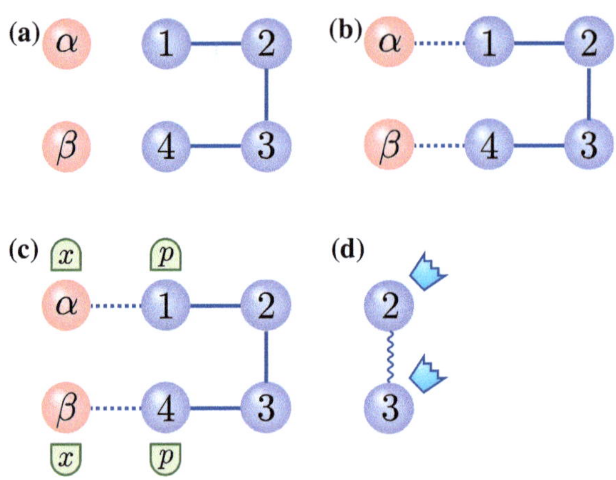

Fig. 7.2 Interpretation of the abstract illustration. **a** Cluster state and input state, **b** Input coupling, **c** Homodyne measurement, **d** Output

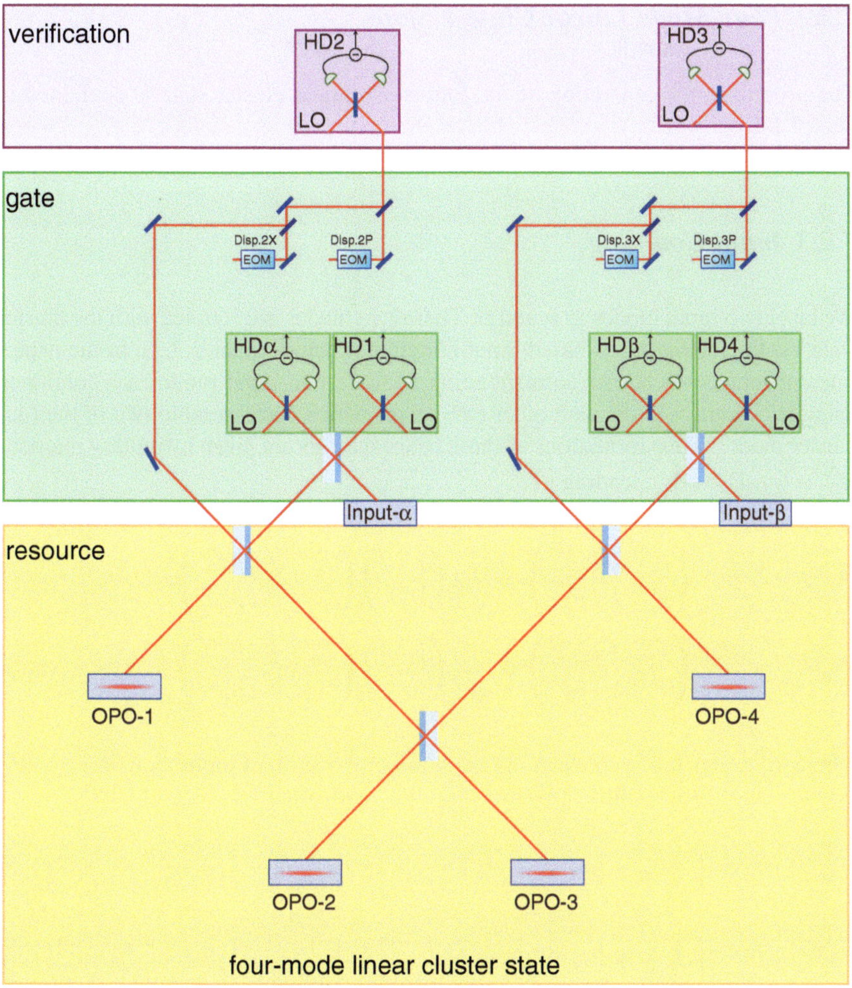

Fig. 7.3 Abstract experimental setup of controlled-Z gate

7.2.1.4 Output

(Figure 7.2d) The unmeasured modes 2 and 3 correspond to the output modes of the one-way quantum computation. We have described the operation by the wavy line.

7.2.2 Four-Mode Linear Cluster State

The procedure for generation of the four-mode linear cluster state is explained in Sect. 6.1.3.

7.2.3 Input Coupling

We label two input modes as α and β. These two modes are coupled with the cluster state via the teleportation-based input-coupling schemes (Sect. 5.3.3). In the experimental setup, two beams corresponding to these two input modes are combined using two beam splitters with other two beams which correspond to two of the four cluster modes. Transformations of these beam splitters are given by unitary matrices $B^{(2)}_{\alpha 1}(\frac{1}{\sqrt{2}})$, $B^{(2)}_{\beta 4}(\frac{1}{\sqrt{2}})$, leading to

$$\begin{pmatrix} \hat{x}^{(b)}_\alpha + i\hat{p}^{(b)}_\alpha \\ \hat{x}^{(b)}_1 + i\hat{p}^{(b)}_1 \end{pmatrix} = \frac{1}{\sqrt{2}} \begin{pmatrix} 1 & -1 \\ 1 & 1 \end{pmatrix} \begin{pmatrix} \hat{x}^{(in)}_\alpha + i\hat{p}^{(in)}_\alpha \\ \hat{x}^{(c)}_1 + i\hat{p}^{(c)}_1 \end{pmatrix},$$
$$\begin{pmatrix} \hat{x}^{(b)}_\beta + i\hat{p}^{(b)}_\beta \\ \hat{x}^{(b)}_4 + i\hat{p}^{(b)}_4 \end{pmatrix} = \frac{1}{\sqrt{2}} \begin{pmatrix} 1 & -1 \\ 1 & 1 \end{pmatrix} \begin{pmatrix} \hat{x}^{(in)}_\beta + i\hat{p}^{(in)}_\beta \\ \hat{x}^{(c)}_4 + i\hat{p}^{(c)}_4 \end{pmatrix}, \tag{7.3}$$

where $\hat{x}^{(in)}_k$ and $\hat{p}^{(in)}_k$ represent the quadrature operators of input mode $k(=\alpha, \beta)$. As a result, the quadrature operators after the input couplings are given by

$$\hat{x}^{(b)}_\alpha + i\hat{p}^{(b)}_\alpha = \left[\frac{1}{\sqrt{2}}\hat{x}_\alpha - \frac{1}{2}\hat{x}_1 - \frac{1}{2\sqrt{5}}\hat{x}_2 - \frac{1}{\sqrt{5}}\hat{p}_3\right] + i\left[\frac{1}{\sqrt{2}}\hat{p}_\alpha - \frac{1}{2}\hat{p}_1 - \frac{1}{2\sqrt{5}}\hat{p}_2 + \frac{1}{\sqrt{5}}\hat{x}_3\right], \tag{7.4}$$

$$\hat{x}^{(b)}_1 + i\hat{p}^{(b)}_1 = \left[\frac{1}{\sqrt{2}}\hat{x}_\alpha + \frac{1}{2}\hat{x}_1 + \frac{1}{2\sqrt{5}}\hat{x}_2 + \frac{1}{\sqrt{5}}\hat{p}_3\right] + i\left[\frac{1}{\sqrt{2}}\hat{p}_\alpha + \frac{1}{2}\hat{p}_1 + \frac{1}{2\sqrt{5}}\hat{p}_2 - \frac{1}{\sqrt{5}}\hat{x}_3\right], \tag{7.5}$$

$$\hat{x}^{(c)}_2 + i\hat{p}^{(c)}_2 = \left[-\frac{1}{\sqrt{2}}\hat{p}_1 + \frac{1}{\sqrt{10}}\hat{p}_2 - \frac{2}{\sqrt{10}}\hat{x}_3\right] + i\left[\frac{1}{\sqrt{2}}\hat{x}_1 - \frac{1}{\sqrt{10}}\hat{x}_2 - \frac{2}{\sqrt{10}}\hat{p}_3\right], \tag{7.6}$$

$$\hat{x}^{(c)}_3 + i\hat{p}^{(c)}_3 = \left[-\frac{2}{\sqrt{10}}\hat{x}_2 + \frac{1}{\sqrt{10}}\hat{p}_3 + \frac{1}{\sqrt{2}}\hat{p}_4\right] + i\left[-\frac{2}{\sqrt{10}}\hat{p}_2 - \frac{1}{\sqrt{10}}\hat{x}_3 - \frac{1}{\sqrt{2}}\hat{x}_4\right], \tag{7.7}$$

$$\hat{x}_4^{(b)} + i\hat{p}_4^{(b)} = \left[\frac{1}{\sqrt{2}}\hat{x}_\beta + \frac{1}{\sqrt{5}}\hat{p}_2 + \frac{1}{2\sqrt{5}}\hat{x}_3 - \frac{1}{2}\hat{x}_4\right] + i\left[\frac{1}{\sqrt{2}}\hat{p}_\beta - \frac{1}{\sqrt{5}}\hat{x}_2 + \frac{1}{2\sqrt{5}}\hat{p}_3 - \frac{1}{2}\hat{p}_4\right], \tag{7.8}$$

$$\hat{x}_\beta^{(b)} + i\hat{p}_\beta^{(b)} = \left[\frac{1}{\sqrt{2}}\hat{x}_\beta - \frac{1}{\sqrt{5}}\hat{p}_2 - \frac{1}{2\sqrt{5}}\hat{x}_3 + \frac{1}{2}\hat{x}_4\right] + i\left[\frac{1}{\sqrt{2}}\hat{p}_\beta + \frac{1}{\sqrt{5}}\hat{x}_2 - \frac{1}{2\sqrt{5}}\hat{p}_3 + \frac{1}{2}\hat{p}_4\right], \tag{7.9}$$

where we have omitted the superscripts (in) for input modes α, β and (r) for resource modes 1, 2, 3, and 4. For example, $\hat{x}_1$ represents $\hat{x}_1 = \hat{x}_1^{(r)} = e^{r_1}\hat{x}_1^{(0)}$.

7.2.4 Measurement

We perform homodyne measurements on modes α, 1, 4, and β. We utilize η *representation* for measurements on modes α and β, while κ *representation* for measurements on modes 1 and 4 (Sect. 5.3.1). Therefore, the measurement observables $\hat{s}_\alpha$, $\hat{s}_\beta$, $\hat{s}_1$, $\hat{s}_4$ for measurements on modes α, 1, 4, and β are

$$\hat{s}_\alpha = \hat{x}_\alpha^{(b)} + \eta_\alpha \hat{p}_\alpha^{(b)}, \tag{7.10}$$

$$\hat{s}_\beta = \hat{x}_\beta^{(b)} + \eta_\beta \hat{p}_\beta^{(b)}, \tag{7.11}$$

$$\hat{s}_1 = \kappa_1 \hat{x}_1^{(b)} + \hat{p}_1^{(b)}, \tag{7.12}$$

$$\hat{s}_4 = \kappa_4 \hat{x}_4^{(b)} + \hat{p}_4^{(b)}. \tag{7.13}$$

7.2.5 Feed-Forward and Operation

Without using the position operators $\hat{x}_k^{(r)}$ for squeezed-state modes $k = 1, 2, 3, 4$, the quadrature operators $\hat{x}_k^{(c)}$ and $\hat{p}_k^{(c)}$ for output modes $k = 2, 3$ can be reformulated as

$$\begin{pmatrix} \hat{x}_2^{(c)} \\ \hat{p}_2^{(c)} \\ \hat{x}_3^{(c)} \\ \hat{p}_3^{(c)} \end{pmatrix} = M_{op} \begin{pmatrix} \hat{x}_\alpha^{(in)} \\ \hat{p}_\alpha^{(in)} \\ \hat{x}_\beta^{(in)} \\ \hat{p}_\beta^{(in)} \end{pmatrix} + M_{disp} \begin{pmatrix} \hat{s}_\alpha \\ \hat{s}_1 \\ \hat{s}_\beta \\ \hat{s}_4 \end{pmatrix} + \begin{pmatrix} -\hat{\delta}_1 \\ \hat{\delta}_2 - \hat{\delta}_4 \\ -\hat{\delta}_4 \\ \hat{\delta}_3 - \hat{\delta}_1 \end{pmatrix}, \tag{7.14}$$

where

$$M_{op} = \begin{pmatrix} \frac{-2\kappa_1}{1-\kappa_1\eta_\alpha} & -\frac{1+\kappa_1\eta_\alpha}{1-\kappa_1\eta_\alpha} & 0 & 0 \\ \frac{1+\kappa_1\eta_\alpha}{1-\kappa_1\eta_\alpha} & \frac{2\eta_\alpha}{1-\kappa_1\eta_\alpha} & \frac{-2\kappa_4}{1-\kappa_4\eta_\beta} & -\frac{1+\kappa_4\eta_\beta}{1-\kappa_4\eta_\beta} \\ 0 & 0 & \frac{-2\kappa_4}{1-\kappa_4\eta_\beta} & -\frac{1+\kappa_4\eta_\beta}{1-\kappa_4\eta_\beta} \\ \frac{-2\kappa_1}{1-\kappa_1\eta_\alpha} & -\frac{1+\kappa_1\eta_\alpha}{1-\kappa_1\eta_\alpha} & \frac{1+\kappa_4\eta_\beta}{1-\kappa_4\eta_\beta} & \frac{2\eta_\beta}{1-\kappa_4\eta_\beta} \end{pmatrix},$$

$$M_{disp} = \sqrt{2} \begin{pmatrix} \frac{\kappa_1}{1-\kappa_1\eta_\alpha} & \frac{1}{1-\kappa_1\eta_\alpha} & 0 & 0 \\ \frac{-1}{1-\kappa_1\eta_\alpha} & \frac{-\eta_\alpha}{1-\kappa_1\eta_\alpha} & \frac{\kappa_4}{1-\kappa_4\eta_\beta} & \frac{1}{1-\kappa_4\eta_\beta} \\ 0 & 0 & \frac{\kappa_4}{1-\kappa_4\eta_\beta} & \frac{1}{1-\kappa_4\eta_\beta} \\ \frac{\kappa_1}{1-\kappa_1\eta_\alpha} & \frac{1}{1-\kappa_1\eta_\alpha} & \frac{-1}{1-\kappa_4\eta_\beta} & \frac{-\eta_\beta}{1-\kappa_4\eta_\beta} \end{pmatrix}. \tag{7.15}$$

Note that the reformulation above holds before the measurements are carried out. On the contrary, when the measurements on modes α, 1, 4, and β have been carried out indeed, the quadratures of the output modes become

$$\begin{pmatrix} \hat{x}_2^{(cm)} \\ \hat{p}_2^{(cm)} \\ \hat{x}_3^{(cm)} \\ \hat{p}_3^{(cm)} \end{pmatrix} = M_{op} \begin{pmatrix} \hat{x}_\alpha^{(in)} \\ \hat{p}_\alpha^{(in)} \\ \hat{x}_\beta^{(in)} \\ \hat{p}_\beta^{(in)} \end{pmatrix} + M_{disp} \begin{pmatrix} s_\alpha \\ s_1 \\ s_\beta \\ s_4 \end{pmatrix} + \begin{pmatrix} -\hat{\delta}_1 \\ \hat{\delta}_2 - \hat{\delta}_4 \\ -\hat{\delta}_4 \\ \hat{\delta}_3 - \hat{\delta}_1 \end{pmatrix}, \tag{7.16}$$

where s_α, s_β, s_1, and s_4 represent measurement results of measurement variables $\hat{s}_\alpha$, $\hat{s}_\beta$, $\hat{s}_1$, and $\hat{s}_4$, respectively. $\hat{\delta}_k$ represents the k-th nullifier of the resource cluster state (see Eq. (6.49)). By performing feed-forwards (displacement operations) based on the measurement results:

$$\begin{pmatrix} \hat{x}_2^{(out)} \\ \hat{p}_2^{(out)} \\ \hat{x}_3^{(out)} \\ \hat{p}_3^{(out)} \end{pmatrix} = \begin{pmatrix} \hat{x}_2^{(cm)} \\ \hat{p}_2^{(cm)} \\ \hat{x}_3^{(cm)} \\ \hat{p}_3^{(cm)} \end{pmatrix} - M_{disp} \begin{pmatrix} s_\alpha \\ s_1 \\ s_\beta \\ s_4 \end{pmatrix}, \tag{7.17}$$

we get

$$\begin{pmatrix} \hat{x}_2^{(out)} \\ \hat{p}_2^{(out)} \\ \hat{x}_3^{(out)} \\ \hat{p}_3^{(out)} \end{pmatrix} = M_{op} \begin{pmatrix} \hat{x}_\alpha^{(in)} \\ \hat{p}_\alpha^{(in)} \\ \hat{x}_\beta^{(in)} \\ \hat{p}_\beta^{(in)} \end{pmatrix} + \begin{pmatrix} -\hat{\delta}_1 \\ \hat{\delta}_2 - \hat{\delta}_4 \\ -\hat{\delta}_4 \\ \hat{\delta}_3 - \hat{\delta}_1 \end{pmatrix}. \tag{7.18}$$

This is the input-output relationship which can be achieved by using our experimental setup. The term of M_{op} represents the main operation. On the other hand, the term of $\hat{\delta}_k$ represents error of the operation, which derives from finite level of resource squeezing. In the limit of infinite squeezing $r \to \infty$, each $\hat{\delta}_k$ goes to zero, meaning that the ideal operation is achieved.

We mention that the input-output relationship shown in Eq. (7.18) is formulated by the δ *representation* (Sect. 5.6). Therefore, Eq. (7.18) holds no matter how a cluster state is generated (for example, canonical cluster state or Gaussian cluster state).

7.2.6 Covariance Matrix

We define $V_{\alpha\beta}^{(in)}$ as the covariance matrix of a two-mode input state. We here choose the *xpxp notation* for all covariance matrices (Sect. 3.2.5).

Since the two-mode input state and the resource four-mode linear cluster state has no correlations, the covariance matrix of the output state becomes

$$V_{23}^{(out)} = M_{op} V_{\alpha\beta}^{(in)} M_{op}^T + V_\delta, \tag{7.19}$$

where V_δ represents excess noise derived from the finite level of resource squeezing. It is given by

$$V_\delta = \frac{\hbar}{2} \begin{pmatrix} 2e^{-2r_1} & 0 & 0 & e^{-2r_1} \\ 0 & \frac{5}{2}e^{-2r_3} + \frac{1}{2}e^{-2r_4} & e^{-2r_4} & 0 \\ 0 & e^{-2r_4} & 2e^{-2r_4} & 0 \\ e^{-2r_1} & 0 & 0 & \frac{1}{2}e^{-2r_1} + \frac{5}{2}e^{-2r_2} \end{pmatrix}. \tag{7.20}$$

In the special case of identical squeezing parameters $r_i = r$, it becomes

$$V_\delta = \frac{\hbar}{2} \begin{pmatrix} 2e^{-2r} & 0 & 0 & e^{-2r} \\ 0 & 3e^{-2r} & e^{-2r} & 0 \\ 0 & e^{-2r} & 2e^{-2r} & 0 \\ e^{-2r} & 0 & 0 & 3e^{-2r} \end{pmatrix}. \tag{7.21}$$

In the limit of infinite squeezing $r_i \to \infty$, V_δ goes to the 4×4 zero matrix O, meaning that the ideal operation is achieved.

7.2.7 Decomposition of Operation

The matrix M_{op} can be decomposed into

$$M_{op} = C_Z \begin{pmatrix} M_{tele}(\eta_\alpha, \kappa_1) & O \\ O & M_{tele}(\eta_\beta, \kappa_4) \end{pmatrix}, \tag{7.22}$$

where

$$C_Z = \begin{pmatrix} 1 & 0 & 0 & 0 \\ 0 & 1 & 1 & 0 \\ 0 & 0 & 1 & 0 \\ 1 & 0 & 0 & 1 \end{pmatrix}, \quad M_{tele}(\eta, \kappa) = \begin{pmatrix} \frac{-2\kappa}{1-\kappa\eta} & -\frac{1+\kappa\eta}{1-\kappa\eta} \\ \frac{1+\kappa\eta}{1-\kappa\eta} & \frac{2\eta}{1-\kappa\eta} \end{pmatrix}. \tag{7.23}$$

$M_{tele}(\eta, \kappa)$ is the matrix representing an operation by a teleportation-based input-coupling scheme (Sect. 5.3.3), which is a member of one-mode Gaussian operations. C_Z is the matrix of the controlled-Z gate, which is a member of two-mode Gaussian operations. Therefore, M_{op} represents operations on two single-modes via teleportation-based input-couplings, followed by a controlled-Z gate operation.

7.2.8 Interpretation as Gate Teleportation

Consider a quantum circuit shown in Fig. 7.4.

In this figure, modes α and β represent two input modes, while modes 1, 2, 3, and 4 represent resource modes. Each resource mode is initially in the zero eigenstate of the momentum operator $|p = 0\rangle$, or a p-squeezed vacuum state, which is an approximation of $|p = 0\rangle$. On the other hand, modes α and β can be initially in an arbitrary state. "CZ 1" shows a controlled-Z gate given by $\hat{C}_{Zab} = e^{\frac{i\hat{x}_a\hat{x}_b}{\hbar}}$, while "$BS2$ 1/2" shows a beam splitter $B^{(2)}$ with the energy reflectivity 1/2. Since the operator of the beam splitter depends on the order of target modes, we define that the mode on which "$BS2$ " is drawn in the figure corresponds to the mode i of $B^{(2)}_{ij}$. All measurements are homodyne measurements, while feed-forwards are displacement operations $\hat{X}(s_a, s_b)\hat{Z}(s_a, s_b)$, where s_a, s_b are measurement results. Therefore, Fig. 7.4 shows two teleportation-based input-couplings, followed by a

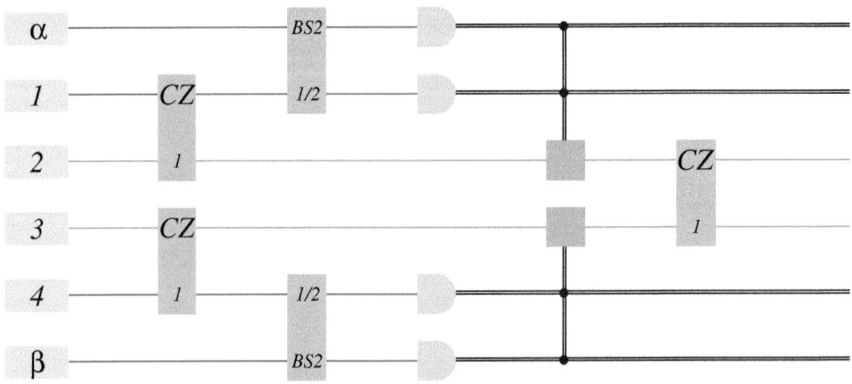

Fig. 7.4 Interpretation as gate teleportation 1

controlled-Z gate operation. It is obvious that the operation achieved by Fig 7.4 is identical to Eq. (7.22):

$$M_{op} = C_Z \begin{pmatrix} M_{tele} & O \\ O & M_{tele} \end{pmatrix}. \tag{7.24}$$

We then consider transformation of the circuit. The essence of this transformation is the same to that of the gate teleportation.

In general, a controlled-Z operator does not commute with a displacement operator. However, by using

$$\hat{C}_{Zij}\hat{Z}_k(s) = \hat{Z}_k(s)\hat{C}_{Zij}, \quad \hat{C}_{Zij}\hat{X}_k(s) = \hat{X}_k(s)\hat{Z}_{k'}(s)\hat{C}_{Zij}, \quad (k,k') = (i,j),(j,i), \tag{7.25}$$

we can change the order of the controlled-Z operator and the displacement operator by adding another displacement operator (Fig. 7.5).

Here, the quantum state acquired by entangling four modes 1, 2, 3, and 4 via three controlled-Z gates is the four-mode linear cluster state. We know that the same cluster state can be generated by using an appropriate network of three beam splitters (Fig. 7.6).

Note that we have omitted phase rotations, for simplicity. Figure 7.6 is nothing but the quantum circuit of this experiment.

In the transformation above, the controlled-Z gate which is originally applied to the output of the two quantum teleportations is moved to the earlier position, and then is applied before the two quantum teleportations. Therefore, this experiment can be considered as a demonstration of a two-mode gate teleportation (Sect. 4.1.2). The transformation from Figs. 7.5 to 7.6 is an example of generation of the same state via a different and easier scheme discussed in Sect. 4.1.2. Since the resource state

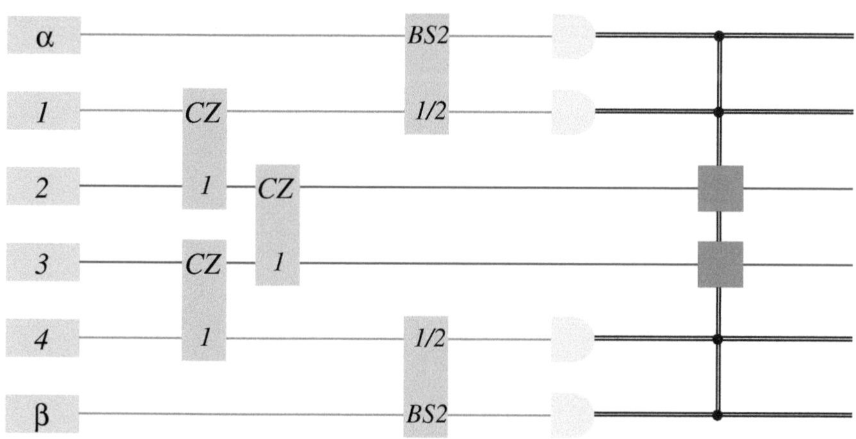

Fig. 7.5 Interpretation as gate teleportation 2

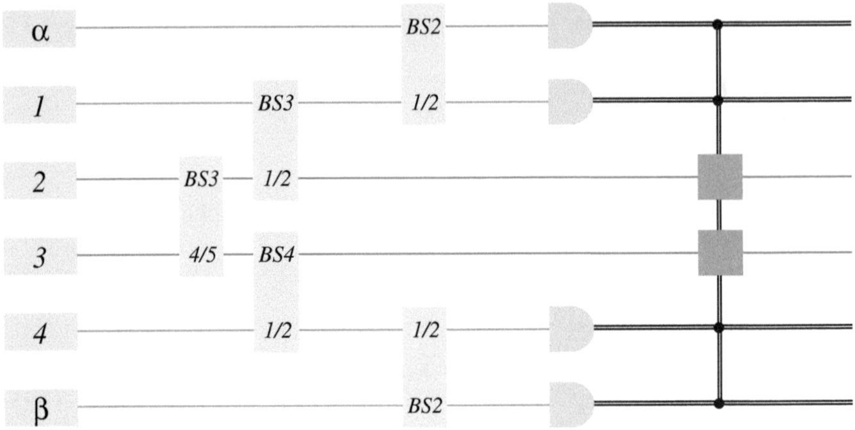

Fig. 7.6 Interpretation as gate teleportation 3

can be generated deterministically (with success rate of 100 %) in this experiment, its transformation does not provide the advantage of success probability. However, we can introduce entanglement distillation, although it is not implemented in this experiment.

7.2.9 Operation We Implement

Although we have discussed the general operation which can be implemented by using our experimental setup, we here discuss the operation we implement in the experiment. In order to focus on the two-mode gate component in the total operation, we choose the single-mode operation components to be the simplest operations. It is achieved by

$$\eta_\alpha = \eta_\beta = \kappa_1 = \kappa_4 = 0, \tag{7.26}$$

leading to four measurement variables

$$\hat{s}_\alpha = \hat{x}_\alpha^{(b)}, \quad \hat{s}_\beta = \hat{x}_\beta^{(b)}, \quad \hat{s}_1 = \hat{p}_1^{(b)}, \quad \hat{s}_4 = \hat{p}_4^{(b)}. \tag{7.27}$$

The matrices of the operation and displacement are given by

$$M_{tele} = \begin{pmatrix} 0 & -1 \\ 1 & 0 \end{pmatrix}, \quad M_{op} = \begin{pmatrix} 0 & -1 & 0 & 0 \\ 1 & 0 & 0 & -1 \\ 0 & 0 & 0 & -1 \\ 0 & -1 & 1 & 0 \end{pmatrix}, \quad M_{disp} = \sqrt{2} \begin{pmatrix} 0 & 1 & 0 & 0 \\ -1 & 0 & 0 & 1 \\ 0 & 0 & 0 & 1 \\ 0 & 1 & -1 & 0 \end{pmatrix}. \tag{7.28}$$

It shows that the single-mode operation components are the Fourier transformations. The feed-forward operation is given by

$$\begin{pmatrix} \hat{x}_2^{(out)} \\ \hat{p}_2^{(out)} \\ \hat{x}_3^{(out)} \\ \hat{p}_3^{(out)} \end{pmatrix} = \begin{pmatrix} \hat{x}_2^{(cm)} \\ \hat{p}_2^{(cm)} \\ \hat{x}_3^{(cm)} \\ \hat{p}_3^{(cm)} \end{pmatrix} - M_{disp} \begin{pmatrix} s_\alpha \\ s_1 \\ s_\beta \\ s_4 \end{pmatrix}, \tag{7.29}$$

or identically

$$\hat{X}_2(-\sqrt{2}s_1)\hat{Z}_2(\sqrt{2}s_\alpha - \sqrt{2}s_4)\hat{X}_3(-\sqrt{2}s_4)\hat{Z}_3(\sqrt{2}s_\beta - \sqrt{2}s_1). \tag{7.30}$$

7.2.10 Variances of Quadrature Operators

The input–output relationship in this experiment is given by

$$\begin{pmatrix} \hat{x}_2^{(out)} \\ \hat{p}_2^{(out)} \\ \hat{x}_3^{(out)} \\ \hat{p}_3^{(out)} \end{pmatrix} = \begin{pmatrix} 0 & -1 & 0 & 0 \\ 1 & 0 & 0 & -1 \\ 0 & 0 & 0 & -1 \\ 0 & -1 & 1 & 0 \end{pmatrix} \begin{pmatrix} \hat{x}_\alpha^{(in)} \\ \hat{p}_\alpha^{(in)} \\ \hat{x}_\beta^{(in)} \\ \hat{p}_\beta^{(in)} \end{pmatrix} + \begin{pmatrix} -\hat{\delta}_1 \\ \hat{\delta}_2 - \hat{\delta}_4 \\ -\hat{\delta}_4 \\ \hat{\delta}_3 - \hat{\delta}_1 \end{pmatrix} = \begin{pmatrix} -\hat{p}_\alpha^{(in)} \\ \hat{x}_\alpha^{(in)} - \hat{p}_\beta^{(in)} \\ -\hat{p}_\beta^{(in)} \\ -\hat{p}_\alpha^{(in)} + \hat{x}_\beta^{(in)} \end{pmatrix} + \begin{pmatrix} -\hat{\delta}_1 \\ \hat{\delta}_2 - \hat{\delta}_4 \\ -\hat{\delta}_4 \\ \hat{\delta}_3 - \hat{\delta}_1 \end{pmatrix}. \tag{7.31}$$

We here assume that both two input modes α and β are initially in vacuum states. Since these two input modes and four squeezed state resources have no correlations with each other, the variances of quadrature operators for output modes $i = 2$ and 3 become

$$\langle \Delta^2 \hat{x}_i^{(out)} \rangle = \frac{\hbar}{2}[1 + 2 \times 10^{-\frac{x}{10}}], \quad \langle \Delta^2 \hat{p}_i^{(out)} \rangle = \frac{\hbar}{2}[2 + 3 \times 10^{-\frac{x}{10}}], \tag{7.32}$$

where we have assumed that all squeezed state resources have the identical squeezing level x.

In the special case of the ideal controlled-Z gate with $x \to \infty$, they become

$$\langle \Delta^2 \hat{x}_i^{(out)} \rangle = \frac{\hbar}{2}, \quad \langle \Delta^2 \hat{p}_i^{(out)} \rangle = \frac{\hbar}{2} \times 2. \tag{7.33}$$

Since the position operators of output modes 2 and 3 become $-\hat{p}_\alpha^{(in)}$ and $-\hat{p}_\beta^{(in)}$, the variance $\langle \Delta^2 \hat{x}_i^{(out)} \rangle$ is identical to that of a vacuum state. On the other hand, because the momentum operators of output modes become $\hat{x}_\alpha^{(in)} - \hat{p}_\beta^{(in)}$ and $-\hat{p}_\alpha^{(in)} + \hat{x}_\beta^{(in)}$, and because two input modes α and β have no correlations, the variance $\langle \Delta^2 \hat{p}_i^{(out)} \rangle$ is identical to twice the variance of a vacuum state.

In another special case of vacuum resource states with $x = 0$, the variances of the output-mode quadratures become

$$\langle \Delta^2 \hat{x}_i^{(out)} \rangle = \frac{\hbar}{2} \times 3, \quad \langle \Delta^2 \hat{p}_i^{(out)} \rangle = \frac{\hbar}{2} \times 5. \tag{7.34}$$

In our experiment, we utilize finite-level squeezed states as resources for one-way quantum computation. As a result, variances of the output modes become between Eqs. (7.33) and (7.34).

7.2.11 Inseparability Criteria

7.2.11.1 Van Loock-Furusawa Criterion

In order to verify entanglement at the output, we utilize van Loock-Furusawa criterion (Sect. 3.7.4).

The input-output relationship is given by

$$\begin{pmatrix} \hat{x}_2^{(out)} \\ \hat{p}_2^{(out)} \\ \hat{x}_3^{(out)} \\ \hat{p}_3^{(out)} \end{pmatrix} = \begin{pmatrix} 0 & -1 & 0 & 0 \\ 1 & 0 & 0 & -1 \\ 0 & 0 & 0 & -1 \\ 0 & -1 & 1 & 0 \end{pmatrix} \begin{pmatrix} \hat{x}_\alpha^{(in)} \\ \hat{p}_\alpha^{(in)} \\ \hat{x}_\beta^{(in)} \\ \hat{p}_\beta^{(in)} \end{pmatrix} = \begin{pmatrix} -\hat{p}_\alpha^{(in)} \\ \hat{x}_\alpha^{(in)} - \hat{p}_\beta^{(in)} \\ -\hat{p}_\beta^{(in)} \\ -\hat{p}_\alpha^{(in)} + \hat{x}_\beta^{(in)} \end{pmatrix}. \tag{7.35}$$

We can guess that $\hat{x}_2^{(out)}$ and $\hat{p}_3^{(out)}$ have positive correlation, while $\hat{x}_3^{(out)}$ and $\hat{p}_2^{(out)}$ have positive correlation. By applying van Loock-Furusawa entanglement criterion, we get the following sufficient condition for entanglement.

7.2.11.2 Sufficient Condition for Entanglement

If the output state satisfies

$$\langle \Delta^2 (g\hat{p}_2^{(out)} - \hat{x}_3^{(out)}) \rangle + \langle \Delta^2 (g\hat{p}_3^{(out)} - \hat{x}_2^{(out)}) \rangle < 2g\hbar, \tag{7.36}$$

for some $g > 0$, the output state is an entangled state.

We assume that two input modes α and β are in vacuum states. $\langle \Delta^2 (g\hat{p}_2^{(out)} - \hat{x}_3^{(out)}) \rangle$ and $\langle \Delta^2 (g\hat{p}_3^{(out)} - \hat{x}_2^{(out)}) \rangle$ are given by

$$\begin{aligned}\langle \Delta^2(g\hat{p}_2^{(out)} - \hat{x}_3^{(out)})\rangle =& \langle \Delta^2(g\hat{p}_3^{(out)} - \hat{x}_2^{(out)})\rangle \\ =& \frac{\hbar}{2}\Big[g^2 + (1-g)^2 + \frac{5}{2}g^2 e^{-2r} + \big(-\frac{1}{\sqrt{2}}g + \sqrt{2}\big)^2 e^{-2r}\Big],\end{aligned} \tag{7.37}$$

where we have assumed that all squeezed resource states have an identical squeezing parameter r. By choosing

$$g = \frac{3}{4}, \tag{7.38}$$

the sufficient condition for entanglement is satisfied with minimum r ($\geqq 0$). It becomes

$$e^{-2r} < \frac{2}{5}. \tag{7.39}$$

Therefore, when the effective squeezing level (Sect. 6.1.1) is higher than $-10\log(2/5)=4.0$ dB, the output state becomes an entangled state.

Note that the van Loock-Furusawa criterion gives us only a sufficient condition, not the necessary and sufficient condition, for entanglement. In other words, dissatisfaction of the van Loock-Furusawa criterion does not mean that the output state is not entangled. Therefore, there exists a possibility that the output state is entangled even if the effective squeezing level is lower than 4.0 dB.

We negate this possibility by acquiring Simon criterion (Sect. 3.7.5) in the next subsection.

7.2.11.3 Simon Criterion

We consider Simon criterion (Sect. 3.7.5), which gives us necessary and sufficient entanglement condition for a two-mode Gaussian state.

We assume that two input modes α and β are in vacuum states, and all squeezed resource states have an identical squeezing parameter r. In this case, the covariance matrix of the output state is given by

$$V_{23}^{(out)} = \frac{\hbar}{2}\left[\begin{pmatrix} 1&0&0&1\\ 0&2&1&0\\ 0&1&1&0\\ 1&0&0&2 \end{pmatrix} + e^{-2r}\begin{pmatrix} 2&0&0&1\\ 0&3&1&0\\ 0&1&2&0\\ 1&0&0&3 \end{pmatrix}\right]. \tag{7.40}$$

By using the Simon criterion, $V_{23}^{(out)}$ represents a covariance matrix of an entangled state if and only if r satisfies

$$e^{-2r} < \frac{2}{5}, \tag{7.41}$$

which is identical to the result of van Loock-Furusawa criterion. Since the Simon criterion gives us the necessary and sufficient condition for entanglement, we find that the van Loock-Furusawa criterion Eq. (7.36) also gives us the optimum condition for entanglement.

In this experiment, we will use the van Loock-Furusawa criterion since entanglement can be detected by using the minimum set of measurement results (Sect. 3.7.7).

7.3 Experimental Setup

7.3.1 From Laser to Cluster State Generation

The resource for this experiment is a four-mode linear cluster state. In order to generate it, we utilize the same experimental setup to that in Chap. 6.

7.3.2 Input States

7.3.2.1 Input States

We utilize vacuum states and coherent states as input states for the controlled-Z gate. Although a vacuum state is also a coherent state with amplitude of zero, we declare that the word "coherent state" in this section shows a coherent state with *non-zero* amplitude.

7.3.2.2 Preparation of Coherent State

We utilize the 1 MHz sideband of the laser beam as the carrier of quantum states. A coherent state at 1 MHz sideband can be generated by phase-modulating the laser beam with the frequency of 1 MHz. In order to enhance interference visibility, we utilize the output beam of the mode cleaning cavity (MCC). The power of each input-mode beam is adjusted to 2 μW. Phase modulation is achieved by using a piezo electric transducer (PZT) (Sect. 6.4.1).

7.3.2.3 Preparation of Vacuum State

Since a vacuum state is a coherent state with amplitude of zero, a vacuum state at 1 MHz sideband can be generated by switching off the phase modulation. Equivalently, it can be achieved by shutting off the laser beam.

7.3.3 Classical Channels and Displacement Operations

7.3.3.1 Terminologies

In this subsection, we mention feed-forward operations (displacement operations) given by Eqs. (7.29) and (7.30). Before we move onto the detailed discussion, we first summarize two terminologies for the feed-forward system.

7.3.3.2 Classical Channel

A classical channel is an electrical circuit which transmits measurement results from homodyne detectors to output modes.

7.3.3.3 Displacement Beam

Since classical channels transmit electrical signals from detectors, we have to encode them to optical beams in order to achieve feed-forwards. Displacement beams are the beams in which electrical signals from the classical channels are encoded.

7.3.3.4 Components

Figure 7.7 shows a basic structure of a feed-forward system.

7.3.3.5 Power Splitter

Electrical signal from a homodyne detector is split into two paths by using a power splitter. One is utilized to acquire the error signal to lock the relative phase between the signal beam and a local oscillator beam in the homodyne detection. The other is transmitted to the classical channel.

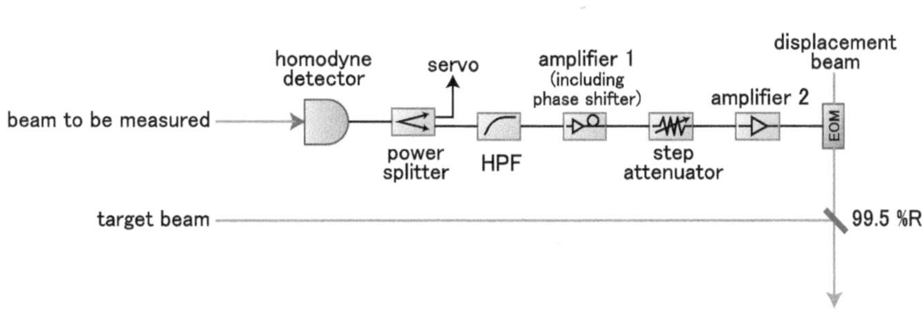

Fig. 7.7 Basic structure of a feed-forward system

7.3.3.6 High Pass Filter

The signal from the power splitter is sent to a high pass filter (HPF). The signal from the homodyne detector contains not only 1 MHz signal of information but modulation signals (derived from phase modulation) of at most 210 kHz which are utilized to lock relative phases. If they are transmitted to the feed-forward target beams, lockings at the later parts of the feed-forward system are disturbed. Therefore, we have to shut out the modulation signals by using a HPF. For that purpose, we use two fifth-order Butterworth filters of cutoff frequency 500 kHz. The modulation signals are reduced to less than -70 dB.

7.3.3.7 Amplifiers, Phase Shifters, and Step Attenuators

We use two amplifiers to increase the power of electrical signals. The amplifier 2 is utilized to operate an EOM, while the amplifier 1 works as a phase shifter as well.

As is mentioned in Sect. 6.3.5, we utilize ± 1 MHz sidebands away from the center frequency of the laser as a carrier of quantum information, while HWHM of these sidebands is 15 kHz. The merit of sideband experiment is that the experimental setup becomes simpler.

In the strict case, two wave packets from the direct optical path (target beam) and from the classical channel path should be combined so that they perfectly coincides with each other. It is also important that these two signals have the same amplitude. Since the classical channel needs electrical cables, its length potentially becomes longer than that of the optical path. Typically, the optical length is to be lengthened by about 10 m with a delay line [2].

On the other hand of the sideband experiment, we can consider that a wave packet which is acquired by shifting one cycle of its inner sine wave is almost equivalent to the original one. In our experiment, by adjusting a phase shifter in the amplifier 1, we can control the timing of these two wave packets so that they are shifted by one cycle. Note that the classical channel is adjusted only to signals of 1 MHz, and thus it is not adjusted to signals of frequencies away from 1 MHz.

In order to adjust the amplitude of electrical signal, we use step attenuators, which enable us to adjust in units of 0.1 dB.

7.3.3.8 Displacement Beam and EOM

The electrical signal transmitted through the classical channel is sent to an EOM, with which it is encoded in a beam named displacement beam. The EOM is adjusted so that phase modulation is applied to the beam.

We define that a_D and $\hat{a}$ show the displacement beam and the target beam. By using a beam splitter with high reflectivity R ($\simeq$99.5 %), the feed-forward operation is described as

$$\hat{a}_{out} = \sqrt{R}\hat{a} + \sqrt{1-R}a_D. \tag{7.42}$$

Since $R \simeq 1$, we can apply the displacement operation of $\sqrt{1-R}a_D$ with almost no information loss of $\hat{a}$.

7.3.3.9 Classical Channels in This Experiment

The displacement operation (feed-forward operation) in this experiment is given by Eqs. (7.29) or (7.30):

$$\hat{X}_2(-\sqrt{2}s_1)\hat{Z}_2(\sqrt{2}s_\alpha - \sqrt{2}s_4)\hat{X}_3(-\sqrt{2}s_4)\hat{Z}_3(\sqrt{2}s_\beta - \sqrt{2}s_1). \tag{7.43}$$

Therefore, we need six paths of classical channels $(1, 2x)$, $(\alpha, 2p)$, $(4, 2p)$, $(4, 3x)$, $(\beta, 3p)$, and $(1, 3p)$, where parentheses show (mode of homodyne detection, destination of feed-forward).

Figure 7.8 shows the schematic of classical channels for the controlled-Z gate experiment.

7.3.4 Locking the Relative Phases

Figure 7.9 shows arrangement of probe beams and phase modulations for the controlled-Z gate experiment. By using Fig. 7.9, we can determine how each relative phase between two beams is locked.

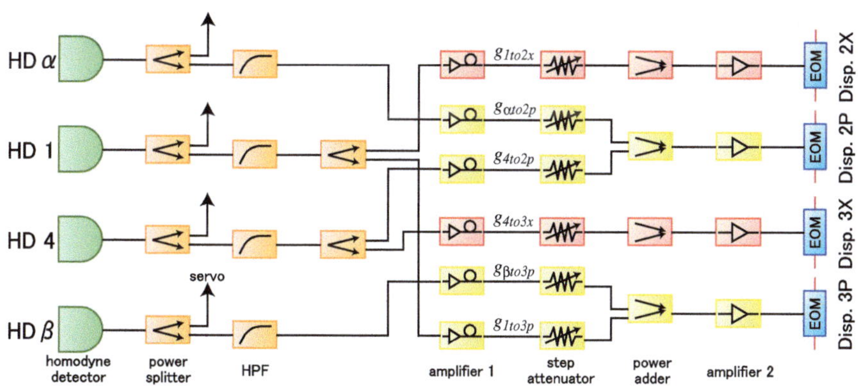

Fig. 7.8 Classical channels for controlled-Z gate experiment

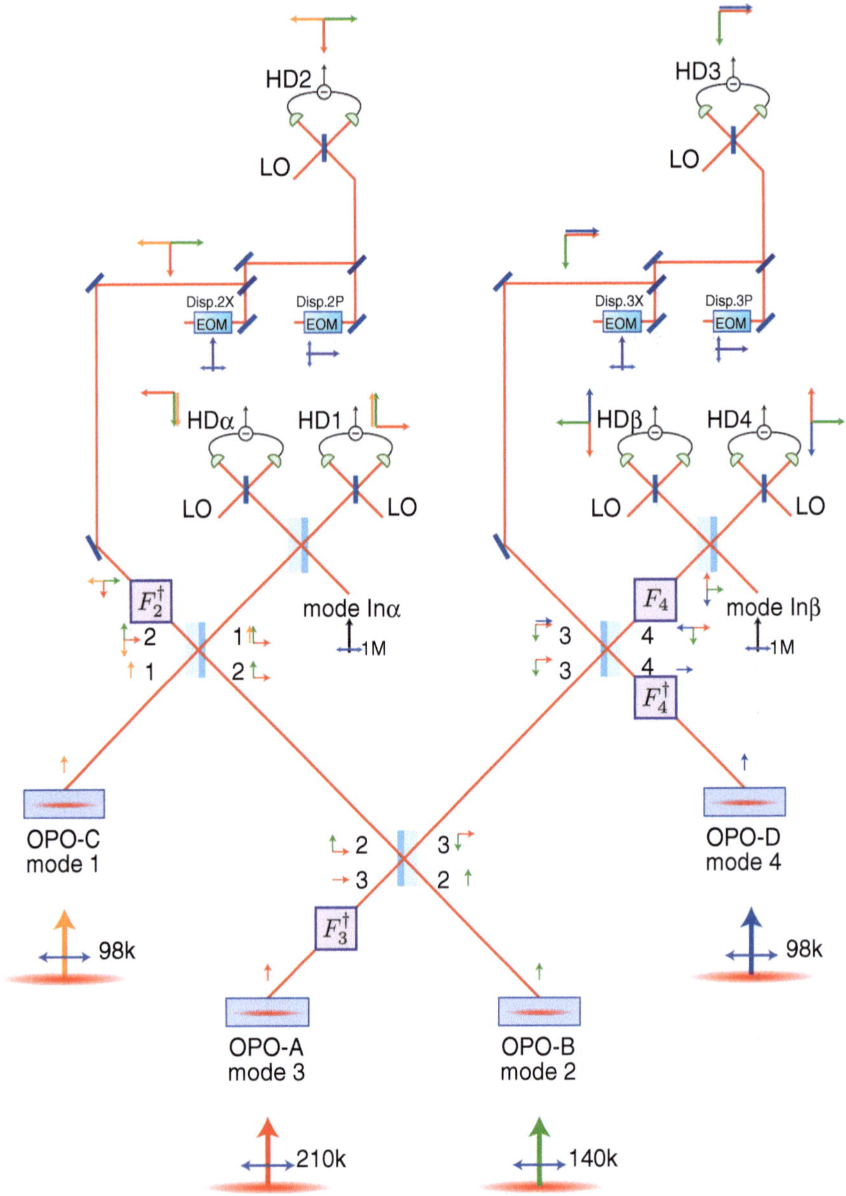

Fig. 7.9 Probe beams and phase modulations for controlled-Z gate experiment

7.3.4.1 Generation of Four-Mode Linear Cluster State

We use the same scheme as Chap. 6 in order to generate a four-mode linear cluster state.

7.3.4.2 Input Coupling

When we use a vacuum state as an input state, we shut off the carrier input-mode beam. As a result, we do not have to lock the phase of the input beam.

When we use a coherent state as an input state, on the other hand, we use the input beam, thus we have to lock the relative phase between the input-mode beam and the probe beams from OPOs. We here take an example where input mode α is to be set to a coherent state with amplitude in x quadrature. Since phase modulation is orthogonal to the carrier input beam, we can set the phase of the coherent-state amplitude to the x quadrature by locking the carrier beam to the p axis. Since the probe beam from OPO-B points to the p axis at the input coupling beam splitter for mode α, this phase locking can be achieved by using the AC locking technique with demodulation signal with the frequency of 140 kHz, which have been utilized to modulate the probe beam of OPO-B.

7.3.4.3 Displacement Beam

Since displacement beams utilize phase modulations similar to the input beams, phase lockings of displacement beams can be achieved in a similar way. Note that the displacement beam Disp2X, for example, should point to the p axis, because phase modulation is orthogonal to the carrier beam.

7.3.4.4 Summary of Phase Locking

Table 7.2 shows phase locking techniques for the controlled-Z gate experiment. Parentheses in Table 7.2 show that their phases are not utilized in experiment.

7.3.5 *Cancellation*

Cancellation refers to procedure to adjust gains and phases of classical channels. In the following, we take an example of the feed-forward from the homodyne measurement on mode 1 to the position quadrature of mode 2. We define g_{1to2x} as the corresponding gain.

We measure $\hat{p}_1^{(b)}$ in the homodyne detection on mode 1. Its measurement result is added to $\hat{x}_2^{(c)}$ with the gain of g_{1to2x}. By using Eqs. (7.5) and (7.6), we get

Table 7.2 Phase lockings for controlled-Z gate

Relative phase	Technique, demodulation frequency
Parametric gain of OPO-A	AC locking, 210 kHz
Parametric gain of OPO-B	AC locking, 140 kHz
Parametric gain of OPO-C	AC locking, 98 kHz
Parametric gain of OPO-D	AC locking, 98 kHz
Beam splitter $B_{23}^{(3)}(\frac{2}{\sqrt{5}})$	DC locking
Beam splitter $B_{12}^{(3)}(\frac{1}{\sqrt{2}})$	AC Locking, 140 kHz
Beam splitter $B_{34}^{(4)}(\frac{1}{\sqrt{2}})$	AC locking, 210 kHz
Input α	AC locking, amplitude x: 140 kHz, amplitude p: 210 kHz
Input β	AC locking, amplitude x: 210 kHz, amplitude p: 140 kHz
Displacement in mode 2	AC locking, displacement x: 210 kHz, displacement p: 140 kHz
Displacement in mode 3	AC locking, displacement x: 140 KHz, displacement p: 210 kHz
Homodyne measurement in mode α	AC locking, x: 210 kHz, (p: 140 kHz)
Homodyne measurement in mode 1	AC locking, (x: 210 kHz), p: 140 kHz
Homodyne measurement in mode 2	AC locking, x: 140 kHz, p: 210 kHz
Homodyne measurement in mode 3	AC locking, x: 210 kHz, p: 140 kHz
Homodyne measurement in mode 4	AC locking, (x: 140 kHz), p: 210 kHz
Homodyne measurement in mode β	AC locking, x: 140 kHz, (p: 210 kHz)

$$\hat{x}_2^{(c)} + g_{1to2x}\hat{p}_1^{(b)} = \left[-\frac{1}{\sqrt{2}}\hat{p}_1 + \frac{1}{\sqrt{10}}\hat{p}_2 - \frac{2}{\sqrt{10}}\hat{x}_3 \right] + g_{1to2x}\left[\frac{1}{\sqrt{2}}\hat{p}_\alpha + \frac{1}{2}\hat{p}_1 + \frac{1}{2\sqrt{5}}\hat{p}_2 - \frac{1}{\sqrt{5}}\hat{x}_3 \right]. \tag{7.44}$$

We apply phase modulation onto the probe beam of OPO-A, leading to a coherent state with the amplitude in $\hat{x}_3$. With measuring the power of $\hat{x}_2$, we adjust the gain and phase of the classical channel so that the coherent signal is not observed. Since

$$-\frac{2}{\sqrt{10}}\hat{x}_3 + g_{1to2x}\left[-\frac{1}{\sqrt{5}}\hat{x}_3 \right] = 0 \iff g_{1to2x} = -\sqrt{2}, \tag{7.45}$$

g_{1to2x} can be adjusted to $g_{1to2x} = -\sqrt{2}$.

The gains and phases of the other classical channels are also adjusted in a similar manner.

Table 7.3 Interference visibilities for controlled-Z gate experiment

Beam splitter	Beams used for adjustment	Visibility (%)
Beam splitter $B_{23}^{(3)}(\frac{2}{\sqrt{5}})$	OPO-A, OPO-B	98.7
Beam splitter $B_{12}^{(3)}(\frac{1}{\sqrt{2}})$	OPO-A, OPO-C	98.9
Beam splitter $B_{34}^{(4)}(\frac{1}{\sqrt{2}})$	OPO-B, OPO-D	98.0
Input coupling α	OPO-A, In-α	98.1
Input coupling β	OPO-B, In-β	97.2
Displacement $2x$	OPO-A, Disp-2X	98.3
Displacement $2p$	OPO-A, Disp-2P	97.9
Displacement $3x$	OPO-B, Disp-3X	97.9
Displacement $3p$	OPO-B, Disp-3P	97.7
Homodyne measurement α	OPO-A, LO-α	98.7
Homodyne measurement 1	OPO-A, LO-1	97.9
Homodyne measurement 2	OPO-A, LO-2	97.5
Homodyne measurement 3	OPO-B, LO-3	98.3
Homodyne measurement 4	OPO-B, LO-4	97.0
Homodyne measurement β	OPO-B, LO-β	96.7

Table 7.4 Parametric gains for controlled-Z gate experiment

OPO name	Parametric gain (G_+)	Phase matching temperature (degrees)
OPO-A	7.0	40.3
OPO-B	8.1	39.7
OPO-C	4.0	40.2
OPO-D	6.8	40.7

7.4 Preparation for Measurement

7.4.1 Visibilities and Parametric Gains

Interference visibilities and parametric gains are shown in Tables 7.3 and 7.4, respectively.

7.4.2 Adjustment of EOM (Purity)

A displacement operation in our experiment is achieved by using a displacement beam which is phase-modulated via an EOM. If the *purity* of the phase modulation by the EOM is not enough, the displacement signal also transmits to the orthogonal quadrature via unwanted amplitude modulation. In order to increase the purity of the phase modulation, we adjust the holding angle of the EOM.

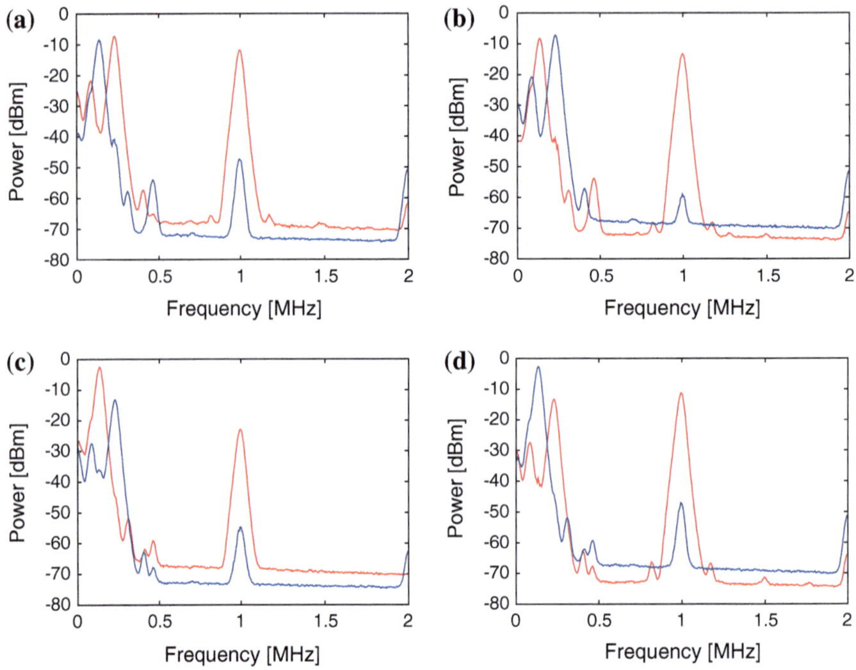

Fig. 7.10 Adjustment of purity for controlled-Z gate experiment. **a** Disp-2X, **b** Disp-2P, **c** Disp-3X, **d** Disp-3P

Table 7.5 Adjustment of purity for controlled-Z gate experiment

EOM	Purity (dB)
Disp-2X	35.7
Disp-2P	46.3
Disp-3X	31.8
Disp-3P	35.9

In order to explain the procedure of its adjustment, we here take an example of Disp-2X.

We use two probe beams from OPO-A and OPO-B, while those from OPO-C and OPO-D are blocked. The relative phases on the first beam splitter $B_{23}^{(3)}(\frac{2}{\sqrt{5}})$ and on the Disp-2X beam splitter are locked in the usual schemes (via the DC locking and AC locking). We impress 1 MHz sine wave signal to the EOM for Disp-2X. If the purity is perfectly adjusted, we will detect the 1 MHz signal when we measure x in the homodyne detection for mode 2 (red traces in Fig. 7.10), while we will detect no 1 MHz signal when we measure p in mode 2 (blue traces in Fig. 7.10). We adjust the holding angle of the EOM so that the 1 MHz signal is minimized when we measure p in mode 2.

Each purity is listed in Table 7.5.

7.4.3 Cancellation

Figure 7.11 shows adjustment results of classical channels (cancellation). Red traces show measurement results when electrical cables for classical channels are disconnected, which correspond to classical channel gains of zero. Thus, the 1 MHz signals are detected which have been initially applied to the probe beams of the OPOs. Blue traces show measurement results when cancellations have been carried out. We find that the 1 MHz signals are almost perfectly canceled out.

Each cancellation level is listed in Table 7.6.

7.4.4 Balance Between Homodyne Detectors

The balance between two homodyne detections for modes 2 and 3 is adjusted in the same way as Chap. 6. We apply phase modulation on the probe beam of OPO-B, which corresponds to mode 2. Figure 7.12 shows adjustment results. Red trace labeled by "HD2p" is a measurement result of p_2, while blue trace labeled by "HD2p-HD3x" is a measurement result of subtraction of p_2 and x_3. By defining g_{23} as the gain for x_3, we get

$$\hat{p}_2^{(c)} - g_{23}\hat{x}_3^{(c)} = -\frac{1}{\sqrt{10}}\hat{x}_2 - g_{23}\left[-\frac{2}{\sqrt{10}}\hat{x}_2\right] = 0 \iff g_{23} = \frac{1}{2}, \tag{7.46}$$

where we have utilized Eqs. (7.6) and (7.7). Thus, g_{23} can be adjusted to $g_{23} = \frac{1}{2}$. In order to acquire the balance between two homodyne detections, the signal from homodyne detection 3 should be amplified with the gain of 2 in amplitude, or equivalently with the gain of 4 in power. For this purpose, we reduce the amount of attenuation by 6.0 dB. Figure 7.12 shows that we have achieved cancellation of 46.7 dB.

7.5 Measurement Results and Their Analysis

7.5.1 Power of Output with Vacuum Inputs

In this subsection, we show measurement results of output-mode quadrature variances when two input modes are in vacuum states. We then estimate effective squeezing level (Sect. 6.1.1).

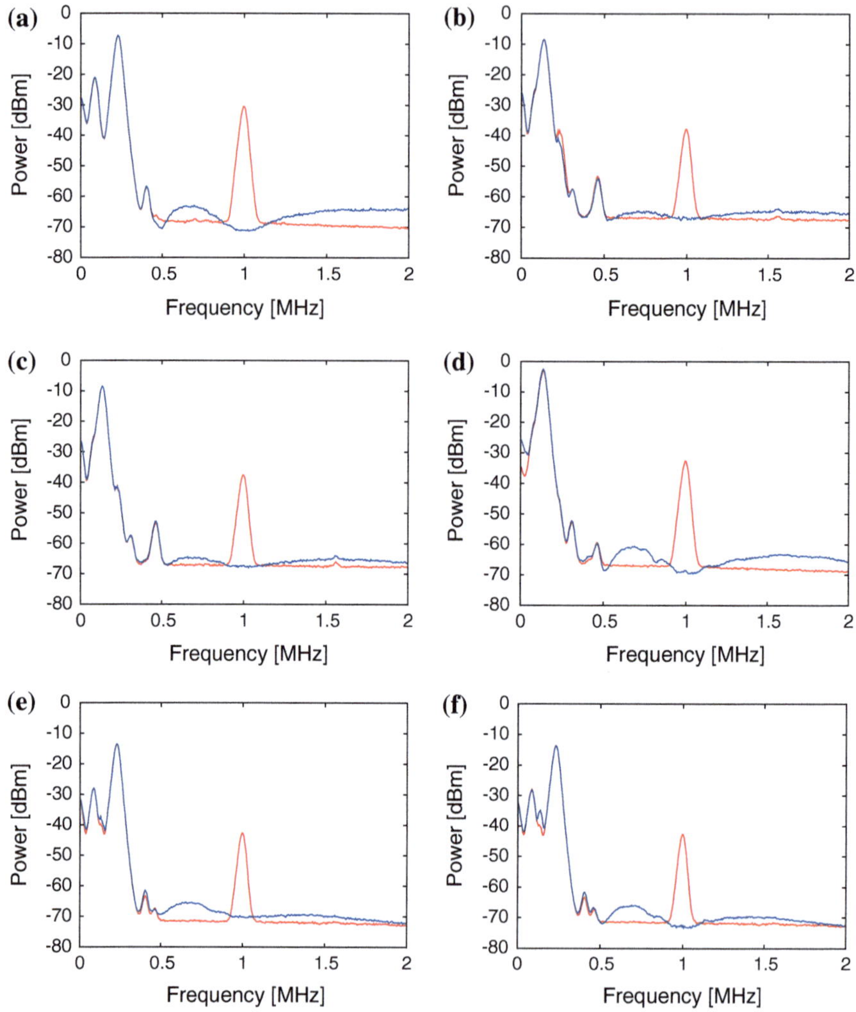

Fig. 7.11 Adjustment of classical channels (cancellation) for controlled-Z gate experiment. **a** g_{1to2x}, **b** $g_{\alpha to2p}$, **c** g_{4to2p}, **d** g_{4to3x}, **e** $g_{\beta to3p}$, **f** g_{1to3p}

Table 7.6 Adjustment of classical channels (cancellation) for controlled-Z gate experiment

Feed-forward	Cancellation level (dB)
g_{1to2x}	40.5
$g_{\alpha to2p}$	28.8
g_{4to2p}	29.6
g_{4to3x}	35.8
$g_{\beta to3p}$	27.2
g_{1to3p}	29.7

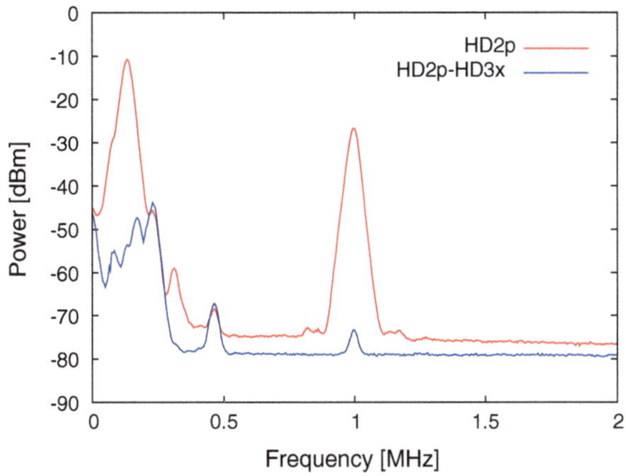

Fig. 7.12 Balance between homodyne detections

7.5.1.1 Method

We generate four squeezed vacuum states by using four pumped OPOs. Two input-mode beams are blocked because we will use vacuum states as input states. We lock all relative phases on beam splitters except for input couplings. We measure both quadratures x and p in the output-mode homodyne detections for mode 2 and 3. The signal from the detectors is sent to and analyzed by a spectrum analyzer. The setting of the spectrum analyzer is listed in the following.

- Center Frequency (CF): 1 MHz.
- Resolution BandWidth (RBW): 30 kHz.
- Video BandWidth (VBW): 300 Hz.
- Sweep Time: 0.2 s.
- Points: 401 points.
- Average: 20 times.
- Span: Zero Span.

7.5.1.2 Measurement Results

Figure 7.13 shows measurement results. Red traces show measurement results when inputs of homodyne detections are shut off. They correspond to the shot noise level (SNL). Orange traces show measurement results of output-mode quadrature variances.

Each result is listed in Table 7.7.

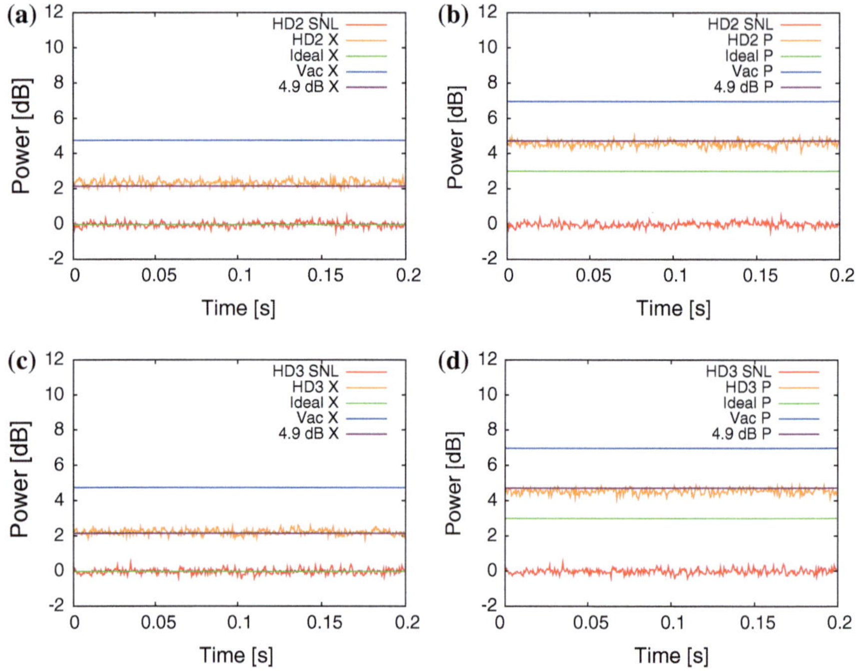

Fig. 7.13 Measurement results with vacuum inputs. **a** $\hat{x}_2$, **b** $\hat{p}_2$ **c** $\hat{x}_3$ **d** $\hat{p}_3$

Table 7.7 Measurement results with vacuum inputs

Measurement variable	Result (dB)
$\langle \Delta^2 \hat{x}_2 \rangle$	2.37 ± 0.02
$\langle \Delta^2 \hat{p}_2 \rangle$	4.58 ± 0.02
$\langle \Delta^2 \hat{x}_3 \rangle$	2.25 ± 0.02
$\langle \Delta^2 \hat{p}_3 \rangle$	4.56 ± 0.02

7.5.1.3 Effective Squeezing Level

We estimate the effective squeezing level (Sect. 6.1.1). For simplicity, we assume that all squeezed states have the identical squeezing level x.

From Eq. 7.32, we get

$$\langle \Delta^2 \hat{x}_i \rangle = \frac{\hbar}{2}\left[1 + 2 \times 10^{-\frac{x}{10}}\right], \quad \langle \Delta^2 \hat{p}_i \rangle = \frac{\hbar}{2}\left[2 + 3 \times 10^{-\frac{x}{10}}\right], \tag{7.47}$$

for $i = 2, 3$. When an ideal controlled-Z gate is accessible, or identically $x \to \infty$, they become $\langle \Delta^2 \hat{x}_i \rangle = \frac{\hbar}{2}$, $\langle \Delta^2 \hat{p}_i \rangle = \frac{\hbar}{2} \times 2$, leading to 0.0 dB and 3.0 dB relative to the SNL (green lines in Fig. 7.13). When no squeezed resource states are available

($x = 0$), they become $\langle \Delta^2 \hat{x}_i \rangle = \frac{\hbar}{2} \times 3$, $\langle \Delta^2 \hat{p}_i \rangle = \frac{\hbar}{2} \times 5$, leading to 4.8 dB and 7.0 dB relative to the SNL (blue lines in Fig. 7.13).

In experiment, we utilize finite-level squeezed states as resources for the controlled-Z gate. The larger the squeezing level is, the lower the variances become. We can estimate the effective squeezing level by using these measurement results. We get $x = 4.9$ dB, which is almost the same as the effective squeezing level of the resource cluster state.

7.5.2 Power of Output with Coherent Inputs

In this subsection, we show measurement results of output-mode quadrature powers when we use coherent states as input states, which enable us to verify input-output relationship.

7.5.2.1 Measurement Results of Input States

We measure the power of both input coherent states in modes α and β.

We here take an example of mode α. The beam of mode α is split into two paths by the input-coupling beam splitter. They are transmitted to homodyne detection systems for modes α and 1. We can acquire the power of the input state by adding two measurement results of these homodyne detections.

Figure 7.14 shows measurement results. Red traces show the SNL. Green traces show measurement results when input-mode beams are blocked. Since the OPOs generate squeezed vacuum states, these results are larger than the SNL. Blue traces show measurement results when input-mode beams are also made effective. The differences between blue traces and green traces correspond to the powers of the input coherent states.

Each power is listed in Table 7.9.

As a result, we get 21.55 dB for mode α and 21.23 dB for mode β, respectively, compared to the SNL. All errors are less than ± 0.02 dB, including in Table 7.9.

7.5.2.2 Measurement Result of Output Power with Coherent Inputs

Figures 7.15, 7.16, 7.17 and 7.18 show measurement results of output powers with coherent input states. We have four types of coherent input states: α_x, α_p, β_x, and β_p, and four measurement variables: x_2, p_2, x_3, and p_3. Thus, we have $4 \times 4 = 16$ measurement results. In these figures, red traces show measurement results of SNL, while green traces show measurement results of output-mode powers.

Each power is listed in Table 7.8. All error bars are less than ± 0.02 dB.

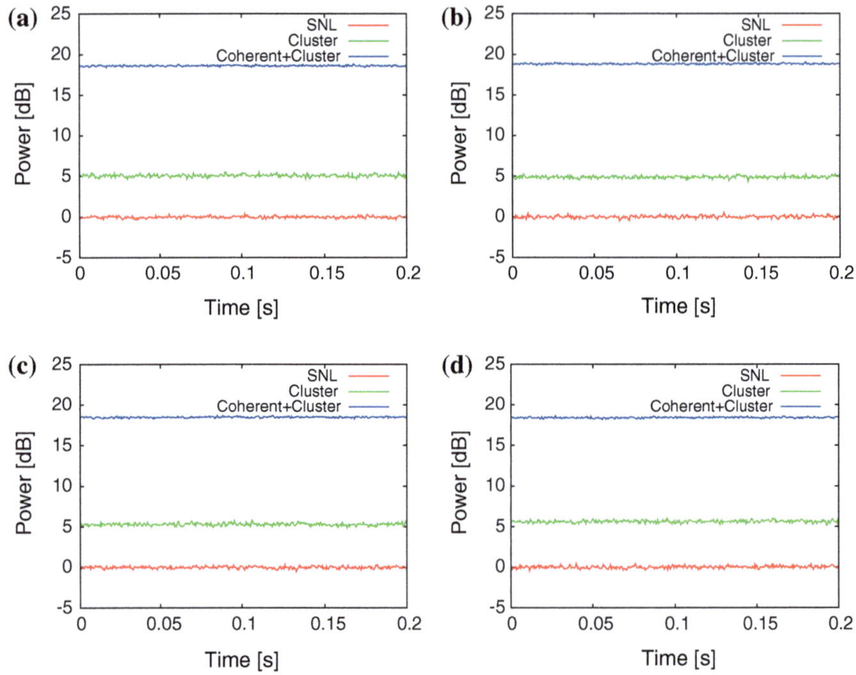

Fig. 7.14 Measurement results of input states. **a** Input α with HDα, **b** Input α with HD1,**c** Input β with HD β, **d** Input β with HD4

Table 7.8 Measurement results of quadrature operator powers with coherent-state inputs (dB)

Input mode and phase \ Output mode and phase	$\hat{x}_2$	$\hat{p}_2$	$\hat{x}_3$	$\hat{p}_3$
Vacuum inputs in both α and β	2.37	4.58	2.25	4.56
Amplitude in x of α	2.75	21.36	2.10	4.88
Amplitude in p of α	20.83	4.60	2.20	21.43
Amplitude in x of β	2.06	4.62	2.63	21.09
Amplitude in p of β	2.36	20.93	20.55	4.87

Table 7.10 shows theoretical predictions. We have here assumed that the resource squeezing level is 4.9 dB, and the input coherent-state amplitudes are 21.55 dB for mode α and 21.23 dB for mode β, respectively.

Figure 7.19 shows graphs of Tables 7.8 and 7.10. Note that they are not covariance matrices.

7.5.2.3 Analysis

We take an example of the coherent state input with amplitude in x of mode α (InAx). The input-output relationship Eq. (7.35) shows that $\hat{x}_\alpha$ is transmitted to $\hat{p}_2$.

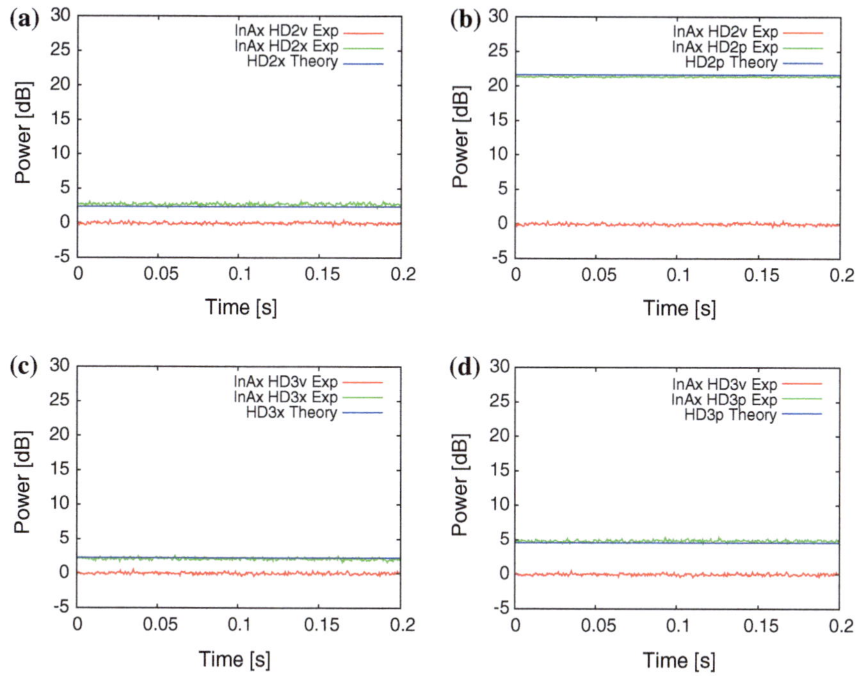

Fig. 7.15 Measurement results of quadrature operator powers with coherent-state input α_x. **a** $2x$, **b** $2p$, **c** $3x$, **d** $3p$

Corresponding to this transmission, we have obtained 21.36 dB. Since the absolute value of the matrix element is 1, it is almost the same as the power of the input state amplitude.[1] In a similar manner, we have observed coherent-state signals which correspond to matrix elements in Eq. (7.35). Experimental results agree well with theoretical predictions.

The difference between experimental result and theoretical value is 0.2 dB on average, and 0.8 dB at maximum. We can consider the following two reasons for this error.

- Cancellation was not perfect since propagation losses and visibilities were not symmetrical.
- Visibilities had become worse with time.

[1] The sign of each matrix element cannot be verified through Table 7.10. Concurrence of signs is indirectly proven by satisfaction of entanglement criterion.

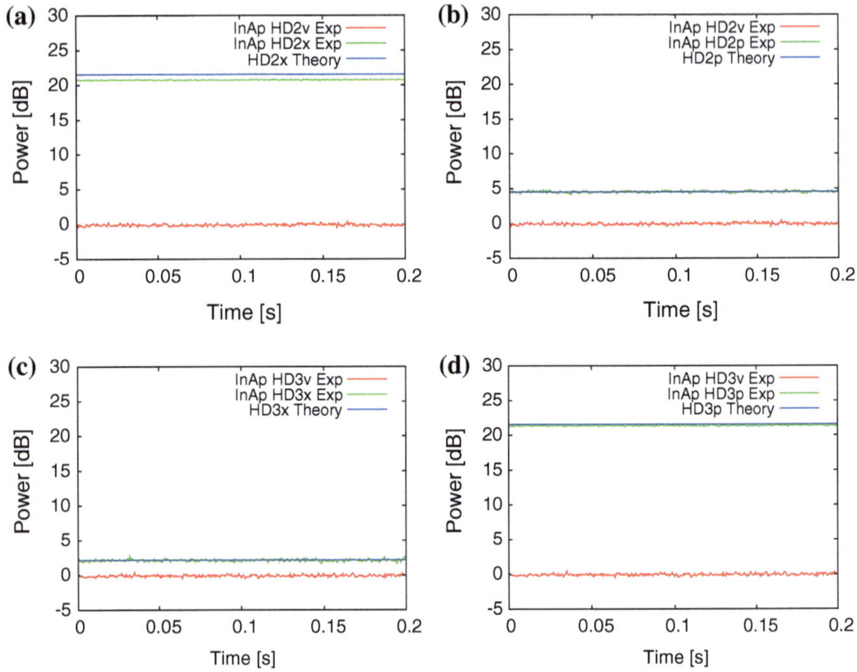

Fig. 7.16 Measurement results of quadrature operator powers with coherent-state input α_p. **a** $2x$, **b** $2p$, **c** $3x$, **d** $3p$

7.5.3 Entanglement at Output (via van Loock–Furusawa Criterion)

In this subsection, we verify entanglement at the output by using the van Loock–Furusawa criterion.

7.5.3.1 Measurement Results and Analysis

Figure 7.20 shows measurement results of $\langle \Delta^2(g\hat{p}_2^{(out)} - \hat{x}_3^{(out)})\rangle + \langle \Delta^2(g\hat{p}_3^{(out)} - \hat{x}_2^{(out)})\rangle$ with several gains g, where we have used the system of units with $\hbar = \dfrac{1}{2}$.

Measurement results with 0 dB resources are acquired by using vacuum states, instead of squeezed states, as resources. Measurement results with squeezed state resources agree well with theoretical predictions with 4.9 dB squeezing.

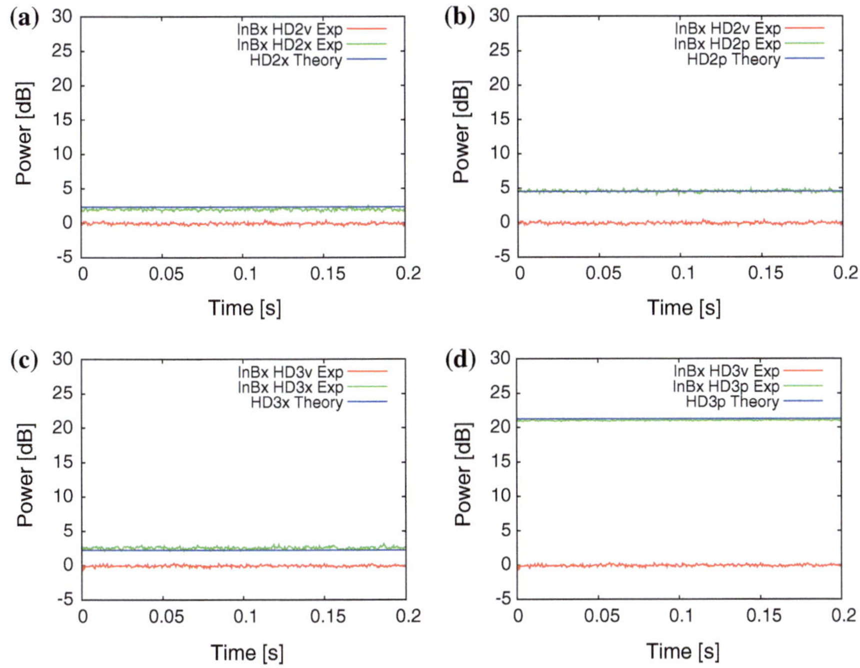

Fig. 7.17 Measurement results of quadrature operator powers with coherent-state input β_x. **a** $2x$, **b** $2p$, **c** $3x$, **d** $3p$

By choosing $g = 0.63, 0.75, 0.89$, we get

$$\langle \Delta^2(g\hat{p}_2^{(out)} - \hat{x}_3^{(out)})\rangle + \langle \Delta^2(g\hat{p}_3^{(out)} - \hat{x}_2^{(out)})\rangle = 0.609 \pm 0.002 < 0.63 \ (g = 0.63), \tag{7.48}$$

$$\langle \Delta^2(g\hat{p}_2^{(out)} - \hat{x}_3^{(out)})\rangle + \langle \Delta^2(g\hat{p}_3^{(out)} - \hat{x}_2^{(out)})\rangle = 0.689 \pm 0.002 < 0.75 \ (g = 0.75), \tag{7.49}$$

$$\langle \Delta^2(g\hat{p}_2^{(out)} - \hat{x}_3^{(out)})\rangle + \langle \Delta^2(g\hat{p}_3^{(out)} - \hat{x}_2^{(out)})\rangle = 0.843 \pm 0.003 < 0.89 \ (g = 0.89), \tag{7.50}$$

from which we find that the measured variances are below the entanglement criteria. Therefore, the output state is entangled.

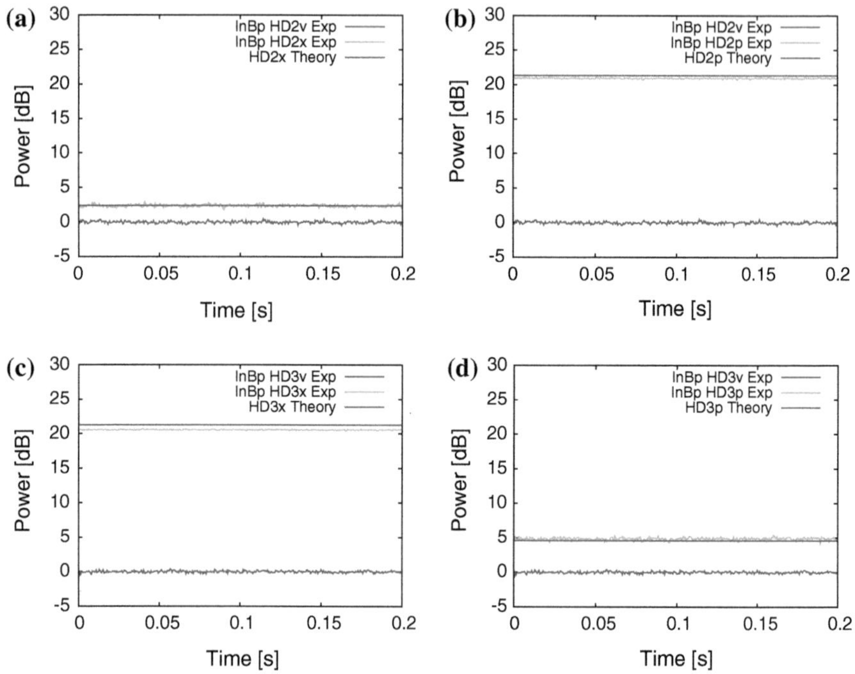

Fig. 7.18 Measurement results of quadrature operator powers with coherent-state input β_p. **a** $2x$, **b** $2p$, **c** $3x$, **d** $3p$

Table 7.9 Measurement results of input states

HD name	Cluster only (dB)	Coherent + Cluster (dB)
HDα	5.08	18.62
HD1	4.89	18.83
HDβ	5.31	18.48
HD4	5.64	18.41

Table 7.10 Theoretical prediction of quadrature operator powers with coherent-state inputs (dB)

Input mode and phase $\diagdown$ Output mode and phase	$\hat{x}_2$	$\hat{p}_2$	$\hat{x}_3$	$\hat{p}_3$
Vacuum inputs in both α and β	2.17	4.73	2.25	4.73
Amplitude in x of α	2.17	21.64	2.17	4.73
Amplitude in p of α	21.60	4.73	2.17	21.64
Amplitude in x of β	2.17	4.73	2.17	21.33
Amplitude in p of β	2.17	21.33	21.29	4.73

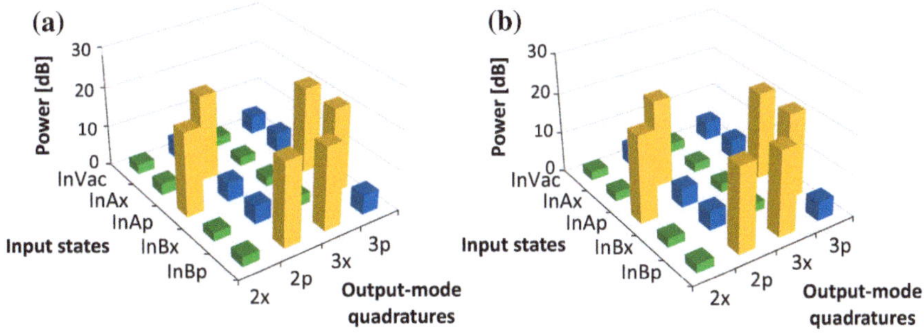

Fig. 7.19 Quadrature operator powers with coherent-state inputs. **a** and **b** correspond to α and β, respectively. **a** Measurement results. **b** Theoretical prediction

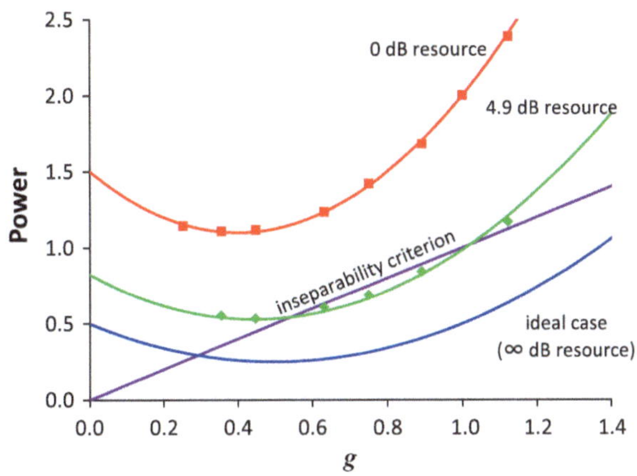

Fig. 7.20 Entanglement at the output

References

1. Ukai, R., Iwata, N., Shimokawa, Y., Armstrong, S.C., Politi, A., Yoshikawa, J., van Loock, P., Furusawa, A.: Demonstration of unconditional one-way quantum computations for continuous variables. Phys. Rev. Lett. **106**, 240504 (2011)
2. Lee, N., Benichi, H., Takeno, Y., Takeda, S., Webb, J., Huntington, E., Furusawa, A.: Teleportation of non-classical wave-packets of light. Science **332**, 330 (2011)

Chapter 8
Experimental Demonstration of Optimum Nonlocal Gate for Continuous Variables

8.1 Optimum Nonlocal Gate Experiment

8.1.1 Operation

In the experimental demonstration of an optimum *nonlocal gate* for continuous-variable one-way quantum computation, we utilize a bipartite entangled state, called the *two-mode cluster state*, as a resource (Sect. 6.1.1). We prepare a two-mode input state independently of the cluster state. The input coupling with the cluster is achieved via the *squeezer-based input-coupling scheme* (Sect. 5.3.4). The two modes to which the input state is transmitted are equivalent to the output modes. By changing the relative phases between signal beams and local oscillator beams in homodyne detections for two input modes, we can implement single-mode Gaussian operations. We choose these measurement bases so that the operation for each single mode becomes the 3.0 dB p-squeezing operation. This is the simplest operation which can be achieved by the experimental setup we use. The input-output relationship in the Heisenberg picture is given by

$$\begin{pmatrix} \hat{x}'_\alpha \\ \hat{p}'_\alpha \\ \hat{x}'_\beta \\ \hat{p}'_\beta \end{pmatrix} = C_Z\left(\frac{1}{2}\right) S_\alpha S_\beta \begin{pmatrix} \hat{x}_\alpha \\ \hat{p}_\alpha \\ \hat{x}_\beta \\ \hat{p}_\beta \end{pmatrix} = S_\alpha S_\beta C_Z(1) \begin{pmatrix} \hat{x}_\alpha \\ \hat{p}_\alpha \\ \hat{x}_\beta \\ \hat{p}_\beta \end{pmatrix}, \tag{8.1}$$

where $\hat{x}_i$ and $\hat{p}_i$ represent quadrature operators of the input mode i, while $\hat{x}'_i$ and $\hat{p}'_i$ represent those of the output mode i. The matrices $C_Z(g)$, S_α, and S_β:

$$C_Z(g) = \begin{pmatrix} 1 & 0 & 0 & 0 \\ 0 & 1 & g & 0 \\ 0 & 0 & 1 & 0 \\ g & 0 & 0 & 1 \end{pmatrix}, \quad S_\alpha = \begin{pmatrix} \sqrt{2} & 0 & 0 & 0 \\ 0 & \frac{1}{\sqrt{2}} & 0 & 0 \\ 0 & 0 & 1 & 0 \\ 0 & 0 & 0 & 1 \end{pmatrix}, \quad S_\beta = \begin{pmatrix} 1 & 0 & 0 & 0 \\ 0 & 1 & 0 & 0 \\ 0 & 0 & \sqrt{2} & 0 \\ 0 & 0 & 0 & \frac{1}{\sqrt{2}} \end{pmatrix} \tag{8.2}$$

represent the controlled-Z gate with gain g, a 3.0 dB p-squeezing operation on mode α, and a 3.0 dB p-squeezing operation on mode β, respectively.

R. Ukai, *Multi-Step Multi-Input One-Way Quantum Information Processing with Spatial and Temporal Modes of Light*, Springer Theses,
DOI 10.1007/978-4-431-55019-8_8

8.1.2 Importance of This Experiment

We summarize importance of this experiment.

8.1.2.1 Gaussian Operation on Two-Mode Input State (Common to the Three Experiments)

This is an experimental demonstration of one-way quantum computation where Gaussian operations on two-mode input states are implemented. By combining the experimental demonstration of one-mode Gaussian operations we have reported earlier [1], we can implement an arbitrary multi-mode Gaussian operations in the framework of one-way quantum computation in principle (Sects. 3.6, 5.5).

This is a common property to the three experiments in this thesis.

8.1.2.2 Nonlocal Gate (Common to the Three Experiments)

This is an experimental demonstration of a nonlocal gate, where target two modes are located at a distance (Sect. 8.3).

This is a common property to the three experiments in this thesis.

8.1.2.3 Optimum Nonlocal Gate (Characteristic of This Experiment)

This is an experimental demonstration of an *optimum* nonlocal controlled-Z gate in the sense that it is implemented using the minimum resource: a bipartite entangled state shared in advance, and a classical channel in each direction (two channels in total) (Sect. 8.3.2). It can also be considered as an experimental demonstration of nonlocal gate with offline scheme (Sect. 4.1.2).

8.1.2.4 Minimum Setup of Two-Mode Gate with Squeezer-Based Input-Coupling Scheme (Characteristic of This Experiment)

Since one mode of a cluster state is consumed during a squeezer-based input coupling (Sect. 5.3.4), we need at least two modes of a cluster state to implement a two-mode operation where a two-mode input state is coupled with the cluster state via two squeezer-based input couplings. Therefore, this is the minimum setup of implementing a two-mode gate with squeezer-based input-coupling scheme.

8.1.3 Comparison of Three Experiments

In order to compare three experiments in this thesis, we summarize the properties of this experiment in a common form. It is given in Table. 8.1.

Table 8.1 Properties of optimum nonlocal gate in a common form

Item	Optimum nonlocal gate experiment
Main feature	Nonlocal gate is implemented with the minimum resource
Resource	Two-mode cluster state
Input coupling scheme	Squeezer-based input-coupling scheme
Excess 3 dB squeezing derived from input coupling	Exists
Operation experimentally demonstrated	$C_Z(\frac{1}{2})S_\alpha S_\beta$, fixed
Entanglement verification at the output	Van Loock-Furusawa, PT symplectic eigenvalue
Logarithmic negativity	$E_N = 0.40 \pm 0.01$

8.2 Theory

8.2.1 Abstract Illustration and Abstract Experimental Setup

Figures 8.1 and 8.2 show an abstract illustration and an abstract setup of the optimum nonlocal gate experiment.

The abstract illustration shown in Fig. 8.1 gives us a brief explanation of the experimental procedure (see Figs. 7.1 and 7.2).

8.2.2 Two-Mode Cluster State

The procedure for generation of the two-mode cluster state is explained in Sect. 6.1.1.

8.2.3 Input Coupling

We label two input modes as α and β. These two modes are coupled with the cluster state via the squeezer-based input-coupling schemes (Sect. 5.3.4). In the experimental setup, two beams corresponding to these two input modes are combined using two beam splitters with other two beams which correspond to two cluster modes. Transformations of these beam splitters are given by unitary matrices $B^{(2)}_{\alpha 1}(\frac{1}{\sqrt{2}})$, $B^{(2)}_{\beta 2}(\frac{1}{\sqrt{2}})$, leading to

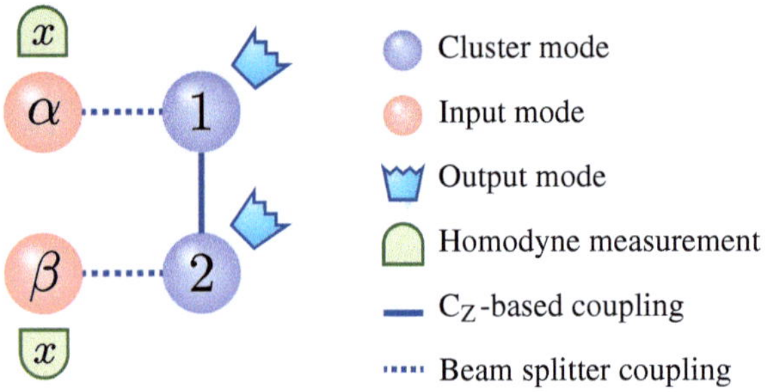

Fig. 8.1 Abstract illustration of optimum nonlocal gate

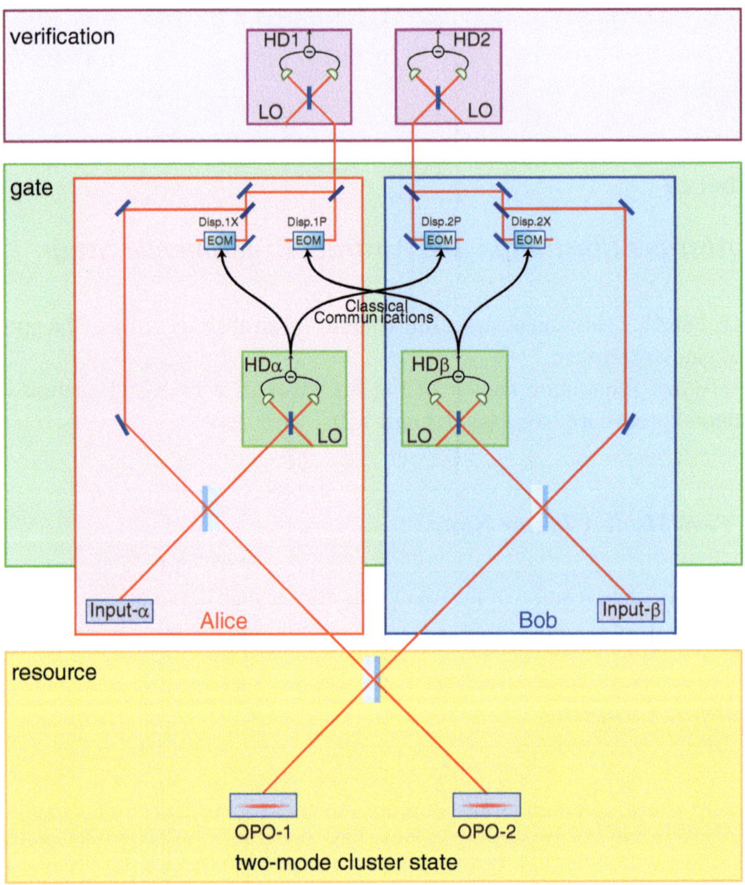

Fig. 8.2 Abstract experimental setup of optimum nonlocal gate

$$\begin{pmatrix} \hat{x}_\alpha^{(b)} + i\hat{p}_\alpha^{(b)} \\ \hat{x}_1^{(b)} + i\hat{p}_1^{(b)} \end{pmatrix} = \frac{1}{\sqrt{2}} \begin{pmatrix} 1 & -1 \\ 1 & 1 \end{pmatrix} \begin{pmatrix} \hat{x}_\alpha^{(in)} + i\hat{p}_\alpha^{(in)} \\ \hat{x}_1^{(c)} + i\hat{p}_1^{(c)} \end{pmatrix}, \quad \begin{pmatrix} \hat{x}_\beta^{(b)} + i\hat{p}_\beta^{(b)} \\ \hat{x}_2^{(b)} + i\hat{p}_2^{(b)} \end{pmatrix} = \frac{1}{\sqrt{2}} \begin{pmatrix} 1 & -1 \\ 1 & 1 \end{pmatrix} \begin{pmatrix} \hat{x}_\beta^{(in)} + i\hat{p}_\beta^{(in)} \\ \hat{x}_2^{(c)} + i\hat{p}_2^{(c)} \end{pmatrix}, \tag{8.3}$$

where $\hat{x}_k^{(in)}$ and $\hat{p}_k^{(in)}$ represent the quadrature operators of input mode $k(=\alpha, \beta)$. As a result, the quadrature operators after the input couplings are given by

$$\hat{x}_\alpha^{(b)} + i\hat{p}_\alpha^{(b)} = \left[\frac{1}{\sqrt{2}}\hat{x}_\alpha + \frac{1}{2}\hat{p}_1 - \frac{1}{2}\hat{x}_2\right] + i\left[\frac{1}{\sqrt{2}}\hat{p}_\alpha - \frac{1}{2}\hat{x}_1 - \frac{1}{2}\hat{p}_2\right], \tag{8.4}$$

$$\hat{x}_1^{(b)} + i\hat{p}_1^{(b)} = \left[\frac{1}{\sqrt{2}}\hat{x}_\alpha - \frac{1}{2}\hat{p}_1 + \frac{1}{2}\hat{x}_2\right] + i\left[\frac{1}{\sqrt{2}}\hat{p}_\alpha + \frac{1}{2}\hat{x}_1 + \frac{1}{2}\hat{p}_2\right], \tag{8.5}$$

$$\hat{x}_2^{(b)} + i\hat{p}_2^{(b)} = \left[\frac{1}{\sqrt{2}}\hat{x}_\beta + \frac{1}{2}\hat{x}_1 - \frac{1}{2}\hat{p}_2\right] + i\left[\frac{1}{\sqrt{2}}\hat{p}_\beta + \frac{1}{2}\hat{p}_1 + \frac{1}{2}\hat{x}_2\right], \tag{8.6}$$

$$\hat{x}_\beta^{(b)} + i\hat{p}_\beta^{(b)} = \left[\frac{1}{\sqrt{2}}\hat{x}_\beta - \frac{1}{2}\hat{x}_1 + \frac{1}{2}\hat{p}_2\right] + i\left[\frac{1}{\sqrt{2}}\hat{p}_\beta - \frac{1}{2}\hat{p}_1 - \frac{1}{2}\hat{x}_2\right], \tag{8.7}$$

where we have omitted the superscripts (in) for input modes α, β and (r) for resource modes 1 and 2. For example, $\hat{x}_1$ represents $\hat{x}_1 = \hat{x}_1^{(r)} = e^{r_1}\hat{x}_1^{(0)}$.

8.2.4 Measurement

We perform homodyne measurements on modes α and β. The relative phases between signal beams and local oscillator beams are set to 0 degree. Therefore, the measurement observables $\hat{s}_\alpha$ and $\hat{s}_\beta$ for measurements on modes α and β are

$$\hat{s}_\alpha = \hat{x}_\alpha^{(b)}, \quad \hat{s}_\beta = \hat{x}_\beta^{(b)}. \tag{8.8}$$

8.2.5 Feed-Forward and Operation

Without using the position operators $\hat{x}_k^{(r)}$ for squeezed-state modes $k = 1, 2$, the quadrature operators $\hat{x}_k^{(b)}$ and $\hat{p}_k^{(b)}$ for output modes $k = 1, 2$ can be reformulated as

$$\begin{pmatrix} \hat{x}_1^{(b)} \\ \hat{p}_1^{(b)} \\ \hat{x}_2^{(b)} \\ \hat{p}_2^{(b)} \end{pmatrix} = M_{op} \begin{pmatrix} \hat{x}_\alpha^{(in)} \\ \hat{p}_\alpha^{(in)} \\ \hat{x}_\beta^{(in)} \\ \hat{p}_\beta^{(in)} \end{pmatrix} + M_{\text{disp}} \begin{pmatrix} \hat{s}_\alpha \\ \hat{s}_\beta \end{pmatrix} + \frac{1}{\sqrt{2}} \begin{pmatrix} 0 \\ \hat{\delta}_1 \\ 0 \\ \hat{\delta}_2 \end{pmatrix}, \tag{8.9}$$

where

$$M_{op} = \begin{pmatrix} \sqrt{2} & 0 & 0 & 0 \\ 0 & \frac{1}{\sqrt{2}} & \frac{1}{\sqrt{2}} & 0 \\ 0 & 0 & \sqrt{2} & 0 \\ \frac{1}{\sqrt{2}} & 0 & 0 & \frac{1}{\sqrt{2}} \end{pmatrix}, \quad M_{\text{disp}} = \begin{pmatrix} -1 & 0 \\ 0 & -1 \\ 0 & -1 \\ -1 & 0 \end{pmatrix}. \tag{8.10}$$

Note that the reformulation above holds before the measurements are carried out. On the contrary, when the measurements on modes α and β have been carried out indeed, the quadratures of the output modes become

$$\begin{pmatrix} \hat{x}_1^{(bm)} \\ \hat{p}_1^{(bm)} \\ \hat{x}_2^{(bm)} \\ \hat{p}_2^{(bm)} \end{pmatrix} = M_{op} \begin{pmatrix} \hat{x}_\alpha^{(in)} \\ \hat{p}_\alpha^{(in)} \\ \hat{x}_\beta^{(in)} \\ \hat{p}_\beta^{(in)} \end{pmatrix} + M_{\text{disp}} \begin{pmatrix} s_\alpha \\ s_\beta \end{pmatrix} + \frac{1}{\sqrt{2}} \begin{pmatrix} 0 \\ \hat{\delta}_1 \\ 0 \\ \hat{\delta}_2 \end{pmatrix}, \tag{8.11}$$

where s_α and s_β represent measurement results of measurement variables $\hat{s}_\alpha$ and $\hat{s}_\beta$, respectively. $\hat{\delta}_k$ represents the k-th nullifier of the resource cluster state (see Eq. 6.6). By performing feed-forwards (displacement operations) based on the measurement results:

$$\begin{pmatrix} \hat{x}_1^{(out)} \\ \hat{p}_1^{(out)} \\ \hat{x}_2^{(out)} \\ \hat{p}_2^{(out)} \end{pmatrix} = \begin{pmatrix} \hat{x}_1^{(bm)} \\ \hat{p}_1^{(bm)} \\ \hat{x}_2^{(bm)} \\ \hat{p}_2^{(bm)} \end{pmatrix} - M_{\text{disp}} \begin{pmatrix} s_\alpha \\ s_\beta \end{pmatrix}, \tag{8.12}$$

or equivalently, by applying displacement operators:

$$\hat{X}_1(s_\alpha)\hat{Z}_1(s_\beta)\hat{X}_2(s_\beta)\hat{Z}_2(s_\alpha), \tag{8.13}$$

we get

$$\begin{pmatrix} \hat{x}_1^{(out)} \\ \hat{p}_1^{(out)} \\ \hat{x}_2^{(out)} \\ \hat{p}_2^{(out)} \end{pmatrix} = M_{op} \begin{pmatrix} \hat{x}_\alpha^{(in)} \\ \hat{p}_\alpha^{(in)} \\ \hat{x}_\beta^{(in)} \\ \hat{p}_\beta^{(in)} \end{pmatrix} + \frac{1}{\sqrt{2}} \begin{pmatrix} 0 \\ \hat{\delta}_1 \\ 0 \\ \hat{\delta}_2 \end{pmatrix}. \tag{8.14}$$

This is the input-output relationship in this experiment. The term of M_{op} represents the main operation. On the other hand, the term of $\hat{\delta}_k$ represents error of the operation, which derives from finite level of resource squeezing. In the limit of infinite squeezing $r \to \infty$, each $\hat{\delta}_k$ goes to zero, meaning that the ideal operation is achieved.

8.2.6 Decomposition of Operation

The matrix M_{op} can be decomposed into

$$M_{op} = C_Z\left(\frac{1}{2}\right)\begin{pmatrix} S & O \\ O & S \end{pmatrix} = \begin{pmatrix} S & O \\ O & S \end{pmatrix} C_Z(1), \tag{8.15}$$

where

$$C_Z(g) = \begin{pmatrix} 1 & 0 & 0 & 0 \\ 0 & 1 & g & 0 \\ 0 & 0 & 1 & 0 \\ g & 0 & 0 & 1 \end{pmatrix}, \quad S = \begin{pmatrix} \sqrt{2} & 0 \\ 0 & \frac{1}{\sqrt{2}} \end{pmatrix}. \tag{8.16}$$

Here, S is the matrix of the 3.0 dB p-squeezing operation, while $C_Z(g)$ represents the controlled-Z gate with the interaction gain g.

8.2.7 Covariance Matrix

We define $V_{\alpha\beta}^{(in)}$ as the covariance matrix of a two-mode input state. We here choose the *xpxp notation* for all covariance matrices (Sect. 3.2.5).

Since the two-mode input state and the resource two-mode cluster state has no correlations, the covariance matrix of the output state becomes

$$V_{12}^{(out)} = M_{op} V_{\alpha\beta}^{(in)} M_{op}^T + V_\delta, \tag{8.17}$$

where V_δ represents excess noise derived from the finite level of resource squeezing. It is given by

$$V_\delta = \frac{\hbar}{2}\begin{pmatrix} 0 & 0 & 0 & 0 \\ 0 & e^{-2r_2} & 0 & 0 \\ 0 & 0 & 0 & 0 \\ 0 & 0 & 0 & e^{-2r_1} \end{pmatrix}. \tag{8.18}$$

In the limit of infinite squeezing $r_i \to \infty$, V_δ goes to the 4×4 zero matrix O, meaning that the ideal operation is achieved.

Finally, we assume that two input modes α and β are initially in vacuum states. The covariance matrix of the output state becomes

$$V_{12}^{(out)} = \frac{\hbar}{2}\left[\begin{pmatrix} 2 & 0 & 0 & 1 \\ 0 & 1 & 1 & 0 \\ 0 & 1 & 2 & 0 \\ 1 & 0 & 0 & 1 \end{pmatrix} + \begin{pmatrix} 0 & 0 & 0 & 0 \\ 0 & e^{-2r_2} & 0 & 0 \\ 0 & 0 & 0 & 0 \\ 0 & 0 & 0 & e^{-2r_1} \end{pmatrix}\right]. \tag{8.19}$$

8.2.8 Inseparability Criteria

In this experimental demonstration of the optimum nonlocal gate, we verify entanglement at the output by using both van Loock-Furusawa criterion (Sect. 3.7.4) and the inseparable criterion with PT symplectic eigenvalue (3.7.5).

8.2.8.1 Van Loock-Furusawa Criterion

By using the input-output relationship Eq. (8.14), we can guess that $\hat{x}_1^{(out)}$ and $\hat{p}_2^{(out)}$ have positive correlation, while $\hat{x}_2^{(out)}$ and $\hat{p}_1^{(out)}$ have positive correlation. By applying van Loock-Furusawa entanglement criterion, we get the following sufficient condition for entanglement.

8.2.8.2 Sufficient Condition for Entanglement

If the output state satisfies

$$\langle \Delta^2(g\hat{p}_1^{(out)} - \hat{x}_2^{(out)})\rangle + \langle \Delta^2(g\hat{p}_2^{(out)} - \hat{x}_1^{(out)})\rangle < 2g\hbar, \tag{8.20}$$

for some $g > 0$, the output state is an entangled state.

We assume that two input modes α and β are in vacuum states. $\langle \Delta^2(g\hat{p}_1^{(out)} - \hat{x}_2^{(out)})\rangle$ and $\langle \Delta^2(g\hat{p}_2^{(out)} - \hat{x}_1^{(out)})\rangle$ are given by

$$\langle \Delta^2(g\hat{p}_1^{(out)} - \hat{x}_2^{(out)})\rangle = \frac{\hbar}{2}\left[\frac{g^2}{2} + \left(\frac{g}{\sqrt{2}} - \sqrt{2}\right)^2 + g^2 e^{-2r_2}\right], \tag{8.21}$$

$$\langle \Delta^2(g\hat{p}_2^{(out)} - \hat{x}_1^{(out)})\rangle = \frac{\hbar}{2}\left[\frac{g^2}{2} + \left(\frac{g}{\sqrt{2}} - \sqrt{2}\right)^2 + g^2 e^{-2r_1}\right]. \tag{8.22}$$

By choosing

$$g = 1, \tag{8.23}$$

the sufficient condition for entanglement is satisfied with minimum r_i (≥ 0). It becomes

$$e^{-2r_1} + e^{-2r_2} < 2. \tag{8.24}$$

Thus, no matter how little the levels of resource squeezed states are, the output state becomes an entangled state.

Note that the van Loock-Furusawa criterion gives us only a sufficient condition, not the necessary and sufficient condition, for entanglement. However, it is obvious that the output state never becomes an entangled state when squeezed states are not utilized as resources. Therefore, we find that Eq. (8.20) is the optimum condition for entanglement.

8.2.8.3 Simon Criterion

We consider Simon criterion (Sect. 3.7.5), which gives us necessary and sufficient entanglement condition for two-mode Gaussian states.

We assume that two input modes α and β are in vacuum states. By using the Simon criterion, we find that Eq. (8.19) shows a covariance matrix of an entangled state if and only if r_i satisfies

$$e^{-2r_1} + e^{-2r_2} < 2, \tag{8.25}$$

which is identical to the result of van Loock-Furusawa criterion.

8.3 Interpretation as Optimum Nonlocal Gate

8.3.1 Nonlocal Gate

We consider Alice and Bob are located at a distance (Fig. 8.3). Each Alice and Bob possesses a one-mode quantum state. We consider that a unitary operation should be applied to these two one-mode quantum states. The quantum gate whose target modes are located at a distance is called a *nonlocal* gate [2–7]. Note that these two modes might be entangled at the beginning, although we have assumed that they are $|\psi_A\rangle$ and $|\psi_B\rangle$ in Fig. 8.3.

Generation of entanglement can be classified into two cases based on whether two quantum states to be entangled are located locally or nonlocally [3]. When they are located locally, an entangling gate on them becomes a member of local unitary operations, and thus it is considered that it can be easily implemented. A theoretical solution for this is to use a medium which has the corresponding interaction Hamiltonian. When they are located nonlocally, a nonlocal entangling gate is required, which cannot be achieved directly with an actual device such as a nonlinear medium.

One possible solution for implementation of a nonlocal gate is to combine a local entangling gate with quantum teleportation. Here, quantum teleportation is a protocol with which an unknown quantum state can be transmitted from one place to another. It consists of the following components: (1) a two-mode entangled state shared by

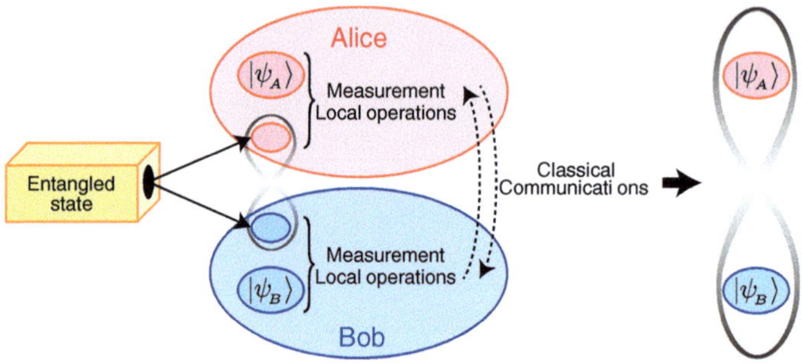

Fig. 8.3 Abstract illustration of nonlocal gate

Alice and Bob in advance; (2) measurements and local unitary operations; and (3) two channels of classical channels with which measurement results are transmitted from one party to another. The processes (2) and (3) are called the Local Operations and Classical Communications (LOCC). By assuming that Bob owns an entangling gate and that they can transmit any quantum state to each other by using quantum teleportation, they can entangle their quantum states by the following procedure. First, Alice transmits her quantum state to Bob via a quantum teleportation. Then, Bob performs the local entangling gate on his state and the state received from Alice. Finally, Bob transmits one of two modes to Alice which was initially owned by her. Note that, two two-mode entangled states and four classical channels are utilized in total since two quantum teleportations are involved.

The fact is, the procedure above is not the optimum way to implement a nonlocal gate. Instead, in this section, we show that a nonlocal controlled-Z gate (with additional local operations) can be implemented by using a two-mode entangled state shared by Alice and Bob in advance as well as a classical channel in each direction (two channels in total).

Before we move onto optimality of a nonlocal gate, we mention an example of its application: distributed quantum computer [8, 9]. It is a device in which total quantum system to be processed is divided into several subsystems. In each subsystem, local processor performs local operations on their quantum states. Processes on quantum states over subsystems are carried out via quantum teleportations and nonlocal entangling gates.

8.3.2 Optimum Nonlocal Gate

The optimum nonlocal gate was first discussed in discrete-variable systems [4]. In this subsection, we extend it to continuous-variable system.

8.3.2.1 Theorem

Assume that each Alice and Bob located at a distance possesses a one-mode quantum state. Assume that they perform a nonlocal controlled-Z gate on these two modes of the quantum states. One maximally entangled pair (an EPR state or a two-mode cluster state) shared in advance and one classical channel in each direction (two channels in total) are necessary and sufficient implements for the nonlocal controlled-Z gate.

8.3.2.2 Proof of Necessity

We consider that they apply a controlled-Z gate with the operator $\hat{C}_Z = e^{\frac{i}{\hbar}\hat{x}_A\hat{x}_B}$.

First, we assume that the one-mode quantum state owned by Alice is $|p=0\rangle_A$, while that by Bob is $|p=0\rangle_B$. Note that there is no correlation between these two states. By applying the nonlocal controlled-Z gate, the two-mode state becomes

$$\hat{C}_Z|p=0\rangle_A|p=0\rangle_B = \frac{1}{2\pi\hbar}e^{\frac{i}{\hbar}\hat{x}_A\hat{x}_B}\int_{-\infty}^{\infty} ds\,|x=s\rangle_A\int_{-\infty}^{\infty} dt\,|x=t\rangle_B \tag{8.26}$$

$$= \frac{1}{2\pi\hbar}\int_{-\infty}^{\infty} ds\,|x=s\rangle_A\int_{-\infty}^{\infty} dt\,e^{\frac{i}{\hbar}st}|x=t\rangle_B \tag{8.27}$$

$$= \frac{1}{\sqrt{2\pi\hbar}}\int_{-\infty}^{\infty} ds\,|x=s\rangle_A|p=s\rangle_B, \tag{8.28}$$

which is a maximally entangled state (two-mode cluster state). Since the amount of entanglement does not increase by local operations and LOCC, Alice and Bob have to share at least one pair of maximally entangled state in advance.

Next, we assume that the one-mode quantum state owned by Alice is $|x=a\rangle_A$, while that by Bob is $|p=0\rangle_B$. By applying the nonlocal controlled-Z gate, the two-mode state becomes

$$\hat{C}_Z|x=a\rangle_A|p=0\rangle_B = \frac{1}{\sqrt{2\pi\hbar}}e^{\frac{i}{\hbar}\hat{x}_A\hat{x}_B}|x=a\rangle_A\int_{-\infty}^{\infty} dt\,|x=t\rangle_B \tag{8.29}$$

$$= \frac{1}{\sqrt{2\pi\hbar}}|x=a\rangle_A\int_{-\infty}^{\infty} dt\,e^{\frac{i}{\hbar}at}|x=t\rangle_B \tag{8.30}$$

$$= |x=a\rangle_A|p=a\rangle_B, \tag{8.31}$$

which can be considered as a transmission of a from Alice to Bob. Since local operations cannot transmit any information, they have to share at least one channel of classical channel from Alice to Bob.

Finally, we assume that the one-mode quantum state owned by Alice is $|p = 0\rangle_A$, while that by Bob is $|x = b\rangle_B$. By applying the nonlocal controlled-Z gate, the two-mode state becomes

$$\hat{C}_Z|p = 0\rangle_A|x = b\rangle_B = \frac{1}{\sqrt{2\pi\hbar}} e^{\frac{i}{\hbar}\hat{x}_A\hat{x}_B} \int_{-\infty}^{\infty} \mathrm{d}s \; |x = s\rangle_A|x = b\rangle_B \tag{8.32}$$

$$= \frac{1}{\sqrt{2\pi\hbar}} \int_{-\infty}^{\infty} \mathrm{d}s \; e^{\frac{i}{\hbar}sb}|x = s\rangle_A|x = b\rangle_B \tag{8.33}$$

$$= |p = b\rangle_A|x = b\rangle_B, \tag{8.34}$$

which can be considered as a transmission of b from Bob to Alice. Since local operations cannot transmit any information, they have to share at least one channel of classical channel from Bob to Alice.

In summary, Alice and Bob have to share at least one pair of maximally entangled state as well as one classical channel in each direction (two channels in total).

8.3.2.3 Proof of Sufficiency

It is sufficient to show an example where a nonlocal controlled-Z gate is implemented by using one maximally entangled pair shared in advance and one classical channel in each direction. Figure 8.2 does this work.

8.4 Experimental Setup

8.4.1 From Laser to Cluster State Generation

The resource for this experiment is a two-mode cluster state. In order to generate it, we utilize the same experimental setup to that in Chap. 6.

8.4.2 Input States

Input states for experimental demonstration of the optimum nonlocal gate are vacuum states and coherent states. We utilize the same experimental setup to that in Sect. 7.3.2.

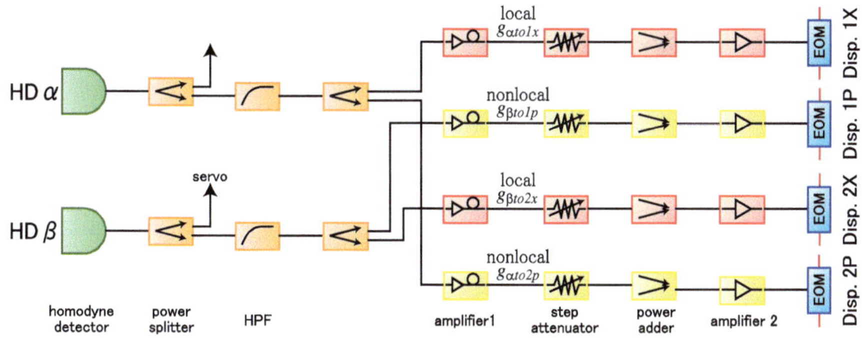

Fig. 8.4 Classical channels for optimum nonlocal gate experiment

8.4.3 Classical Channels and Displacement Operations

The displacement operation (feed-forward operation) in this experiment is given by Eq. (8.12), or equivalently, Eq. (8.13):

$$\hat{X}_1(s_\alpha)\hat{Z}_1(s_\beta)\hat{X}_2(s_\beta)\hat{Z}_2(s_\alpha). \tag{8.35}$$

Therefore, we need four paths of classical channels $(\alpha, 1x)$, $(\alpha, 2p)$, $(\beta, 2x)$, and $(\beta, 1p)$, where parentheses show (mode of homodyne detection, destination of feed-forward).

The schematic of classical channels for the optimum nonlocal gate experiment is shown in Fig. 8.4. In these four classical channels, $(\alpha, 1x)$ and $(\beta, 2x)$, which are shown in red in Fig. 8.4, are local classical channels which work within one parties (Alice or Bob). On the other hand, $(\alpha, 2p)$ and $(\beta, 1p)$, which are shown in yellow, are nonlocal classical channels which transmit measurement results from Alice to Bob and vise versa.

8.4.4 Locking the Relative Phases

Figure 8.5 shows arrangement of probe beams and phase modulations for the optimum nonlocal gate experiment.

8.4.4.1 Summary of Phase Locking

Table 8.2 shows phase locking techniques for the optimum nonlocal gate experiment. Parentheses in Table 8.2 show that their phases are not utilized in experiment.

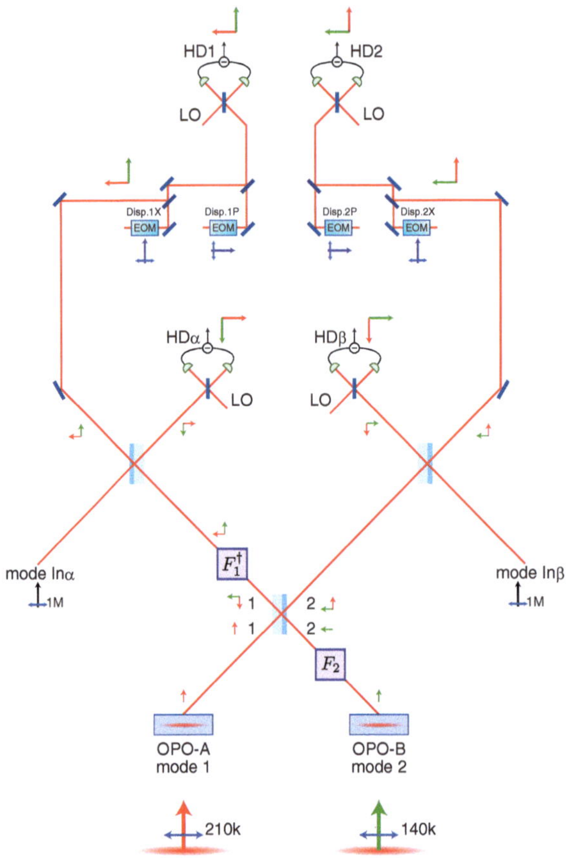

Fig. 8.5 Probe beams and phase modulations for optimum nonlocal gate experiment

8.4.5 Cancellation

We use the same procedure for adjustment of classical channels (cancellation) explained in Sect. 7.3.5.

8.5 Preparation for Measurement

8.5.1 Visibilities and Parametric Gains

Interference visibilities and parametric gains are shown in Tables 8.3 and 8.4, respectively.

Table 8.2 Phase lockings for optimum nonlocal gate

Relative phase	Technique, demodulation frequency
Parametric gain of OPO-A	AC locking, 210 kHz
Parametric gain of OPO-B	AC locking, 140 kHz
Beam splitter $B_{12}^{(1)}(\frac{1}{\sqrt{2}})$	DC locking
Input α	AC locking, amplitude x: 140 kHz, amplitude p: 210 kHz
Input β	AC locking, amplitude x: 210 kHz, amplitude p: 140 kHz
Displacement in mode 1	AC locking, displacement x: 140 kHz, displacement p: 210 kHz
Displacement in mode 2	AC locking, displacement x: 210 kHz, displacement p: 140 kHz
Homodyne measurement in mode α	AC locking, x: 210 kHz, (p: 140 kHz)
Homodyne measurement in mode 1	AC locking, x: 210 kHz, p: 140 kHz
Homodyne measurement in mode 2	AC locking, x: 140 kHz, p: 210 kHz
Homodyne measurement in mode β	AC locking, x: 140 kHz, (p: 210 kHz)

Table 8.3 Interference visibilities for optimum nonlocal gate experiment

Beam splitter	Beams used for adjustment	Visibility (%)
Beam splitter $B_{12}^{(1)}(\frac{1}{\sqrt{2}})$	OPO-A, OPO-B	99.2
Input coupling α	OPO-A, In-α	99.2
Input coupling β	OPO-B, In-β	97.8
Displacement $1x$	OPO-A, Disp-1X	98.5
Displacement $1p$	OPO-A, Disp-1P	97.9
Displacement $2x$	OPO-B, Disp-2X	96.9
Displacement $2p$	OPO-B, Disp-2P	96.6
Homodyne measurement α	OPO-A, LO-α	98.6
Homodyne measurement 1	OPO-A, LO-1	98.0
Homodyne measurement 2	OPO-B, LO-2	96.9
Homodyne measurement β	OPO-B, LO-β	96.3

Table 8.4 Parametric gains for optimum nonlocal gate experiment

OPO name	Parametric gain (G_+)	Phase matching temperature
OPO-A	7.3	40.5°
OPO-B	7.3	40.0°

8.5.2 *Adjustment of EOM (Purity)*

Each purity is listed in Table 8.5.

Table 8.5 Adjustment of purity for optimum nonlocal gate experiment

EOM	Purity (dB)
Disp-1X	34.1
Disp-1P	39.4
Disp-2X	30.8
Disp-2P	37.5

Table 8.6 Adjustment of classical channels (cancellation) for optimum nonlocal gate experiment

Feed-forward	Cancellation level (dB)
$g_{\alpha to 1x}$	39.1
$g_{\alpha to 2p}$	35.5
$g_{\beta to 2x}$	39.9
$g_{\beta to 1p}$	41.3

8.5.3 Cancellation

Each cancellation level is listed in Table 8.6.

8.5.4 Balance Between Homodyne Detections

The balance between two homodyne detections for modes 1 and 2 is adjusted in the same way as Chap. 6. As a result, we have achieved cancellation of 47.8 dB.

8.6 Measurement Results and Their Analysis

8.6.1 Power of Output with Vacuum Inputs

8.6.1.1 Measurement Results

Figure 8.6 shows measurement results. Red traces show measurement results when inputs of homodyne detections are shut off. They correspond to the shot noise level (SNL). Orange traces show measurement results of output-mode quadrature variances.

Each result is listed in Table 8.7.

8.6.1.2 Effective Squeezing Level

We estimate effective squeezing level (Sect. 6.1.1). For simplicity, we assume that all squeezed states have identical squeezing level x.

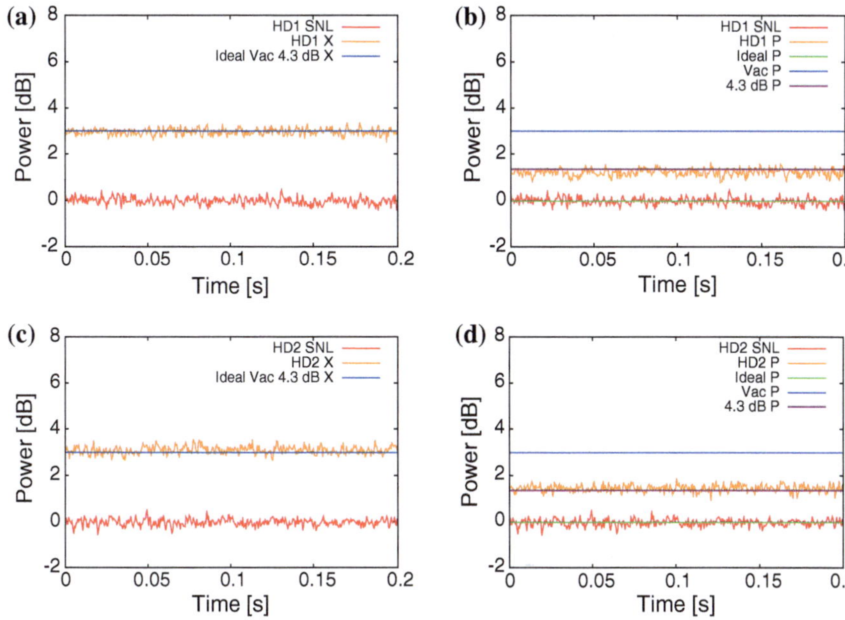

Fig. 8.6 Measurement results with vacuum inputs. **a** $\hat{x}_1$, **b** $\hat{p}_1$, **c** $\hat{x}_2$, **d** $\hat{p}^2$

Table 8.7 Measurement results with vacuum inputs

Measurement variable	Result (dB)
$\langle \Delta \hat{x}_1 \rangle$	2.98 ± 0.02
$\langle \Delta \hat{p}_1 \rangle$	1.24 ± 0.02
$\langle \Delta \hat{x}_2 \rangle$	3.12 ± 0.02
$\langle \Delta \hat{p}_2 \rangle$	1.48 ± 0.02

From Eq. 8.19, we get

$$\langle \Delta^2 \hat{x}_i \rangle = \frac{\hbar}{2} \times 2, \quad \langle \Delta^2 \hat{p}_i \rangle = \frac{\hbar}{2}\left[1 + 10^{-\frac{x}{10}}\right], \tag{8.36}$$

for $i = 1, 2$. Thus, the variance of $\langle \Delta^2 \hat{x}_i \rangle$ is equal to 3.0 dB relative to the SNL, which does not depend on the level of resource squeezed states (blue lines in Fig. 8.6). On the other hand, the variance of $\langle \Delta^2 \hat{p}_i \rangle$ depends on the level of resource squeezed states. In the ideal case of $x \to \infty$, it becomes $\langle \Delta^2 \hat{p}_i \rangle = \frac{\hbar}{2}$, which is identical to the SNL (green lines in Fig. 8.6). When no squeezed resource states are available ($x = 0$), it becomes $\langle \Delta^2 \hat{p}_i \rangle = \frac{\hbar}{2} \times 2$, which is equal to 3.0 dB relative to the SNL (blue lines in Fig. 8.6).

In experiment, we utilize finite-level squeezed states as resources for the optimum nonlocal gate. The larger the squeezing level is, the lower the variance becomes.

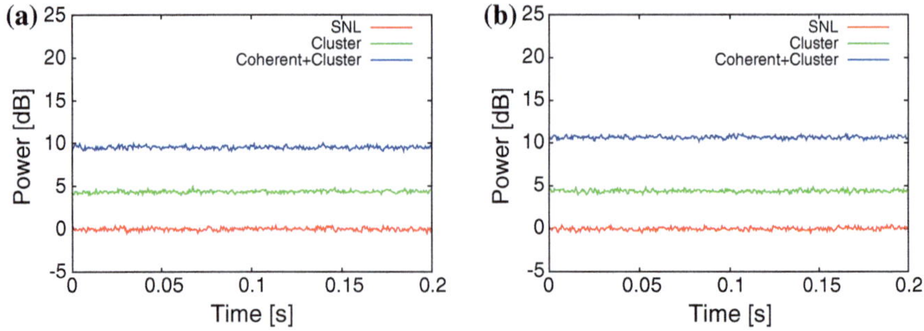

Fig. 8.7 Measurement results of input states. **a** Input α with HD α. **b** Input β with HD β

Table 8.8 Measurement results of input states

HD name	Cluster only (dB)	Coherent + Cluster (dB)
HDα	4.35	9.52
HDβ	4.41	10.68

Therefore, we can estimate the effective squeezing level by using their measurement results. As a result, we get $x = 4.3\,\text{dB}$ as the effective squeezing level.

8.6.2 Power of Output with Coherent Inputs

8.6.2.1 Measurement Results of Input States

We measure the powers of both input coherent states in modes α and β. Different from Chap. 7, we have acquired the powers of the input state by doubling measurement results of single homodyne detections.

Each power is listed in Table 7.8.

As a result, we get 10.96 dB for mode α, while 12.52 dB for mode β, compared to the SNL. All statistical errors are less than ± 0.02 dB, including in Table 8.8.

8.6.2.2 Measurement Result of Output Power

Table 8.9 shows measurement results of powers of output-mode quadratures with vacuum or coherent input states. All error bars are less than ± 0.02 dB.

Table 8.10 shows theoretical predictions. We have here assumed that the resource squeezing level is 4.3 dB, and the input coherent-state amplitudes are 10.96 dB for mode α and 12.52 dB for mode β, respectively.

Table 8.9 Measurement results of quadrature operator powers with coherent-state inputs (dB)

Input mode and phase \ Output mode and phase	$\hat{x}_1$	$\hat{p}_1$	$\hat{x}_2$	$\hat{p}_2$
Vacuum inputs in both α and β	2.98	1.24	3.12	1.48
Amplitude in x of α	14.59	1.44	3.03	8.88
Amplitude in p of α	3.11	8.95	3.03	1.50
Amplitude in x of β	2.83	10.34	16.08	1.53
Amplitude in p of β	2.83	1.56	3.16	10.31

Table 8.10 Theoretical prediction of quadrature operator powers with coherent-state inputs (dB)

Input mode and phase \ Output mode and phase	$\hat{x}_1$	$\hat{p}_1$	$\hat{x}_2$	$\hat{p}_2$
Vacuum inputs in both α and β	3.01	1.37	3.01	1.37
Amplitude in x of α	14.34	1.37	3.01	8.85
Amplitude in p of α	3.01	8.85	3.01	1.37
Amplitude in x of β	3.01	10.11	15.75	1.37
Amplitude in p of β	3.01	1.37	3.01	10.11

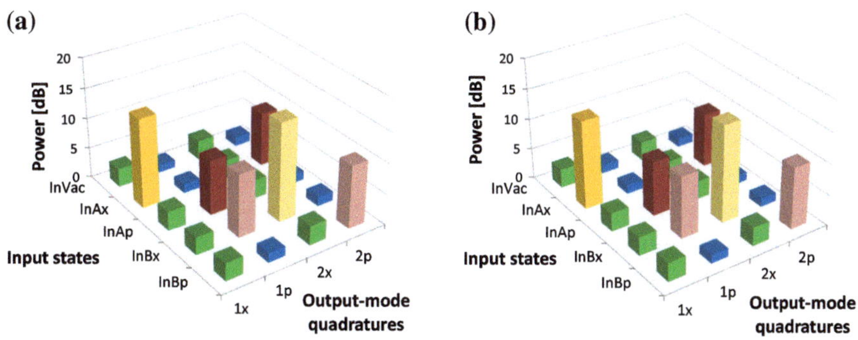

Fig. 8.8 Quadrature operator powers with coherent-state inputs. A and B correspond to α and β, respectively. **a** Measurement result, **b** Theoretical prediction

Figure 8.8 shows graphs of Tables 8.9 and 8.10. Note that they are not covariance matrices.

8.6.2.3 Analysis

We take an example of the coherent state input with amplitude in x of mode α (InAx). The input-output relationship Eq. (8.14) shows that $\hat{x}_\alpha$ is transmitted to $\hat{x}_1$ with the amplitude gain of $\sqrt{2}$, as well as to $\hat{p}_2$ with the amplitude gain of $\frac{1}{\sqrt{2}}$. As a result, we acquired 14.59 and 8.88 dB for measurements of $\hat{x}_1$ and $\hat{p}_2$, which are about twice and a half of the initial power, respectively.

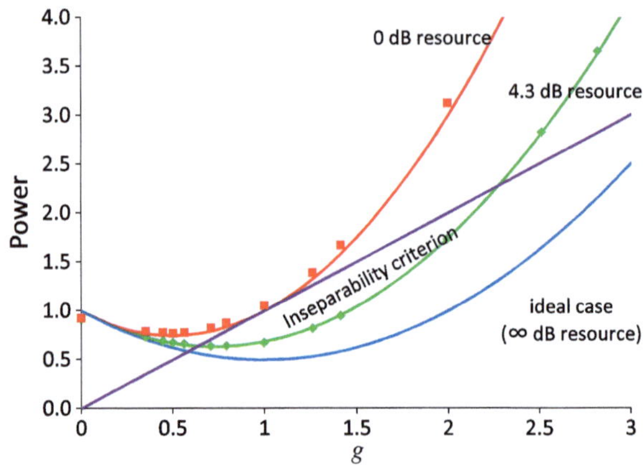

Fig. 8.9 Entanglement at the output

The differences between input and output powers in theoretical values are not precisely 3.0 dB because the power of the input state is not intense enough to neglect the variance of a vacuum state and effect of finite level of resource squeezed states.

8.6.3 Entanglement at Output (via van Loock-Furusawa Criterion)

In this subsection, we verify entanglement at the output by using the van Loock-Furusawa criterion.

8.6.3.1 Measurement Results and Analysis

Figure 8.9 shows measurement results of $\langle\Delta^2(g\hat{p}_1^{(out)} - \hat{x}_2^{(out)})\rangle + \langle\Delta^2(g\hat{p}_2^{(out)} - \hat{x}_1^{(out)})\rangle$ with several gains g, where we have used the system of units with $\hbar = \frac{1}{2}$.

Measurement results with 0 dB resources are acquired by using vacuum states, instead of squeezed states, as resources. Measurement results with squeezed state resources agree well with theoretical predictions with 4.3 dB squeezed states.

If a measurement result is less than $2g\hbar = g$ $(\hbar = 1/2)$ with some gain g, the output state is entangled. We find that the entanglement criterion is certainly satisfied with $g = 0.71, \ldots, 2.0$. By choosing the optimum gain $g = 1$, we get

$$\langle\Delta^2(g\hat{p}_1^{(out)} - \hat{x}_2^{(out)})\rangle + \langle\Delta^2(g\hat{p}_2^{(out)} - \hat{x}_1^{(out)})\rangle = 0.673 \pm 0.002 < 1 \ (g = 1). \tag{8.37}$$

Table 8.11 Measurement results of variances

Measurement Variable	Variance (dB) (Relative to vacuum inputs)	Measurement Variable	Variance (dB) (Relative to vacuum inputs)
$\hat{x}_1$	3.05 ± 0.02	$\hat{x}_2$	2.98 ± 0.02
$\hat{p}_1$	1.05 ± 0.02	$\hat{p}_2$	1.34 ± 0.02
$(\hat{x}_1 + \hat{p}_1)/\sqrt{2}$	2.27 ± 0.02	$(\hat{x}_2 + \hat{p}_2)/\sqrt{2}$	2.08 ± 0.02
$(\hat{x}_1 - \hat{p}_1)/\sqrt{2}$	2.13 ± 0.02	$(\hat{x}_2 - \hat{p}_2)/\sqrt{2}$	2.36 ± 0.02
$\hat{x}_1 + \hat{x}_2$	3.11 ± 0.02	$\hat{p}_1 + \hat{x}_2$	3.93 ± 0.02
$\hat{x}_1 - \hat{x}_2$	2.76 ± 0.02	$\hat{p}_1 - \hat{x}_2$	-1.56 ± 0.02
$\hat{x}_1 + \hat{p}_2$	4.36 ± 0.02	$\hat{p}_1 + \hat{p}_2$	1.04 ± 0.02
$\hat{x}_1 - \hat{p}_2$	-1.54 ± 0.02	$\hat{p}_1 - \hat{p}_2$	1.46 ± 0.02

8.6.4 Covariance Matrix and Entanglement at Output (via PT Symplectic Eigenvalue)

We verify entanglement at the output by using PT symplectic eigenvalue. We here use the system of units with $\hbar = \frac{1}{2}$.

8.6.4.1 Measurement Results of Variances

Table 8.11 shows measurement results of variances which are utilized to acquire the covariance matrix of the output state.

As a result, we get the covariance matrix V:

$$V = \begin{pmatrix} 0.504 & 0.007 & 0.019 & 0.253 \\ 0.007 & 0.318 & 0.222 & -0.016 \\ 0.019 & 0.222 & 0.497 & -0.014 \\ 0.253 & -0.016 & -0.014 & 0.340 \end{pmatrix}, \tag{8.38}$$

where the statistical error in each element is less than ± 0.002. Figure 8.10 shows its graph.

8.6.4.2 Variables Acquired by Covariance Matrix

Table 8.12 shows several variables (Sects. 3.2.6, 3.7.5) which are acquired by the covariance matrix of the output state.

Since all determinants of the k-th order principal submatrices are positive: $\det V_k > 0$ $(k = 1, \ldots, 4)$, and the symplectic eigenvalue ν_- is $\nu_- \geq \frac{\hbar}{2}$, the symmetric matrix V certainly shows a covariance matrix of a physical state.

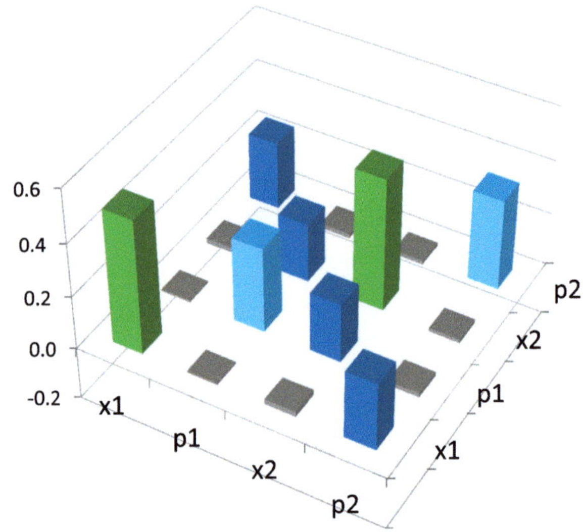

Fig. 8.10 Covariance matrix

Table 8.12 Variables acquired by covariance matrix

Variable	Result
Determinant of 1st order principal submatrix: det $V_1 = V_{11}$	0.504 ± 0.002
Determinant of 2nd order principal submatrix: det $V_2 = \det A$	0.1604 ± 0.0008
Determinant of 3rd order principal submatrix: det V_3	0.0548 ± 0.0006
Determinant of 4th order principal submatrix: det $V_4 = \det V$	0.0116 ± 0.0002
Symplectic eigenvalue: ν_-	0.314 ± 0.002
Symplectic eigenvalue: ν_+	0.343 ± 0.002
PT symplectic eigenvalue: $\tilde{\nu}_-$	0.167 ± 0.002
PT symplectic eigenvalue: $\tilde{\nu}_+$	0.643 ± 0.002
Logarithmic negativity: E_N	0.402 ± 0.007

In addition to this, since the PT symplectic eigenvalue $\tilde{\nu}_-$ satisfies $\tilde{\nu}_- < \frac{\hbar}{2}$, the physical state is an entangled state. We get the logarithmic negativity of $E_N = 0.402 \pm 0.007$.

We can estimate effective squeezing level (Sect. 6.1.1) from the logarithmic negativity, leading to $x^{(e)} = 4.1\,\text{dB}$. It is almost equivalent to the effective squeezing level which is estimated by the variances when both input modes are in vacuum states.

References

1. Ukai, R., Iwata, N., Shimokawa, Y., Armstrong, S.C., Politi, A., Yoshikawa, J., van Loock, P., Furusawa, A.: Demonstration of unconditional one-way quantum computations for continuous variables. Phys. Rev. Lett. **106**, 240504 (2011)
2. Gottesman, D.: The Heisenberg representation of quantum computers, Group22. In: Corney, S.P., Delbourgo, R., Jarvis, P.D. (eds.) Proceedings of the XXII International Colloquium on Group Theoretical Methods in Physics, pp. 32–43. International Press, Cambridge, MA. arXiv:9807006 [quant-ph] (1999)
3. Zhou, X., Leung, D.W., Chuang, I.L.: Methodology for quantum logic gate construction. Phys. Rev. A **62**, 052316 (2000)
4. Eisert, J., Jacobs, K., Papadopoulos, P., Plenio, B.: Optimal local implementation of nonlocal quantum gates. Phys. Rev. A **62**, 052317 (2000)
5. Huang, Y.-F., Ren, X.-F., Zhang, Y.-S., Duan, L.M.G., Guo, G.C.: Experimental teleportation of a quantum Controlled-NOT gate. Phys. Rev. Lett. **93**, 240501 (2004)
6. Filip, R., Marek, P., Andersen, U.L.: Measurement-induced continuous-variable quantum interactions. Phys. Rev. A **71**, 042308 (2005)
7. Gao, W.-B., Goebel, A.M., Lu, C.-Y., Dai, H.-N., Wagenknecht, C., Zhang, Q., Zhao, B., Peng, C.-Z., Chen, Z.g-B., Chen, Y.-A., Pan, J.-W.: Teleportation-based realization of an optical quantum two-qubit entangling gate. Proc. Natl. Acad. Sci. USA **107**, 20869 (2010)
8. Cirac, J.I., Ekert, A.K., Huelga, S.F., Machiavello, C.: Distributed quantum computation over noisy channels. Phys. Rev. A **59**, 4249 (1999)
9. Lo, H.K.: Classical-communication cost in distributed quantum-information processing: a generalization of quantum-communication complexity. Phys. Rev. A **62**, 012313 (2000)

Chapter 9
Experimental Demonstration of Gain-Tunable Entangling Gate for Continuous Variables

9.1 Gain-Tunable Entangling Gate Experiment

9.1.1 Operation

In the experimental demonstration of a *gain-tunable entangling gate* (T_Z gate) for continuous-variable one-way quantum computation, we utilize a three-partite entangled state, called the *three-mode linear cluster state*, as a resource (Sect. 6.1.2). We prepare a two-mode input state independently of the cluster state. The input coupling with the cluster is achieved via the *squeezer-based input-coupling scheme* (Sect. 5.3.4). The two modes to which the input state is transmitted are equivalent to the output modes. By changing the relative phases between signal beams and local oscillator beams in homodyne detections for two input modes, we can implement single-mode Gaussian operations. We choose these measurement bases so that the operation for each single mode becomes the 3.0 dB p-squeezing operation. This is the simplest operation which can be achieved by the experimental setup we use. The remaining mode 2 is measured by a homodyne detection with relative phase θ, which enables us to control the operation on the two-mode input state. The input–output relationship in the Heisenberg picture is given by

$$\begin{pmatrix} \hat{x}'_\alpha \\ \hat{p}'_\alpha \\ \hat{x}'_\beta \\ \hat{p}'_\beta \end{pmatrix} = T_Z\Big(-\frac{g}{2}\Big) S_\alpha S_\beta \begin{pmatrix} \hat{x}_\alpha \\ \hat{p}_\alpha \\ \hat{x}_\beta \\ \hat{p}_\beta \end{pmatrix}, \tag{9.1}$$

where $\hat{x}_i$ and $\hat{p}_i$ represent quadrature operators of the input mode i, while $\hat{x}'_i$ and $\hat{p}'_i$ represent those of the output mode i. The matrices $T_Z(g)$, S_α, and S_β:

$$T_Z(g) = \begin{pmatrix} 1 & 0 & 0 & 0 \\ g & 1 & g & 0 \\ 0 & 0 & 1 & 0 \\ g & 0 & g & 1 \end{pmatrix}, \quad S_\alpha = \begin{pmatrix} \sqrt{2} & 0 & 0 & 0 \\ 0 & \frac{1}{\sqrt{2}} & 0 & 0 \\ 0 & 0 & 1 & 0 \\ 0 & 0 & 0 & 1 \end{pmatrix}, \quad S_\beta = \begin{pmatrix} 1 & 0 & 0 & 0 \\ 0 & 1 & 0 & 0 \\ 0 & 0 & \sqrt{2} & 0 \\ 0 & 0 & 0 & \frac{1}{\sqrt{2}} \end{pmatrix} \tag{9.2}$$

R. Ukai, *Multi-Step Multi-Input One-Way Quantum Information Processing with Spatial and Temporal Modes of Light*, Springer Theses,
DOI 10.1007/978-4-431-55019-8_9

show the gain-tunable entangling gate (T_Z gate), a 3.0 dB p-squeezing operation on mode α, and a 3.0 dB p-squeezing operation on mode β. Here, $g = \tan\theta$. Therefore, the operation is controlled by the relative phase θ in homodyne detection for mode 2.

9.1.2 Importance of This Experiment

We summarize importance of this experiment.

9.1.2.1 Gaussian Operation on Two-Mode Input State (Common to the Three Experiments)

This is an experimental demonstration of one-way quantum computation where Gaussian operations on two-mode input states are implemented. By combining the experimental demonstration of one-mode Gaussian operations we have reported earlier [1], we can implement an arbitrary multi-mode Gaussian operations in the framework of one-way quantum computation in principle (Sects. 3.6 and 5.5).

This is a common property to the three experiments in this thesis.

9.1.2.2 Nonlocal Gate (Common to the Three Experiments)

This is an experimental demonstration of a nonlocal gate, where target two modes are located at a distance (Sect. 8.3).

This is a common property to the three experiments in this thesis.

9.1.2.3 Gain-Tunable Entangling Gate (Characteristic of this Experiment)

In the input–output relationship, S_α and S_β show fixed quantum state manipulations on single-mode systems. On the other hand, $T_Z(g)$ becomes a single-mode operation only when $g = 0$, while it becomes a quantum state manipulation on two-mode systems when $g \neq 0$. In addition, the larger the $|g|$ is, the larger the interaction gain becomes. Therefore, our experimental setup has the ability of the on–off switching as well as the interaction-gain tuning. The gain $g = \tan\theta$ is determined by the relative phase θ in homodyne detection for mode 2, thus it exhibits the characteristic of one-way quantum computation that operations are determined by measurement bases.

Table 9.1 Properties of gain-tunable gate in a common form

Item	Gain-tunable gate experiment
Main feature	Interaction gain is tunable
Resource	Three-mode linear cluster state
Input coupling scheme	Squeezer-based input-coupling scheme
Excess 3 dB squeezing derived from input coupling	Exists
Operation experimentally demonstrated	$T_Z(-\frac{g}{2})S_\alpha S_\beta$, tunable
Entanglement verification at the output	PT symplectic eigenvalue
Logarithmic negativity	$E_N = 0, \ldots, 0.39 \pm 0.05$

9.1.2.4 Three-Mode Connection Gate (Characteristic of this Experiment)

The core of gain-tunable operation in this experiment is equivalent to that of three-mode connection gate [2]. Thus, our experiment can be considered as an experimental demonstration of the three-mode connection gate.

9.1.3 Comparison of Three Experiments

In order to compare three experiments in this thesis, we summarize the properties of this experiment in a common form. It is given in Table 9.1.

9.2 Theory

9.2.1 Abstract Illustration and Abstract Experimental Setup

Figures 9.1 and 9.2 show an abstract illustration and an abstract setup of the gain-tunable gate experiment.

The abstract illustration shown in Fig. 9.1 gives us a brief explanation of the experimental procedure (see Figs. 7.1 and 7.2).

9.2.2 Three-Mode Linear Cluster State

The procedure for generation of the three-mode linear cluster state is explained in Sect. 6.1.2.

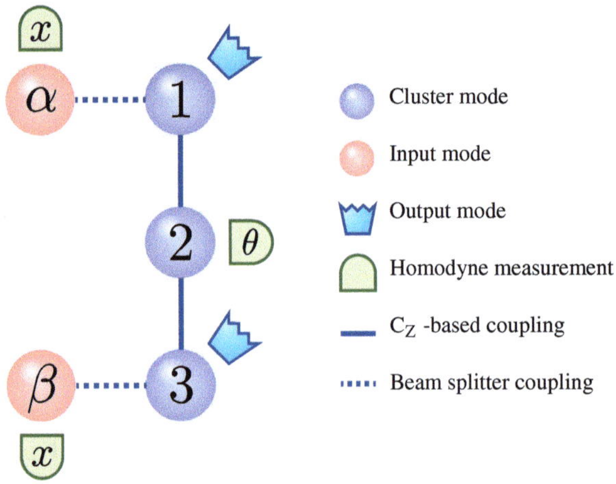

Fig. 9.1 Abstract illustration of gain-tunable gate

9.2.3 Input Coupling

We label two input modes as α and β. These two modes are coupled with the cluster state via the squeezer-based input-coupling schemes (Sect. 5.3.4). In the experimental setup, two beams corresponding to these two input modes are combined using two beam splitters with other two beams which correspond to two of the three cluster modes. Transformations of these beam splitters are given by unitary matrices $B^{(2)}_{\alpha 1}(\frac{1}{\sqrt{2}})$, $B^{(2)}_{\beta 3}(\frac{1}{\sqrt{2}})$, leading to

$$\begin{pmatrix} \hat{x}^{(b)}_\alpha + i\hat{p}^{(b)}_\alpha \\ \hat{x}^{(b)}_1 + i\hat{p}^{(b)}_1 \end{pmatrix} = \frac{1}{\sqrt{2}}\begin{pmatrix} 1 & -1 \\ 1 & 1 \end{pmatrix}\begin{pmatrix} \hat{x}^{(in)}_\alpha + i\hat{p}^{(in)}_\alpha \\ \hat{x}^{(c)}_1 + i\hat{p}^{(c)}_1 \end{pmatrix}, \tag{9.3}$$
$$\begin{pmatrix} \hat{x}^{(b)}_\beta + i\hat{p}^{(b)}_\beta \\ \hat{x}^{(b)}_3 + i\hat{p}^{(b)}_3 \end{pmatrix} = \frac{1}{\sqrt{2}}\begin{pmatrix} 1 & -1 \\ 1 & 1 \end{pmatrix}\begin{pmatrix} \hat{x}^{(in)}_\beta + i\hat{p}^{(in)}_\beta \\ \hat{x}^{(c)}_3 + i\hat{p}^{(c)}_3 \end{pmatrix},$$

where $\hat{x}^{(in)}_k$ and $\hat{p}^{(in)}_k$ represent the quadrature operators of input mode $k(=\alpha, \beta)$. As a result, the quadrature operators after the input couplings are given by

$$\begin{aligned} \hat{x}^{(b)}_\alpha + i\hat{p}^{(b)}_\alpha &= \left[\frac{1}{\sqrt{2}}\hat{x}_\alpha - \frac{1}{2}\hat{x}_1 + \frac{1}{\sqrt{6}}\hat{p}_2 + \frac{1}{2\sqrt{3}}\hat{x}_3\right] \\ &\quad + i\left[\frac{1}{\sqrt{2}}\hat{p}_\alpha - \frac{1}{2}\hat{p}_1 - \frac{1}{\sqrt{6}}\hat{x}_2 + \frac{1}{2\sqrt{3}}\hat{p}_3\right], \\ \hat{x}^{(b)}_1 + i\hat{p}^{(b)}_1 &= \left[\frac{1}{\sqrt{2}}\hat{x}_\alpha + \frac{1}{2}\hat{x}_1 - \frac{1}{\sqrt{6}}\hat{p}_2 - \frac{1}{2\sqrt{3}}\hat{x}_3\right] \end{aligned} \tag{9.4}$$

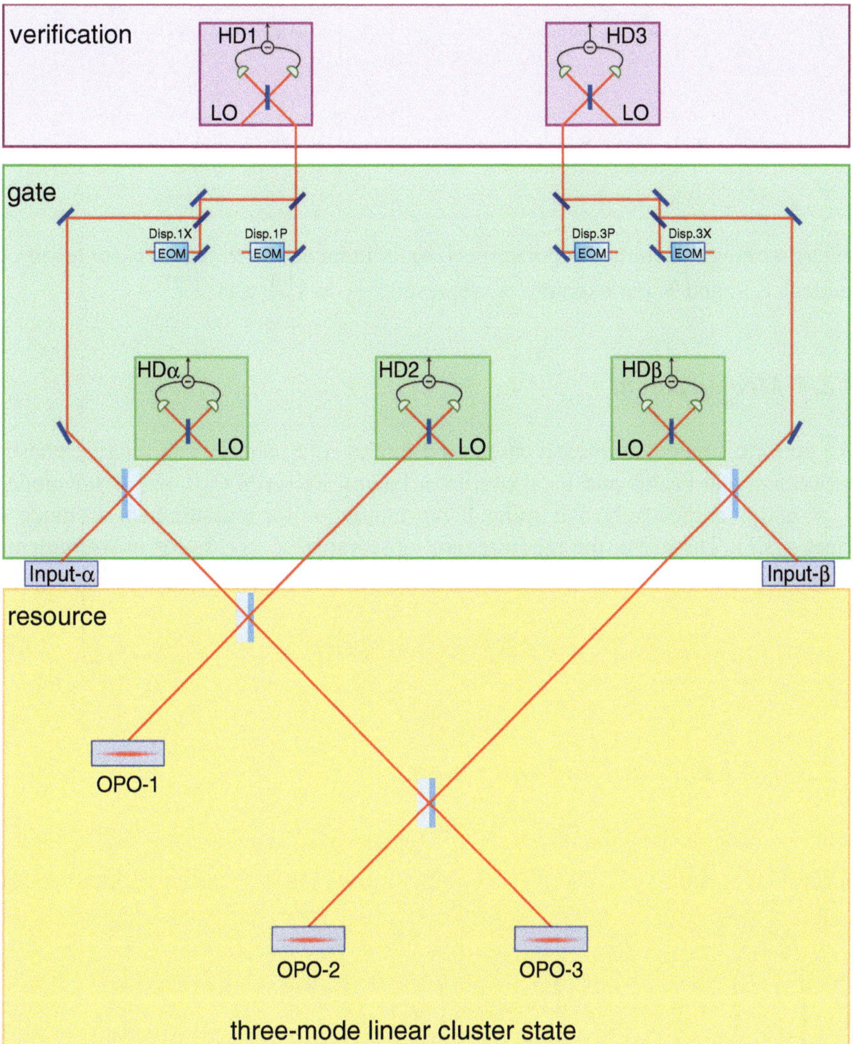

Fig. 9.2 Abstract experimental setup of gain-tunable gate

$$+ i\left[\frac{1}{\sqrt{2}}\hat{p}_\alpha + \frac{1}{2}\hat{p}_1 + \frac{1}{\sqrt{6}}\hat{x}_2 - \frac{1}{2\sqrt{3}}\hat{p}_3\right], \tag{9.5}$$

$$\hat{x}_2^{(c)} + i\,\hat{p}_2^{(c)} = \left[-\frac{1}{\sqrt{2}}\hat{p}_1 + \frac{1}{\sqrt{3}}\hat{x}_2 - \frac{1}{\sqrt{6}}\hat{p}_3\right] + i\left[\frac{1}{\sqrt{2}}\hat{x}_1 + \frac{1}{\sqrt{3}}\hat{p}_2 + \frac{1}{\sqrt{6}}\hat{x}_3\right], \tag{9.6}$$

$$\hat{x}_3^{(b)} + i\hat{p}_3^{(b)} = \left[\frac{1}{\sqrt{2}}\hat{x}_\beta - \frac{1}{\sqrt{6}}\hat{p}_2 + \frac{1}{\sqrt{3}}\hat{x}_3\right] + i\left[\frac{1}{\sqrt{2}}\hat{p}_\beta + \frac{1}{\sqrt{6}}\hat{x}_2 + \frac{1}{\sqrt{3}}\hat{p}_3\right], \tag{9.7}$$

$$\hat{x}_\beta^{(b)} + i\hat{p}_\beta^{(b)} = \left[\frac{1}{\sqrt{2}}\hat{x}_\beta + \frac{1}{\sqrt{6}}\hat{p}_2 - \frac{1}{\sqrt{3}}\hat{x}_3\right] + i\left[\frac{1}{\sqrt{2}}\hat{p}_\beta - \frac{1}{\sqrt{6}}\hat{x}_2 - \frac{1}{\sqrt{3}}\hat{p}_3\right], \tag{9.8}$$

where we have the omitted superscripts (in) for input modes α, β and (r) for resource modes[1] 1, 2, and 3. For example, $\hat{x}_1$ represents $\hat{x}_1 = \hat{x}_1^{(r)} = e^{r_1}\hat{x}_1^{(0)}$.

9.2.4 Measurement

We perform homodyne measurements on modes α, β, and 2. The relative phases between signal beams and local oscillator beams are set to 0, 0, and θ for modes α, β, and 2, respectively. We utilize θ *representation* for measurement on mode 2 (Sect. 5.3.1). Therefore, the measurement observables $\hat{s}_\alpha, \hat{s}_\beta, \hat{s}_2$ for measurements on modes α, β, and 2 are

$$\hat{s}_\alpha = \hat{x}_\alpha^{(b)}, \quad \hat{s}_\beta = \hat{x}_\beta^{(b)}, \quad \hat{s}_2 = \cos\theta\,\hat{x}_2^{(c)} + \sin\theta\,\hat{p}_2^{(c)}. \tag{9.9}$$

9.2.5 Feed-Forward and Operation

Without using the position operators $\hat{x}_k^{(r)}$ for squeezed-state modes $k = 1, 2, 3$, the quadrature operators $\hat{x}_k^{(b)}$ and $\hat{p}_k^{(b)}$ for output modes $k = 1, 3$ can be reformulated as

$$\begin{pmatrix} \hat{x}_1^{(b)} \\ \hat{p}_1^{(b)} \\ \hat{x}_3^{(b)} \\ \hat{p}_3^{(b)} \end{pmatrix} = M_{op} \begin{pmatrix} \hat{x}_\alpha^{(in)} \\ \hat{p}_\alpha^{(in)} \\ \hat{x}_\beta^{(in)} \\ \hat{p}_\beta^{(in)} \end{pmatrix} + M_{disp} \begin{pmatrix} \hat{s}_\alpha \\ \hat{s}_\beta \\ \hat{s}_2 \end{pmatrix} + \begin{pmatrix} 0 \\ \frac{1}{\sqrt{2}}\hat{\delta}_1 - \frac{1}{\sqrt{2}}\hat{\delta}_2\tan\theta \\ 0 \\ \frac{1}{\sqrt{2}}\hat{\delta}_3 - \frac{1}{\sqrt{2}}\hat{\delta}_2\tan\theta \end{pmatrix}, \tag{9.10}$$

where

$$M_{op} = \begin{pmatrix} \sqrt{2} & 0 & 0 & 0 \\ -\frac{\tan\theta}{\sqrt{2}} & \frac{1}{\sqrt{2}} & -\frac{\tan\theta}{\sqrt{2}} & 0 \\ 0 & 0 & \sqrt{2} & 0 \\ -\frac{\tan\theta}{\sqrt{2}} & 0 & -\frac{\tan\theta}{\sqrt{2}} & \frac{1}{\sqrt{2}} \end{pmatrix}, \quad M_{disp} = \begin{pmatrix} -1 & 0 & 0 \\ \tan\theta & \tan\theta & \frac{1}{\sqrt{2}\cos\theta} \\ 0 & -1 & 0 \\ \tan\theta & \tan\theta & \frac{1}{\sqrt{2}\cos\theta} \end{pmatrix}. \tag{9.11}$$

[1] Note that all resource modes are initially in p-squeezed states. This condition simplifies theoretical discussion of one-way quantum computation.

Note that the reformulation above holds before the measurements are carried out. On the contrary, when the measurements on modes α, β, and 2 have been carried out indeed, the quadratures of the output modes become

$$\begin{pmatrix} \hat{x}_1^{(bm)} \\ \hat{p}_1^{(bm)} \\ \hat{x}_3^{(bm)} \\ \hat{p}_3^{(bm)} \end{pmatrix} = M_{op} \begin{pmatrix} \hat{x}_\alpha^{(in)} \\ \hat{p}_\alpha^{(in)} \\ \hat{x}_\beta^{(in)} \\ \hat{p}_\beta^{(in)} \end{pmatrix} + M_{disp} \begin{pmatrix} s_\alpha \\ s_\beta \\ s_2 \end{pmatrix} + \begin{pmatrix} 0 \\ \frac{1}{\sqrt{2}}\hat{\delta}_1 - \frac{1}{\sqrt{2}}\hat{\delta}_2 \tan\theta \\ 0 \\ \frac{1}{\sqrt{2}}\hat{\delta}_3 - \frac{1}{\sqrt{2}}\hat{\delta}_2 \tan\theta \end{pmatrix}, \tag{9.12}$$

where s_α, s_β, and s_2 represent measurement results of measurement variables $\hat{s}_\alpha$, $\hat{s}_\beta$, and $\hat{s}_2$, respectively. $\hat{\delta}_k$ represents the kth nullifier of the resource cluster state (see Eq. (6.42)). By performing feed-forwards (displacement operations) based on the measurement results:

$$\begin{pmatrix} \hat{x}_1^{(out)} \\ \hat{p}_1^{(out)} \\ \hat{x}_3^{(out)} \\ \hat{p}_3^{(out)} \end{pmatrix} = \begin{pmatrix} \hat{x}_1^{(bm)} \\ \hat{p}_1^{(bm)} \\ \hat{x}_3^{(bm)} \\ \hat{p}_3^{(bm)} \end{pmatrix} - M_{disp} \begin{pmatrix} s_\alpha \\ s_\beta \\ s_2 \end{pmatrix}, \tag{9.13}$$

or equivalently, by applying displacement operators:

$$\hat{X}_1(s_\alpha)\hat{Z}_1\left(-\tan\theta s_\alpha - \tan\theta s_\beta - \frac{1}{\sqrt{2}\cos\theta}s_2\right)\hat{X}_3(s_\beta)\hat{Z}_3\left(-\tan\theta s_\alpha - \tan\theta s_\beta - \frac{1}{\sqrt{2}\cos\theta}s_2\right), \tag{9.14}$$

we get

$$\begin{pmatrix} \hat{x}_1^{(out)} \\ \hat{p}_1^{(out)} \\ \hat{x}_3^{(out)} \\ \hat{p}_3^{(out)} \end{pmatrix} = M_{op} \begin{pmatrix} \hat{x}_\alpha^{(in)} \\ \hat{p}_\alpha^{(in)} \\ \hat{x}_\beta^{(in)} \\ \hat{p}_\beta^{(in)} \end{pmatrix} + \begin{pmatrix} 0 \\ \frac{1}{\sqrt{2}}\hat{\delta}_1 - \frac{1}{\sqrt{2}}\hat{\delta}_2 \tan\theta \\ 0 \\ \frac{1}{\sqrt{2}}\hat{\delta}_3 - \frac{1}{\sqrt{2}}\hat{\delta}_2 \tan\theta \end{pmatrix}. \tag{9.15}$$

This is the input–output relationship in this experiment. The term of M_{op} represents the main operation. On the other hand, the term of $\hat{\delta}_k$ represents error of the operation, which derives from finite level of resource squeezing. In the limit of infinite squeezing $r \to \infty$, each $\hat{\delta}_k$ goes to zero, meaning that the ideal operation is achieved.

9.2.6 Decomposition of Operation

The matrix M_{op} can be decomposed into

$$M_{op} = T_Z\left(-\frac{1}{2}\eta\right)\begin{pmatrix} S & O \\ O & S \end{pmatrix} = C_Z\left(-\frac{1}{2}\eta\right)\begin{pmatrix} D(-\frac{1}{2}\eta) & O \\ O & D(-\frac{1}{2}\eta) \end{pmatrix}\begin{pmatrix} S & O \\ O & S \end{pmatrix}, \quad \eta = \tan\theta, \tag{9.16}$$

where

$$T_Z(g) = \begin{pmatrix} 1 & 0 & 0 & 0 \\ g & 1 & g & 0 \\ 0 & 0 & 1 & 0 \\ g & 0 & g & 1 \end{pmatrix}, \quad C_Z(g) = \begin{pmatrix} 1 & 0 & 0 & 0 \\ 0 & 1 & g & 0 \\ 0 & 0 & 1 & 0 \\ g & 0 & 0 & 1 \end{pmatrix}, \quad D(g) = \begin{pmatrix} 1 & 0 \\ g & 1 \end{pmatrix}, \quad S = \begin{pmatrix} \sqrt{2} & 0 \\ 0 & \frac{1}{\sqrt{2}} \end{pmatrix}. \tag{9.17}$$

Here, S is the matrix of a 3.0 dB p-squeezing operation, while $D(g)$ is the matrix of a quadratic phase gate. They are single-mode operations. The matrix $C_Z(g)$ is a controlled-Z gate with interaction gain g, while $T_Z(g)$ is the matrix of the gain-tunable gate (T_Z gate) with interaction gain g. They are two-mode operations.

9.2.7 Squeezer-Based Input Coupling and Gaussian Parallelism

As is clearly shown in Eq. (9.16), the operation can be decomposed into two sub-operations: single-mode operations S derived from the squeezer-based input coupling schemes (Sect. 5.3.4) and a two-mode operation $T_Z(g)$ derived from the measurement on mode 2 of the three-mode linear cluster state. In this subsection, we show interpretation of this experiment as a demonstration of one-way quantum computation which starts with the squeezer-based input-coupling scheme.

The initial states are a three-mode linear cluster state labeled by 1, 2, and 3, and a two-mode input state labeled by α and β (Figs. 9.3a and 9.4). Although we have assumed that the cluster state is a canonical cluster state (or the ideal cluster state; they are generated by combining three p-squeezed states or $|p = 0\rangle$ using controlled-Z gates) in Fig. 9.4 for simplicity, the following discussion can be directly applied to the case of a Gaussian cluster state (Sect. 5.6).

We perform two instances of squeezer-based input-coupling schemes (Sect. 5.3.4). Here, a squeezer-based input-coupling is described in Fig. 9.6a.

The input mode α is coupled with the cluster mode 1 with a beam splitter $B_{\alpha 1}^{(2)}(\frac{1}{\sqrt{2}})$ (Figs. 9.3b and 9.5). Next, the mode α is measured by a homodyne detection with a measurement variable of x. We then perform a feed-forward operation $\hat{X}_1(s)$ on mode 1 with the amplitude gain of 1 and another feed-forward operation $\hat{Z}_2(\sqrt{2}s)$ on mode 2 with the amplitude gain of $\sqrt{2}$, where s represents the measurement result of the homodyne detection.

Figures 9.3c and 9.7 show the abstract illustration and the quantum circuit after the input couplings. Note that the input coupling of mode β is also performed by using the cluster mode 3.

The output quantum state of the input coupling circuit shown in Fig. 9.6a is equivalent to the output state of the circuit in Fig. 9.6b, where a 3.0 dB p-squeezing operation (shown by S) is first performed on mode α, and then a controlled-Z gate with the

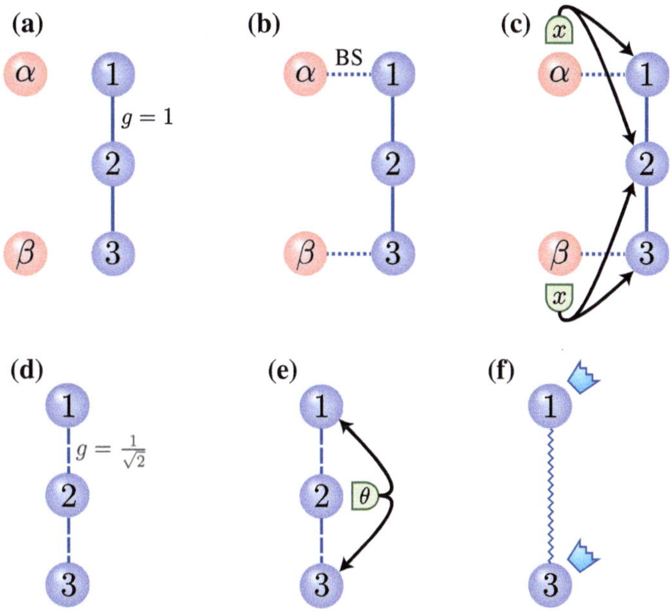

Fig. 9.3 Interpretation as demonstration of quantum gate with squeezer-based input-coupling. **a** Initial state. **b** Input coupling. **c** Measurement and feed-forward. **d** Intermediate state. **e** Measurement and feed-forward. **f** Output state

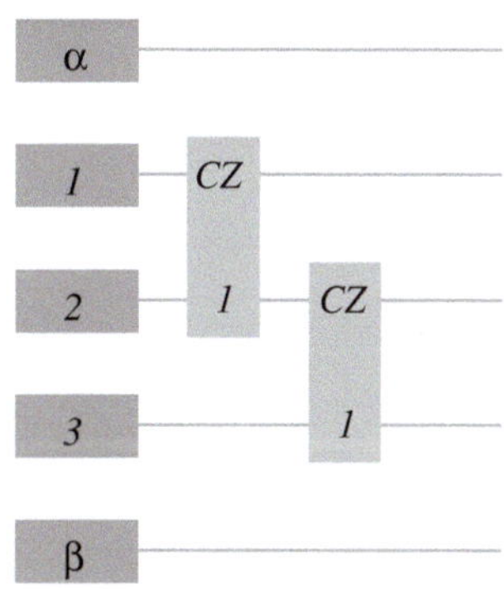

Fig. 9.4 Interpretation as demonstration of quantum gate with squeezer-based input-coupling 1

gain of $g = \frac{1}{\sqrt{2}}$ is applied to modes α and 2. Thus, the result of Fig. 9.7 is equivalent to that of Fig. 9.8 (Fig. 9.3d).

We here consider a quantum circuit shown in Fig. 9.9.

In Fig. 9.9, modes A and B show input modes, while mode 2 shows a resource mode in $|p = 0\rangle$, or a p-squeezed state. First, two pairs of modes $(A, 2)$ and $(B, 2)$ are coupled with controlled-Z gates with the gains of g. Next, a homodyne measurement on mode 2 with the relative phase t is performed, leading to a measurement result s. Finally, feed-forward operations $\hat{Z}_A(-sg \sec t)\hat{Z}_B(-sg \sec t)$ is carried out with the measurement result s. The output modes are modes A and B. The input–output

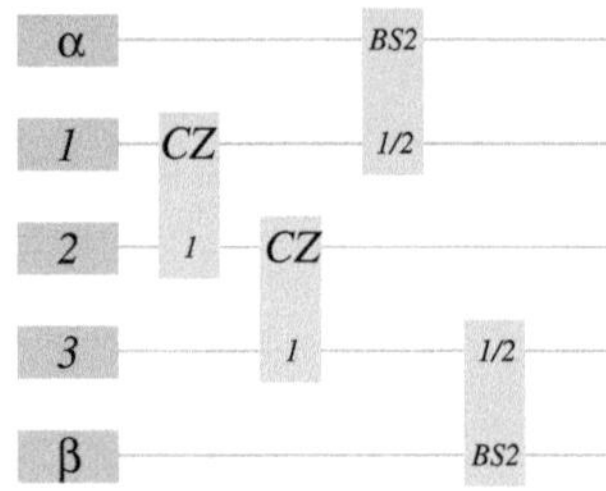

Fig. 9.5 Interpretation as demonstration of quantum gate with squeezer-based input-coupling 2

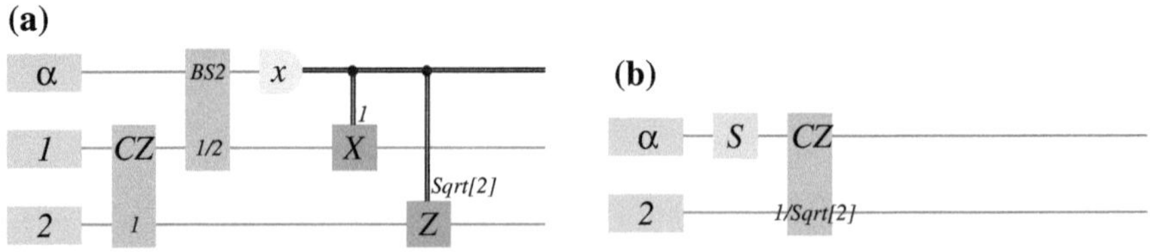

Fig. 9.6 **a** Quantum circuit of squeezer-based input-coupling. **b** Equivalent quantum circuit of squeezer-based input-coupling

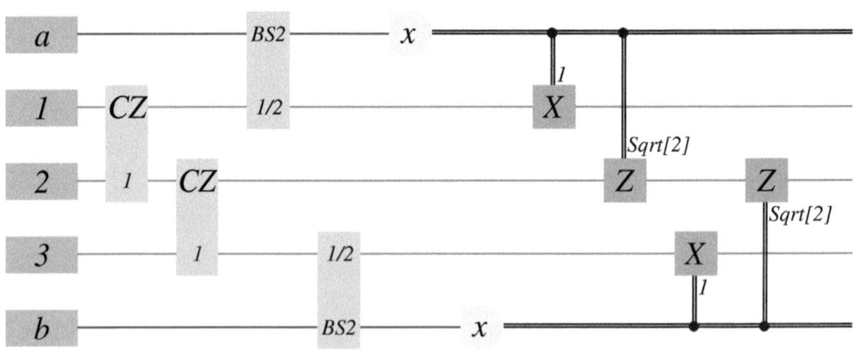

Fig. 9.7 Interpretation as demonstration of quantum gate with squeezer-based input-coupling 3

relationship in the Heisenberg picture is given by

$$\begin{pmatrix} \hat{x}'_A \\ \hat{p}'_A \\ \hat{x}'_B \\ \hat{p}'_B \end{pmatrix} = T_Z(-g^2 \tan t) \begin{pmatrix} \hat{x}_A \\ \hat{p}_A \\ \hat{x}_B \\ \hat{p}_B \end{pmatrix}, \tag{9.18}$$

which is equivalent to that of the gain-tunable entangling gate (T_Z gate).

By applying the operation shown in Fig. 9.9 on the output state of Fig. 9.7 with $g = \frac{1}{\sqrt{2}}$ and $t = \theta$, we get Figs. 9.3e, f and 9.10.

In Fig. 9.10, the measurement on mode 2 is carried out *after* the feed-forward operations on mode 2. These measurement and feed-forward operations can be swapped

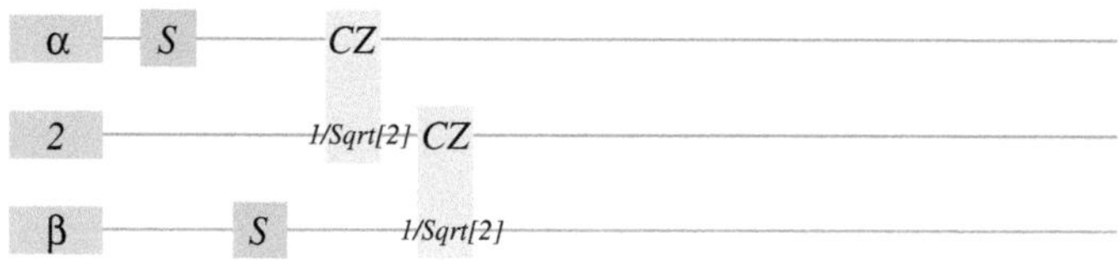

Fig. 9.8 Interpretation as demonstration of quantum gate with squeezer-based input-coupling 4

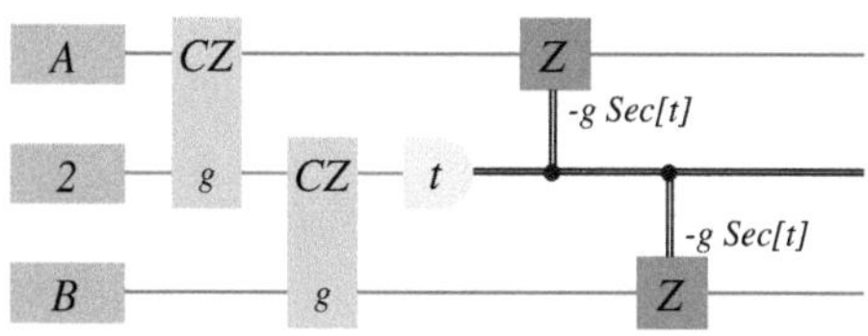

Fig. 9.9 T_Z gate

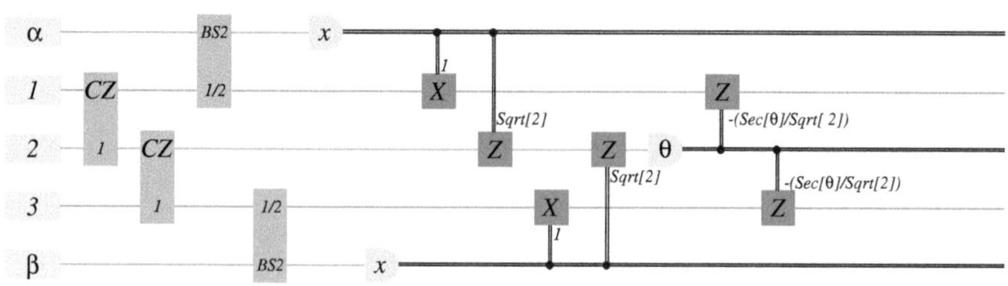

Fig. 9.10 Interpretation as demonstration of quantum gate with squeezer-based input-coupling 5

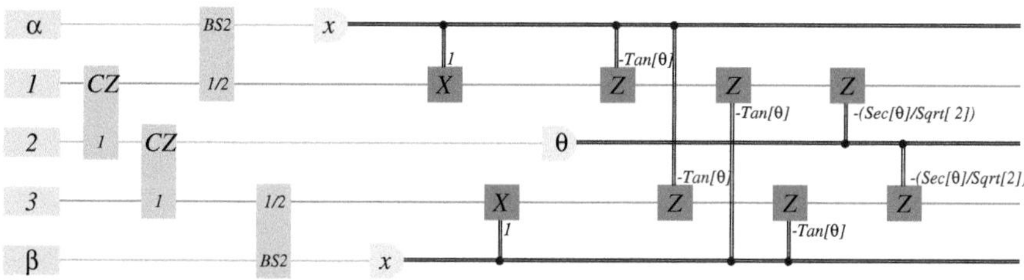

Fig. 9.11 Interpretation as demonstration of quantum gate with squeezer-based input-coupling 6

by changing feed-forward operations on modes 1 and 3 (Fig. 9.11). Note that the measurement on mode 2 is not changed.

Figure 9.11 is nothing but the quantum circuit of our gain-tunable entangling gate experiment. The operation achieved by Fig. 9.11 is given by

$$T_Z\left(-g^2 \tan t\right)\begin{pmatrix} S & O \\ O & S \end{pmatrix} = T_Z\left(-\frac{1}{2}\eta\right)\begin{pmatrix} S & O \\ O & S \end{pmatrix}, \quad \eta = \tan\theta, \tag{9.19}$$

which is identical to Eq. (9.16).

In Fig. 9.10, the measurement on mode 2 is carried out *after* both previous measurements and feed-forwards are finished. On the contrary, Fig. 9.11 shows that all three measurements can be carried out *simultaneously*, and then all feed-forward operations can be performed *simultaneously*. In the transformation of the quantum circuit, all measurement variables are not changed, although feed-forward operations are changed to other Pauli group operations (displacement operations). Therefore, the transformation above can be considered as an example of Gaussian Parallelism (Sect. 4.2.4).

9.2.8 Covariance Matrix

We define $V_{\alpha\beta}^{(in)}$ as the covariance matrix of a two-mode input state. We here choose the *xpxp notation* for all covariance matrices (Sect. 3.2.5).

Since the two-mode input state and the resource three-mode linear cluster state has no correlations, the covariance matrix of the output state becomes

$$V_{13}^{(out)} = M_{op} V_{\alpha\beta}^{(in)} M_{op}^{T} + V_{\delta}, \tag{9.20}$$

where V_{δ} represents excess noise derived from the finite level of resource squeezing. It is given by

$$V_{\delta} = \frac{\hbar}{2} \begin{pmatrix} 0 & 0 & 0 & 0 \\ 0 & e^{-2r_1} + \frac{3}{2}\eta^2 e^{-2r_2} & 0 & \frac{1}{2}e^{-2r_1} + \frac{3}{2}\eta^2 e^{-2r_2} \\ 0 & 0 & 0 & 0 \\ 0 & \frac{1}{2}e^{-2r_1} + \frac{3}{2}\eta^2 e^{-2r_2} & 0 & \frac{1}{4}e^{-2r_1} + \frac{3}{2}\eta^2 e^{-2r_2} + \frac{3}{4}e^{-2r_3} \end{pmatrix}. \tag{9.21}$$

In the special case of identical squeezing parameters $r_i = r$, it becomes

$$V_{\delta} = \frac{\hbar}{2} \begin{pmatrix} 0 & 0 & 0 & 0 \\ 0 & \left(1 + \frac{3}{2}\eta^2\right) e^{-2r} & 0 & \left(\frac{1}{2} + \frac{3}{2}\eta^2\right) e^{-2r} \\ 0 & 0 & 0 & 0 \\ 0 & \left(\frac{1}{2} + \frac{3}{2}\eta^2\right) e^{-2r} & 0 & \left(1 + \frac{3}{2}\eta^2\right) e^{-2r} \end{pmatrix}. \tag{9.22}$$

In the limit of infinite squeezing $r_i \to \infty$, V_{δ} goes to the 4×4 zero matrix O, meaning that the ideal operation is achieved.

Finally, we assume that two input modes α and β are initially in vacuum states. The covariance matrix of the output state becomes

$$V_{13}^{(out)} = \frac{\hbar}{2}\begin{pmatrix} 2 & -\eta & 0 & -\eta \\ -\eta & \eta^2+\frac{1}{2} & -\eta & \eta^2 \\ 0 & -\eta & 2 & -\eta \\ -\eta & \eta^2 & -\eta & \eta^2+\frac{1}{2} \end{pmatrix} + V_\delta. \tag{9.23}$$

9.2.9 Inseparability Criteria

In this experimental demonstration of gain-tunable entangling gate, we verify entanglement at the output by using the inseparable criterion with PT symplectic eigenvalue (Sect. 3.7.5).

9.2.9.1 van Loock-Furusawa Criterion, Duan Criterion

van Loock-Furusawa criterion and Duan criterion (Sect. 3.7.4) are not suitable for verification of entanglement at the output in this experiment. In the following, we briefly mention the reason of this.

From the covariance matrix of the output state shown in Eq. (9.23), we find that its $(1, 2)$ and $(3, 4)$ entries have non-zero values when $\eta \neq 0$ $(\theta \neq 0)$. Thus each position operator $\hat{x}_k$ of mode k has correlation with the momentum operator $\hat{p}_k$ of the same mode k. This inner correlation derives from the quadratic phage gates $D(g)$ in Eq. (9.16). Note that this is not the case with the other experiments in this thesis (Chaps. 7 and 8).

Consider entanglement detection of a two-mode quantum state with correlations within single modes. In order to acquire the appropriate Duan criterion for this state, transformations of its covariance matrix into a standard form must be performed. As a result, the measurement variables have both position and momentum operators of each mode. Although such measurements can be achieved by controlling relative phases θ so that $\theta \neq 0, 90$, it leads to complication of experimental setup. In addition, the appropriate measurement angles depend on the level of resource squeezing. The situation is the same to van Loock-Furusawa criterion. Thus, van Loock-Furusawa criterion and Duan criterion are not suitable in this experiment.

9.2.9.2 Simon Criterion, PT Symplectic Eigenvalue Criterion

We can get the entanglement criterion for our experiment by using the covariance matrix of the output state shown in Eq. (9.23). In the following, we show the entanglement criterion with its PT symplectic eigenvalue, which is equivalent to the well-known Simon criterion.

We assume that all resource squeezed states have identical squeezing parameter r. The one of two PT symplectic eigenvalues is given by

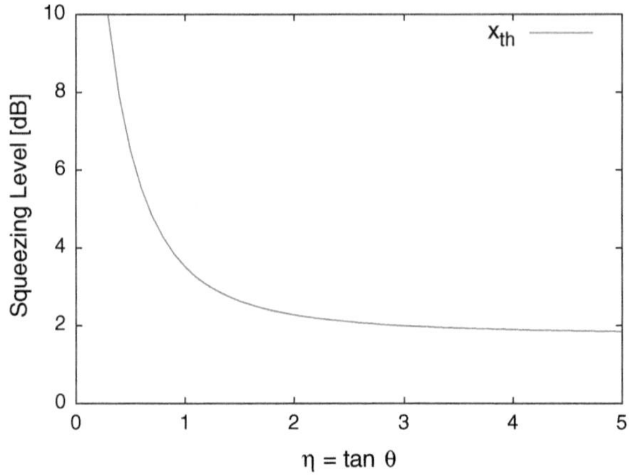

Fig. 9.12 Threshold squeezing level x_{th} (dB) of entanglement for operation with $\eta = \tan\theta$

$$\tilde{\nu}_- = \frac{\hbar}{2}\sqrt{2\eta^2 + 1 + (3\eta^2 + 2)e^{-2r} - \sqrt{(2 + 3e^{-2r})^2\eta^4 + (6e^{-4r} + 8e^{-2r} + 4)\eta^2 + e^{-4r}}}. \tag{9.24}$$

The inseparability criterion is given by $\tilde{\nu}_- < \frac{\hbar}{2}$, leading to

$$\eta \neq 0 \text{ and } \left(10^{\frac{x}{10}}\right)^2 = e^{4r} > \frac{6\eta^2 + 3}{4\eta^2}. \tag{9.25}$$

In the case of $\eta = 0$, since the input–output relationship given by Eq. (9.16) shows single-mode operations, the output state is always separable. In the ideal case of $r \to \infty$, the output state becomes inseparable when $\eta \neq 0$. In the realistic case of finite-level squeezing, the output state might not be entangled even when $\eta \neq 0$. This is because the covariance matrix of the output state has the additional term V_δ which shows error in one-way quantum computation.

Consider an operation with η. The output state becomes an entangled state if the effective squeezing level $x^{(e)}$ (Sect. 6.1.1) satisfies

$$x^{(e)} > x_{th}, \quad x_{th} = 10\log_{10}\sqrt{\frac{6\eta^2 + 3}{4\eta^2}}. \tag{9.26}$$

9.3 Experimental Setup

9.3.1 From Laser to Cluster State Generation

The resource for this experiment is a three-mode linear cluster state. In order to generate it, we utilize the same experimental setup to that in Chap. 6.

9.3.2 Input States

Input states for experimental demonstration of the gain-tunable entangling gate are vacuum states and coherent states. We utilize the same experimental setup to that in Sect. 7.3.2.

9.3.3 Classical Channels and Displacement Operations

The displacement operation (feed-forward operation) in this experiment is given by Eq. (9.13), or equivalently, Eq. (9.14):

$$\hat{X}_1(s_\alpha)\hat{Z}_1\left(-\tan\theta s_\alpha - \tan\theta s_\beta - \frac{1}{\sqrt{2}\cos\theta}s_2\right)\hat{X}_3(s_\beta)\hat{Z}_3\left(-\tan\theta s_\alpha - \tan\theta s_\beta - \frac{1}{\sqrt{2}\cos\theta}s_2\right). \tag{9.27}$$

Therefore, we need eight paths of classical channels $(\alpha, 1x)$, $(\alpha, 1p)$, $(\beta, 1p)$, $(2, 1p)$, $(\beta, 3x)$, $(\alpha, 3p)$, $(\beta, 3p)$, and $(2, 3p)$, where parentheses show (mode of homodyne detection, destination of feed-forward).

The schematic of classical channels for the gain-tunable entangling gate experiment is shown in Fig. 9.13.

9.3.4 Locking the Relative Phases

Figure 9.14 shows arrangement of probe beams and phase modulations for the gain-tunable entangling gate experiment.

9.3.4.1 Oblique Locking

The measurement variable in mode 2 is given by $\hat{x}\cos\theta + \hat{p}\sin\theta$, where θ is determined by the gain of the operation $\eta = \tan\theta$. It is achieved by a homodyne detection with the relative phase θ. Since the local oscillator beam points to an angle other than 0° or 90° in phase space, we name the locking of this relative phase as *oblique locking*.

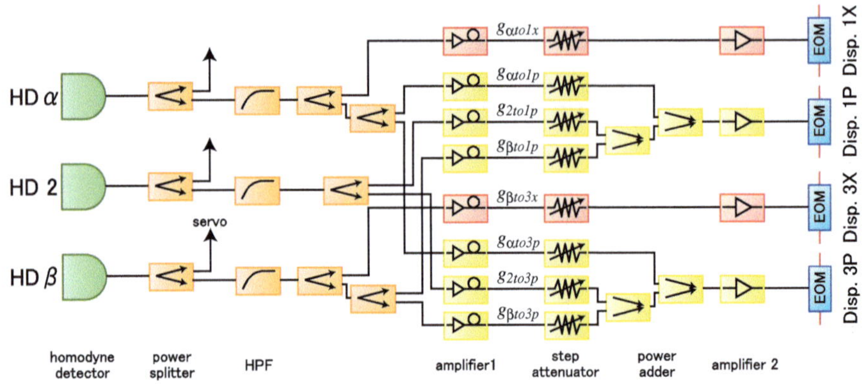

Fig. 9.13 Classical channels for gain-tunable entangling gate experiment

In the homodyne detection in mode 2, we can adjust the relative phase θ between the signal beam and the local oscillator beam so that $\theta = 0$, by using demodulation signal of 210 kHz (Fig. 9.14 and Table 9.2). Similarly, θ becomes $\theta = 90$ by using demodulation signal of 140 kHz. We assume that error signals for these lockings are given by $V_x \sin\phi$ and $-V_p \cos\phi$, and that the phase is locked to the angle ϕ_0 with $V_e(\phi_0) = 0$ and $\left.\frac{\mathrm{d}V_e(\phi)}{\mathrm{d}\phi}\right|_{\phi=\phi_0} > 0$, where $V_e(\phi)$ shows an error signal. By adding these two error signals, we get a new signal:

$$V_x \sin\phi - V_p \cos\phi = \sqrt{V_x^2 + V_p^2}\sin(\phi - \theta), \quad \theta = \tan^{-1}\frac{V_p}{V_x}. \tag{9.28}$$

Thus, oblique locking with an arbitrary θ can be achieved by using it as the error signal, where the ratio between V_x and V_p is appropriately adjusted (Fig. 9.15).

9.3.4.2 Summary of Phase Locking

Table 9.2 shows phase locking techniques for the gain-tunable entangling gate experiment. Parentheses in Table 9.2 show that their phases are not utilized in experiment.

9.3.5 Cancellation

In the following, we take two examples of adjustment of classical channels.

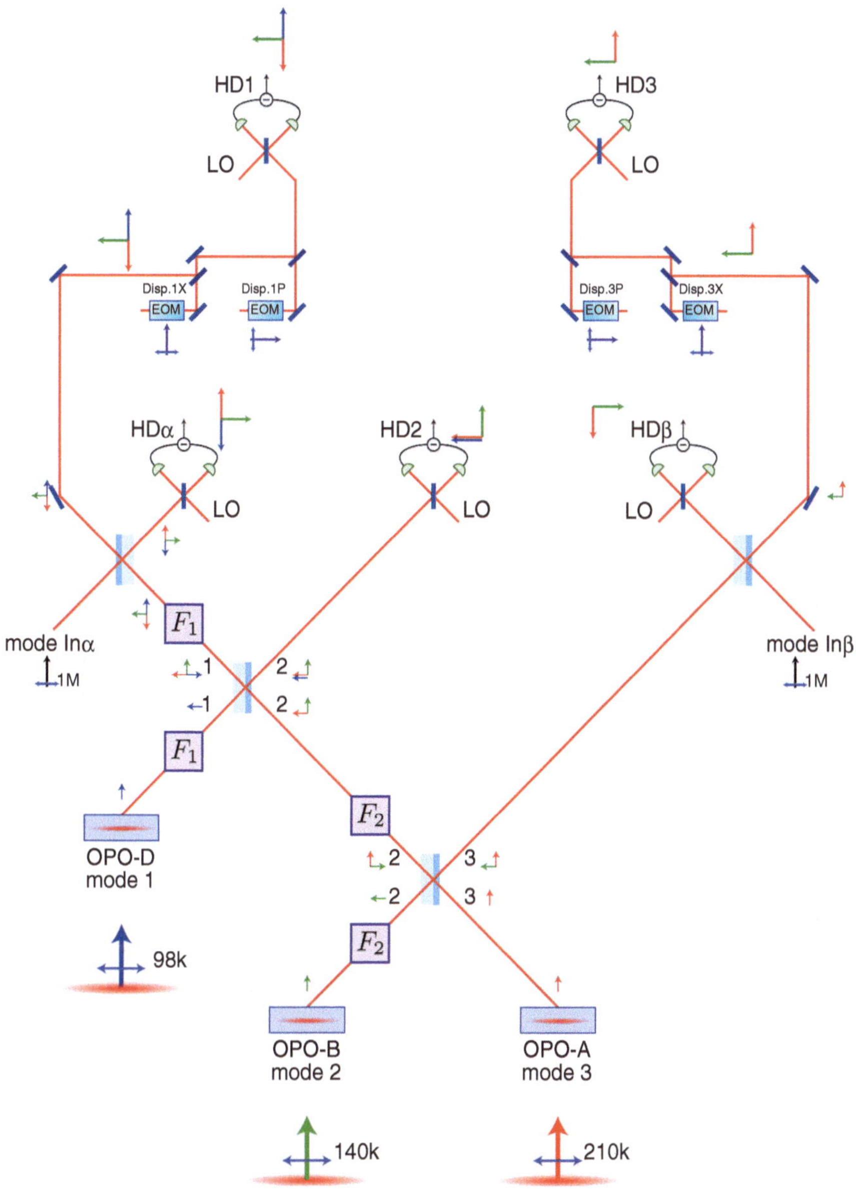

Fig. 9.14 Probe beams and phase modulations for gain-tunable entangling gate experiment

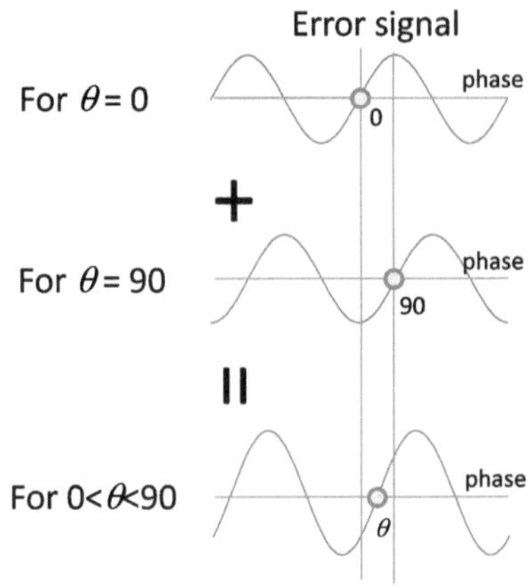

Fig. 9.15 Oblique locking

Table 9.2 Phase lockings for gain-tunable entangling gate

Relative phase	Technique, demodulation frequency
Parametric gain of OPO-A	AC locking, 210 kHz
Parametric gain of OPO-B	AC locking, 140 kHz
Parametric gain of OPO-D	AC locking, 98 kHz
Beam splitter $B_{23}^{(1)}(\sqrt{\frac{2}{3}})$	DC locking
Beam splitter $B_{12}^{(1)}(\frac{1}{\sqrt{2}})$	AC locking, 210 kHz
Input α	AC locking, amplitude x: 210 kHz, amplitude p: 140 kHz
Input β	AC locking, amplitude x: 210 kHz, amplitude p: 140 kHz
Displacement in mode 1	AC locking, displacement x: 210 kHz, displacement p: 140 kHz
Displacement in mode 3	AC locking, displacement x: 210 kHz, displacement p: 140 kHz
Homodyne measurement in mode α	AC locking, x: 140 kHz, (p: 210 kHz)
Homodyne measurement in mode 1	AC locking, x: 140 kHz, p: 210 kHz
Homodyne measurement in mode 2	Oblique (AC) locking, x: 210 kHz, p: 140 kHz
Homodyne measurement in mode 3	AC locking, x: 140 kHz, p: 210 kHz
Homodyne measurement in mode β	AC locking, x: 140 kHz, (p: 210 kHz)

9.3.5.1 Adjustment of g_{2to1p}

We define g_{2to1p} as the gain of classical channel from homodyne 2 to the momentum operator of mode 1.

We apply phase modulation onto the probe beam of OPO-B, leading to a coherent state with amplitude in $\hat{x}_2$. We measure x in the homodyne measurement on mode 2. Its measurement result is added to p_1 with the gain of g_{2to1p}. With measuring the power of $\hat{p}_1$, we adjust the gain and phase of the classical channel so that the coherent signal is not observed. Since

$$\frac{1}{\sqrt{6}}\hat{x}_2 + g_{2to1p}\left[\frac{1}{\sqrt{3}}\hat{x}_2\right] = 0 \Longleftrightarrow g_{2to1p} = -\frac{1}{\sqrt{2}}, \tag{9.29}$$

g_{2to1p} can be adjusted to $g_{2to1p} = -\frac{1}{\sqrt{2}}$.

9.3.5.2 Adjustment of $g_{\alpha to1p}$

We define $g_{\alpha to1p}$ as the gain of the classical channel from homodyne α to the momentum operator of mode 1. The gain $g_{\alpha to1p}$ is indirectly adjusted by using g_{2to1p}.

We apply phase modulation onto the probe beam of OPO-A, leading to a coherent state with amplitude in $\hat{x}_3$. Modes α and 2 are measured by homodyne detections with the measurement variables $\hat{x}_\alpha$ and $\hat{p}_2$. Their measurement results are added electrically. With measuring the power of the added signal, we adjust the gain and phase of the classical channel so that the coherent signal is not observed. Since

$$g_{2to1p}\left[\frac{1}{\sqrt{6}}\hat{x}_3\right] + g_{\alpha to1p}\left[\frac{1}{2\sqrt{3}}\hat{x}_3\right] = 0 \Longleftrightarrow g_{\alpha to1p} = -\sqrt{2}g_{2to1p} = 1, \tag{9.30}$$

$g_{\alpha to1p}$ can be adjusted to $g_{\alpha to1p} = 1$.

9.3.5.3 Adjustment to the Proper Gains

The gains of classical channels are now adjusted to

$$g_{\alpha to1p} = 1, \quad g_{2to1p} = -\frac{1}{\sqrt{2}}. \tag{9.31}$$

On the other hand, these gains should be

$$g_{\alpha to1p} = -\tan\theta, \quad g_{2to1p} = -\frac{1}{\sqrt{2}\cos\theta}, \tag{9.32}$$

where $\eta = \tan\theta$. Thus, $g_{\alpha to1p}$ and g_{2to1p} should be amplified with amplitude gains of $-\tan\theta$ and $\frac{1}{\cos\theta}$, respectively. Therefore, we have to adjust these gains in the following way:

- $g_{\alpha to 1p}$: change the polarity switch in *Amplifier 1*, and then increase the power attenuation by $-20 \log \tan \theta$ dB.
- g_{2to1p}: increase the power attenuation by $-20 \log \frac{1}{\cos \theta}$ dB.

Gains and phases of the other classical channels are also adjusted in a similar manner.

9.4 Digital Locking System

9.4.1 Background

As is mentioned in Sect. 6.4, locking of relative phase is achieved by using the *DC locking* system or the *AC locking* system. Each feed-back loop contains a servo controller, called the *DC servo* or *AC servo*. In all traditional experiments in our laboratory, including the experimental demonstrations of the controlled-Z gate and the optimum nonlocal gate, servo controllers are implemented by using analog circuits. For example, the controlled-Z gate experiment involves 25 independent servo controllers to lock relative phases.

Consider a case where a relative-phase locking is dropped because, for example, the length of a piezo electric transducer (PZT) becomes out of its specifications. Although feed-back loops work independently of each other, optical beams of feed-back target are in common. Thus, when a relative-phase locking is dropped, other phase locking loops may be disturbed. As a result, we have to relock all relative phases where common optical beams are involved.

A possible solution for this is to use an extensibility of a digital locking system. In this system, an electrical signal from a detector is first digitized with an analog to digital converter (ADC). The digitized signal is then processed by a digital filter. Finally, a feed-back signal is transmitted to a PZT after the filtered signal is converted to an analog signal with a digital to analog converter (DAC). Cooperation of all feed-back servos can be implemented by a computer program.

Including the automatic relocking, we can consider that a digital locking system has the following advantages:

- Automatic relocking system can be easily implemented.
- Sophisticated filter can be easily encoded.
- Cooperation with sample-and-hold system can be easily achieved.
- Cooperation with measurement apparatuses can be easily achieved.
- Control system can be easily updated.
- Control system can be easily duplicated.

Table 9.3 Interference visibilities for gain-tunable entangling gate experiment

Beam splitter	Beams used for adjustment	Visibility (%)
Beam splitter $B_{23}^{(1)}(\sqrt{\frac{2}{3}})$	OPO-A, OPO-B	98.1
Beam splitter $B_{12}^{(1)}(\frac{1}{\sqrt{2}})$	OPO-B, OPO-D	97.9
Input coupling α	OPO-B, In-α	97.4
Input coupling β	OPO-A, In-β	96.9
Displacement $1x$	OPO-B, Disp-1X	97.2
Displacement $1p$	OPO-B, Disp-1P	98.0
Displacement $3x$	OPO-A, Disp-3X	97.9
Displacement $3p$	OPO-A, Disp-3P	98.1
Homodyne measurement α	OPO-B, LO-α	97.2
Homodyne measurement 1	OPO-B, LO-1	96.1
Homodyne measurement 2	OPO-B, LO-2	98.1
Homodyne measurement 3	OPO-A, LO-3	97.7
Homodyne measurement β	OPO-A, LO-β	95.9

9.5 Digital Locking System in This Experiment

In the experimental demonstration of the gain-tunable entangling gate, we utilize a digital locking system [3].

The chassis is NI PXI-1033 (National Instruments), which can be controlled by a personal computer. We utilize five PXI modules: NI PXI-7853R $\times$ 4 and NI PXI-7854R $\times$ 1 (National Instruments). Each module has eight analog input ports (up to 750 kHz independent sampling), eight analog output ports (up to 1 MHz independent update), and 96 digital input/output ports (up to 40 MHz). Feed-back servos and their controllers are programed on Field Programmable Gate Array (FPGA) modules by using NI LabVIEW (Laboratory Virtual Instrumentation Engineering Workbench).

For more details, see Ref. [3].

9.6 Preparation for Measurement

9.6.1 Visibilities and Parametric Gains

Interference visibilities and parametric gains are shown in Tables 9.3 and 9.4, respectively.

Table 9.4 Parametric gains for gain-tunable entangling gate experiment

OPO name	Parametric gain (G_+)	Phase matching temperature (°)
OPO-A	5.0	41.3
OPO-B	5.9	40.0
OPO-D	5.8	41.0

Table 9.5 Adjustment of purity for gain-tunable entangling gate experiment

EOM	Purity (dB)
Disp-1X	32.5
Disp-1P	31.2
Disp-3X	32.8
Disp-3P	28.3

Table 9.6 Adjustment of classical channels (cancellation) for gain-tunable entangling gate experiment

Feed-forward	Cancellation level (dB)
$g_{\alpha to1x}$	26.5
$g_{\alpha to1p}$	34.8
$g_{\beta to1p}$	38.8
g_{2to1p}	26.3
$g_{\beta to3x}$	32.4
$g_{\alpha to3p}$	34.5
$g_{\beta to3p}$	39.3
g_{2to3p}	27.6

9.6.2 Adjustment of EOM (Purity)

Each purity is listed in Table 9.5.

9.6.3 Cancellation

Each cancellation level is listed in Table 9.6.

9.6.4 Balance Between Homodyne Detections

The balance between two homodyne detections for modes 1 and 3 is adjusted in the same way as Chap. 6. As a result, we have achieved cancellation of 33.6 dB.

Table 9.7 Measurement results of input states

HD name	Cluster only (dB)	Coherent+Cluster (dB)
HDα	1.94	11.11
HD1	1.80	11.56
HDβ	5.46	14.80
HD3	5.37	14.20

9.7 Measurement Results and Their Analysis

In this section, we show measurement results of the gain-tunable entangling gate experiment. We choose the following entangling gains η:

$$\eta = \tan\theta = 0, \frac{1}{5}, \frac{1}{2}, \frac{1}{\sqrt{2}}, 1, \sqrt{2}, 2, \tag{9.33}$$

which can be achieved by relative phases θ:

$$\theta = 0.0°, 11.3°, 26.6°, 35.3°, 45.0°, 54.7°, 63.4°, \tag{9.34}$$

in homodyne detection for mode 2.

9.7.1 Measurement Results with Vacuum and Coherent Inputs

9.7.1.1 Measurement Results of Input States

We measure the powers of both input coherent states in modes α and β (Fig. 9.16). For this purpose, we utilize the same scheme to Sect. 7.5.2, where two homodyne detections are engaged in measurement for each input mode.

Each power is listed in Table 9.7.

As a result, we get 13.83 dB for mode α and 16.95 dB for mode β, respectively, compared to the SNL. All errors are less than ± 0.02 dB, including in Table 9.7.

9.7.1.2 Measurement Result of Output States (Values)

In Tables 9.8, 9.9, 9.10, 9.11, 9.12, 9.13 and 9.14, we show measurement results of powers of the output-mode quadratures with vacuum or coherent input states. We have a vacuum input case and four cases of coherent input states: α_x, α_p, β_x, and β_p. For measurements of the output states, we have four measurement variables: x_2, p_2, x_3, and p_3. We choose seven operations with interaction gains $\eta = 0, \frac{1}{5}, \frac{1}{2}, \frac{1}{\sqrt{2}}, 1, \sqrt{2}$, and 2. Thus, we have $5 \times 4 \times 7 = 140$ measurement results in total.

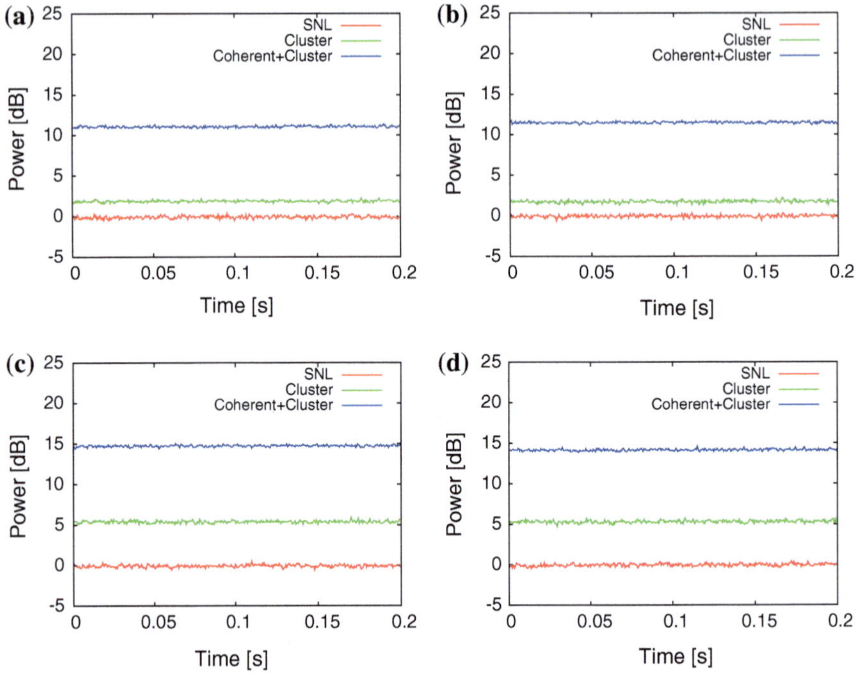

Fig. 9.16 Measurement results of input states. **a** Input α with HD α. **b** Input α with HD 1. **c** Input β with HD β. **d** Input β with HD 3

Table 9.8 Measurement results (M) and theoretical values (T) with $\eta = 0$ (dB, error bars of measurement results are less than 0.06 dB)

Input	Output							
	$\hat{x}_1$(M)	$\hat{p}_1$(M)	$\hat{x}_3$(M)	$\hat{p}_3$(M)	$\hat{x}_1$(T)	$\hat{p}_1$(T)	$\hat{x}_3$(T)	$\hat{p}_3$(T)
Vacuum	2.87	−0.61	2.63	−1.04	3.01	−0.68	3.01	−0.68
Coherent x_α	17.44	0.15	3.10	−0.73	17.02	−0.68	3.01	−0.68
Coherent p_α	3.55	11.63	2.72	−1.13	3.01	11.12	3.01	−0.68
Coherent x_β	2.81	−0.63	20.75	0.20	3.01	−0.68	20.04	−0.68
Coherent p_β	2.76	−0.65	4.28	14.21	3.01	−0.68	3.01	14.08

Table 9.9 Measurement results and theoretical values with $\eta = \frac{1}{5}$ (dB, error bars of measurement results are less than 0.05 dB)

Input	Output							
	$\hat{x}_1$(M)	$\hat{p}_1$(M)	$\hat{x}_3$(M)	$\hat{p}_3$(M)	$\hat{x}_1$(T)	$\hat{p}_1$(T)	$\hat{x}_3$(T)	$\hat{p}_3$(T)
Vacuum	2.66	−0.47	2.96	−0.50	3.01	−0.38	3.01	−0.38
Coherent x_α	17.59	1.18	2.94	1.02	17.02	1.46	3.01	1.46
Coherent p_α	4.33	11.83	2.97	−0.35	3.01	11.14	3.01	−0.38
Coherent x_β	2.80	2.19	20.48	3.86	3.01	2.80	20.04	2.80
Coherent p_β	2.74	−0.24	4.20	14.27	3.01	−0.38	3.01	14.09

In these tables, theoretical values are listed as well. We have here assumed that the resource squeezing level is 4.5 dB, and the input coherent-state amplitudes are 13.83 dB for mode α and 16.95 dB for mode β, respectively (Figs. 9.15 and 9.16).

9.7.1.3 Measurement Result of Output States (Graphs Categorized by Operations)

In Figs. 9.17, 9.18, 9.19, 9.20, 9.21, 9.22 and 9.23, we show graphs of measurement results and theoretical values, where data are categorized by operations. Note that they are not covariance matrices.

9.7.1.4 Measurement Result of Output States (Graphs Categorized by Input States)

Figure 9.24 shows graphs of measurement results and theoretical values, where data are categorized by input states. Note that measurement results and their theoretical curves are shown in the same colors.

9.7.1.5 Analysis

Case where input states are changed with $\eta = 0$

When $\eta = 0$ (Fig. 9.17), the operations are two single-mode 3.0 dB p-squeezing operations (see Eq. (9.11)). Thus, when we use a coherent state with amplitude in x of input mode α, we detect signal of the coherent state in x of output mode 1, where its power is increased by about 3.0 dB. On the other hand, when we use a coherent state with amplitude in p of input mode α, we detect signal of the coherent state in p of output mode 1, where its power is decreased by about 3.0 dB. The differences between input and output powers in theoretical values are not precisely 3.0 dB because the power of the input state is not intense enough to neglect the variance of a vacuum state. The discussion above can be applied to the set of input mode β and output mode 3 as well.

Case where operations are changed with vacuum input states

Next, we consider the case where operations are changed with vacuum input states (Fig. 9.24a). We find from the operation matrix Eq. (9.11) and the diagonal elements of covariance matrix Eq. (9.23) that the position operators of output modes 1 and 3 do not depend on the interaction gain η which determines the operation. Since they become $\sqrt{2}\hat{x}_\alpha$ and $\sqrt{2}\hat{x}_\beta$, the variances become 3.0 dB relative to the SNL. On the other hand, the variances of momentum operators of modes 1 and 3 increase depending on the interaction gain η.

Case where input states are changed with operations fixed

We consider the case where input states are changed, while operation is fixed (in Figs. 9.17, 9.18, 9.19, 9.20, 9.21, 9.22 and 9.23). We have already mentioned the

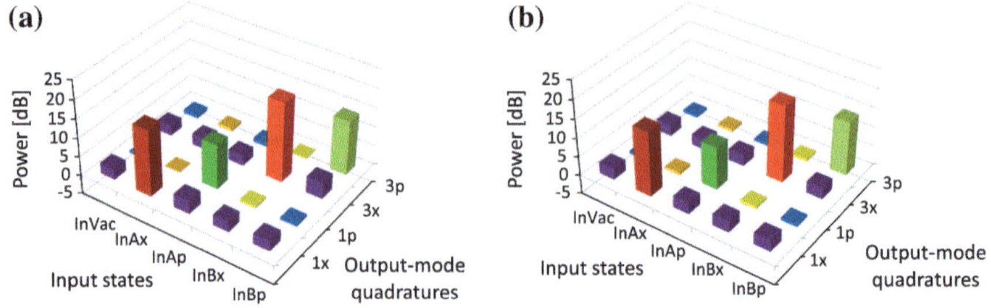

Fig. 9.17 Measurement results and theoretical values with $\eta = 0$. **a** Measurement results. **b** Theoretical values

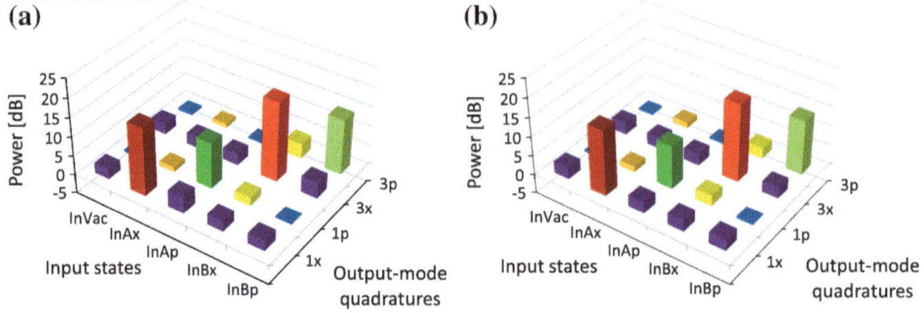

Fig. 9.18 Measurement results and theoretical values with $\eta = \frac{1}{5}$. **a** Measurement results. **b** Theoretical values

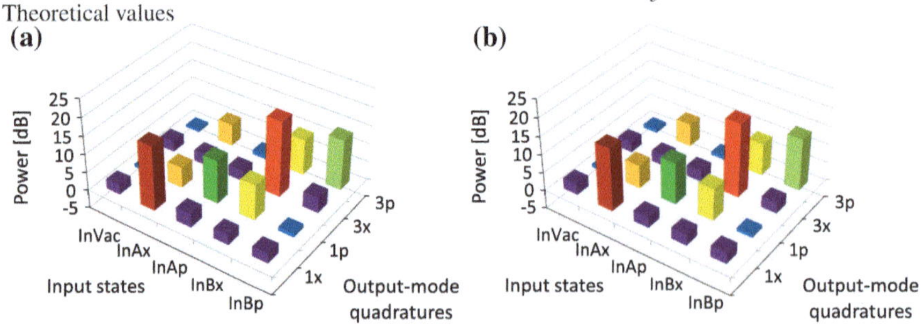

Fig. 9.19 Measurement results and theoretical values with $\eta = \frac{1}{2}$. **a** Measurement results. **b** Theoretical values

case where both input modes are vacuum states. For example, by changing the input state in mode α from a vacuum state to a coherent state with amplitude in x, powers of $\hat{x}_1$, $\hat{p}_1$, and $\hat{p}_3$ increase because $\hat{x}_\alpha$ is transmitted to these quadratures (Eq. 9.11). On the other hand, $\hat{x}_\alpha$ is not transmitted to $\hat{x}_3$, thus the power of $\hat{x}_3$ is not changed.

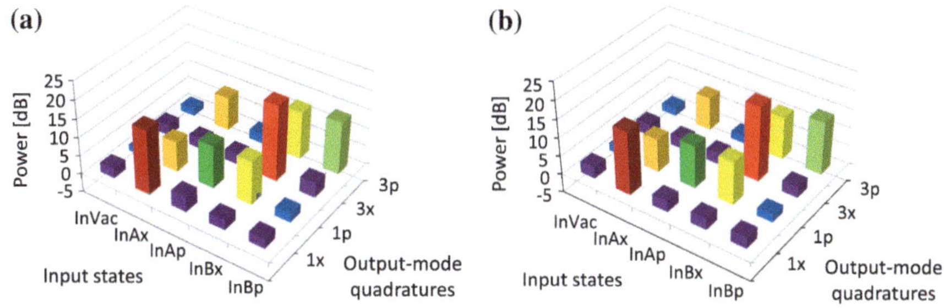

Fig. 9.20 Measurement results and theoretical values with $\eta = \frac{1}{\sqrt{2}}$. **a** Measurement results. **b** Theoretical values

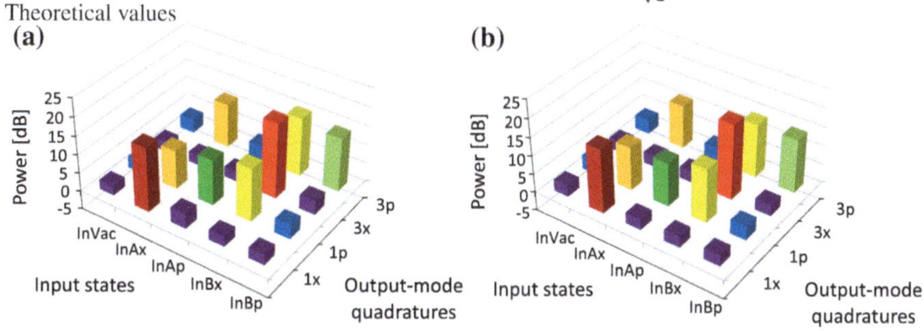

Fig. 9.21 Measurement results and theoretical values with $\eta = 1$. **a** Measurement results. **b** Theoretical values

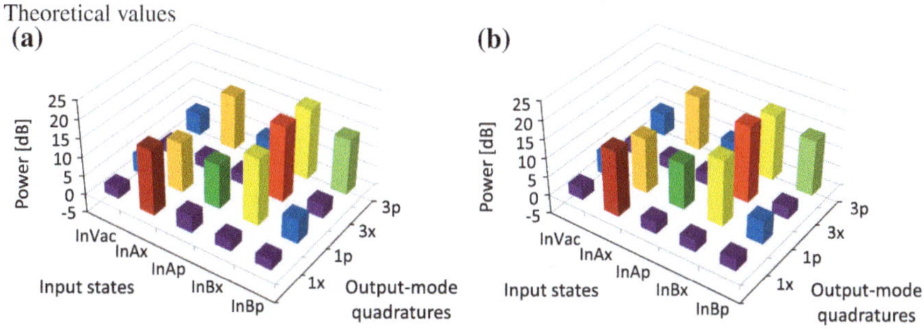

Fig. 9.22 Measurement results and theoretical values with $\eta = \sqrt{2}$. **a** Measurement results. **b** Theoretical values

Case where operations are changed with input states fixed

Finally, we consider the case where operations are changed, while input states are fixed (from Fig. 9.24b to e). Behavior of measurement results can be categorized into the following five cases.

x_1 with InAx, x_3 with InBx (shown in rufous and red in figures)

Because the coherent-state signals are transmitted to these output-mode quadratures with fixed gains, and because the variances of the quadratures do not depend on the operations when both input modes are in vacuum states, the powers of these quadratures become constants.

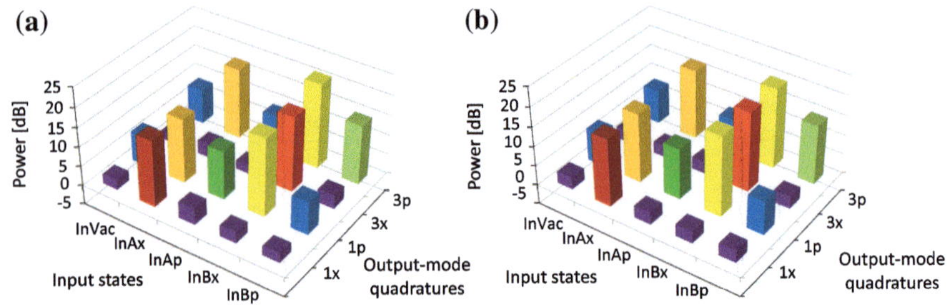

Fig. 9.23 Measurement results and theoretical values with $\eta = 2$. **a** Measurement results. **b** Theoretical values

x_3 with InAx, x_1 and x_3 with InAp, x_1 with InBx, x_1 and x_3 with InBp (shown in purple in figures)

Because the coherent-state signals are not transmitted to these output-mode quadratures, and because the variances of the quadratures do not depend on the operations when both input modes are in vacuum states, the powers of these quadratures become constants.

p_3 with InAp, p_1 with InBp (shown in blue in figures)

Although the coherent-state signals are not transmitted to these output-mode quadratures, the variances of the quadratures depend on the operations when both input modes are in vacuum states. Thus, the variances of these quadratures increase depending on the interaction gain η.

p_1 and p_3 with InAx, p_1 and p_3 with InBx (shown in orange and yellow in figures)

Because the coherent-state signals are transmitted to these output-mode quadratures with operation-dependent gains, and because the variances of the quadratures depend on the operations when both input modes are in vacuum states, the variances of these quadratures increase depending on the interaction gain η.

p_1 with InAp, p_3 with InBp (shown in green and yellow-green in figures)

Although the coherent-state signals are transmitted to these output-mode quadratures with fixed gains, the variances of the quadratures depend on the operations when both input modes are in vacuum states. Thus, the variances of these quadratures increase slightly depending on the interaction gain η.

In all cases, the experimental results agree well with their theoretical predictions.

9.7.2 Covariance Matrices and Entanglement at Output (Via PT Symplectic Eigenvalues)

We verify entanglement at the output by using PT symplectic eigenvalues. We here use the system of units with $\hbar = \frac{1}{2}$.

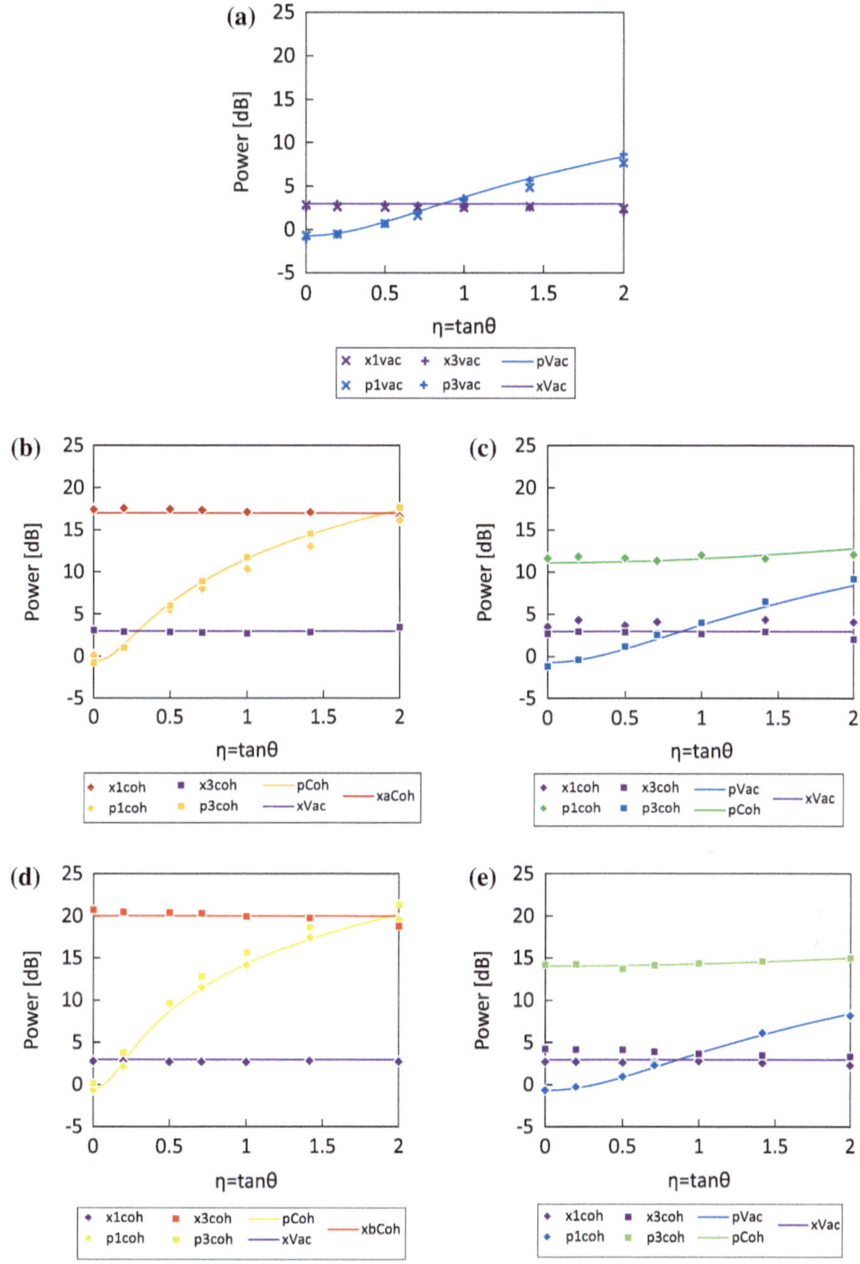

Fig. 9.24 Measurement result of output states (graphs categorized by input states). **a** Vacuum inputs in both α and β (InVac). **b** Amplitude in x of α (InAx). **c** Amplitude in p of α (InAp). **d** Amplitude in x of β (InBx). **e** Amplitude in p of β (InBp)

9.7.2.1 Measurement Results

The procedure for measurement of covariance matrices is explained in Chap. 6. We get the following covariance matrices of the output states with vacuum input states.

$$\eta\text{: (measurement result), (theoretical value with 4.5 dB resource), (ideal case).} \tag{9.35}$$

$$0\text{:} \begin{pmatrix} 0.49 & 0.00 & 0.00 & 0.00 \\ 0.00 & 0.21 & 0.00 & 0.04 \\ 0.00 & 0.00 & 0.50 & 0.00 \\ 0.00 & 0.04 & 0.00 & 0.21 \end{pmatrix}, \begin{pmatrix} 0.50 & 0.00 & 0.00 & 0.00 \\ 0.00 & 0.21 & 0.00 & 0.04 \\ 0.00 & 0.00 & 0.50 & 0.00 \\ 0.00 & 0.04 & 0.00 & 0.21 \end{pmatrix}, \begin{pmatrix} 0.50 & 0.00 & 0.00 & 0.00 \\ 0.00 & 0.13 & 0.00 & 0.00 \\ 0.00 & 0.00 & 0.50 & 0.00 \\ 0.00 & 0.00 & 0.00 & 0.13 \end{pmatrix}. \tag{9.36}$$

$$\frac{1}{5}\text{:} \begin{pmatrix} 0.49 & -0.03 & 0.00 & -0.05 \\ -0.03 & 0.22 & -0.04 & 0.05 \\ 0.00 & -0.04 & 0.50 & -0.05 \\ -0.05 & 0.05 & -0.05 & 0.22 \end{pmatrix}, \begin{pmatrix} 0.50 & -0.05 & 0.00 & -0.05 \\ -0.05 & 0.23 & -0.05 & 0.06 \\ 0.00 & -0.05 & 0.50 & -0.05 \\ -0.05 & 0.06 & -0.05 & 0.23 \end{pmatrix}, \begin{pmatrix} 0.50 & -0.05 & 0.00 & -0.05 \\ -0.05 & 0.14 & -0.05 & 0.01 \\ 0.00 & -0.05 & 0.50 & -0.05 \\ -0.05 & 0.01 & -0.05 & 0.14 \end{pmatrix}. \tag{9.37}$$

$$\frac{1}{2}\text{:} \begin{pmatrix} 0.50 & -0.10 & 0.01 & -0.12 \\ -0.10 & 0.27 & -0.11 & 0.11 \\ 0.01 & -0.11 & 0.50 & -0.12 \\ -0.12 & 0.11 & -0.12 & 0.28 \end{pmatrix}, \begin{pmatrix} 0.50 & -0.13 & 0.00 & -0.13 \\ -0.13 & 0.31 & -0.13 & 0.14 \\ 0.00 & -0.13 & 0.50 & -0.13 \\ -0.13 & 0.14 & -0.13 & 0.31 \end{pmatrix}, \begin{pmatrix} 0.50 & -0.13 & 0.00 & -0.13 \\ -0.13 & 0.19 & -0.13 & 0.06 \\ 0.00 & -0.13 & 0.50 & -0.13 \\ -0.13 & 0.06 & -0.13 & 0.19 \end{pmatrix}. \tag{9.38}$$

$$\frac{1}{\sqrt{2}}\text{:} \begin{pmatrix} 0.50 & -0.16 & 0.00 & -0.17 \\ -0.16 & 0.35 & -0.15 & 0.17 \\ 0.00 & -0.15 & 0.49 & -0.17 \\ -0.17 & 0.17 & -0.17 & 0.37 \end{pmatrix}, \begin{pmatrix} 0.50 & -0.18 & 0.00 & -0.18 \\ -0.18 & 0.41 & -0.18 & 0.24 \\ 0.00 & -0.18 & 0.50 & -0.18 \\ -0.18 & 0.24 & -0.18 & 0.41 \end{pmatrix}, \begin{pmatrix} 0.50 & -0.18 & 0.00 & -0.18 \\ -0.18 & 0.25 & -0.18 & 0.13 \\ 0.00 & -0.18 & 0.50 & -0.18 \\ -0.18 & 0.13 & -0.18 & 0.25 \end{pmatrix}. \tag{9.39}$$

$$1\text{:} \begin{pmatrix} 0.50 & -0.24 & -0.01 & -0.23 \\ -0.24 & 0.50 & -0.21 & 0.32 \\ -0.01 & -0.21 & 0.49 & -0.23 \\ -0.23 & 0.32 & -0.23 & 0.50 \end{pmatrix}, \begin{pmatrix} 0.50 & -0.25 & 0.00 & -0.25 \\ -0.25 & 0.60 & -0.25 & 0.43 \\ 0.00 & -0.25 & 0.50 & -0.25 \\ -0.25 & 0.43 & -0.25 & 0.60 \end{pmatrix}, \begin{pmatrix} 0.50 & -0.25 & 0.00 & -0.25 \\ -0.25 & 0.38 & -0.25 & 0.25 \\ 0.00 & -0.25 & 0.50 & -0.25 \\ -0.25 & 0.25 & -0.25 & 0.38 \end{pmatrix}. \tag{9.40}$$

$$\sqrt{2}\text{:} \begin{pmatrix} 0.48 & -0.32 & -0.01 & -0.29 \\ -0.32 & 0.80 & -0.30 & 0.58 \\ -0.01 & -0.30 & 0.47 & -0.31 \\ -0.29 & 0.58 & -0.31 & 0.75 \end{pmatrix}, \begin{pmatrix} 0.50 & -0.35 & 0.00 & -0.35 \\ -0.35 & 0.98 & -0.35 & 0.81 \\ 0.00 & -0.35 & 0.50 & -0.35 \\ -0.35 & 0.81 & -0.35 & 0.98 \end{pmatrix}, \begin{pmatrix} 0.50 & -0.35 & 0.00 & -0.35 \\ -0.35 & 0.62 & -0.35 & 0.50 \\ 0.00 & -0.35 & 0.50 & -0.35 \\ -0.35 & 0.50 & -0.35 & 0.62 \end{pmatrix}. \tag{9.41}$$

$$2\text{:} \begin{pmatrix} 0.50 & -0.53 & 0.01 & -0.48 \\ -0.53 & 1.43 & -0.44 & 1.18 \\ 0.01 & -0.44 & 0.49 & -0.49 \\ -0.48 & 1.18 & -0.49 & 1.37 \end{pmatrix}, \begin{pmatrix} 0.50 & -0.50 & 0.00 & -0.50 \\ -0.50 & 1.75 & -0.50 & 1.58 \\ 0.00 & -0.50 & 0.50 & -0.50 \\ -0.50 & 1.58 & -0.50 & 1.75 \end{pmatrix}, \begin{pmatrix} 0.50 & -0.50 & 0.00 & -0.50 \\ -0.50 & 1.13 & -0.50 & 1.00 \\ 0.00 & -0.50 & 0.50 & -0.50 \\ -0.50 & 1.00 & -0.50 & 1.13 \end{pmatrix}. \tag{9.42}$$

In these equations, we have listed in the form of "η: measurement result, theoretical value with 4.5 dB resource squeezed state, ideal case with infinitely squeezed state". Error bars of measurement results with $\eta = 2$ are less than ± 0.02 dB, while those with the other η are less than ± 0.01 dB.

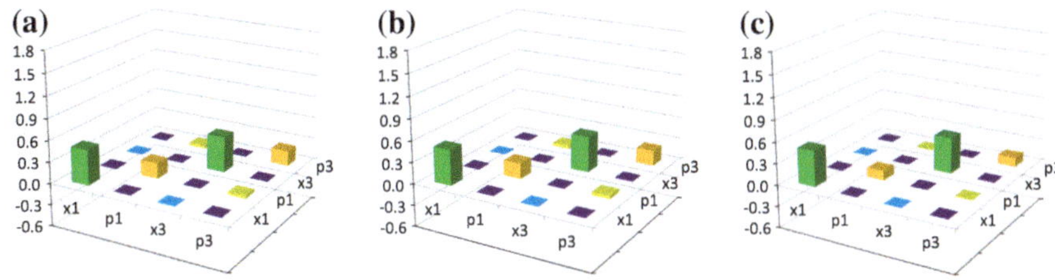

Fig. 9.25 Covariance matrices with $\eta = 0$. **a** Measurement result. **b** Theory (4.5 dB resource). **c** Theory (∞ dB resource)

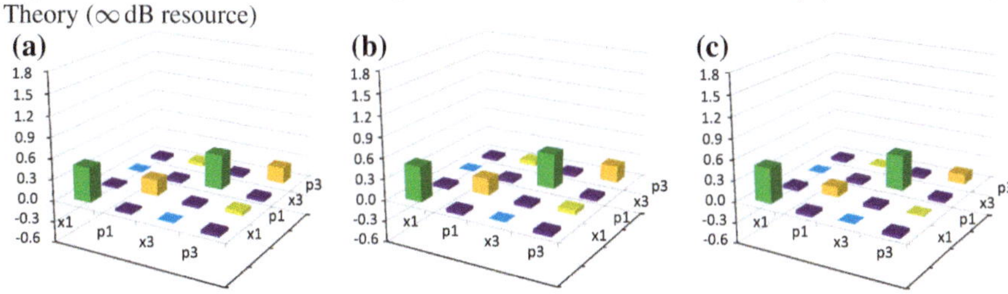

Fig. 9.26 Covariance matrices with $\eta = \frac{1}{5}$. **a** Measurement result. **b** Theory (4.5 dB resource). **c** Theory (∞ dB resource)

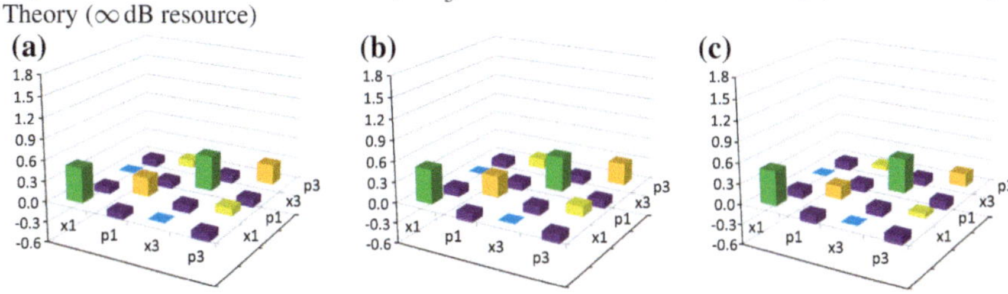

Fig. 9.27 Covariance matrices with $\eta = \frac{1}{2}$. **a** Measurement result. **b** Theory (4.5 dB resource). **c** Theory (∞ dB resource)

In Figs. 9.25, 9.26, 9.27, 9.28, 9.29, 9.30 and 9.31, we show their graphs. Note that the purple elements have negative values.

9.7.2.2 Analysis

Case where $\eta = 0$

When $\eta = 0$, the operations are two single-mode 3.0 dB p-squeezing operations (see Eq. (9.16)).

In the ideal case with infinitely squeezed state resources, the variances of $\hat{x}$ of the output modes are given by $\frac{\hbar}{2} \times 2 = \frac{1}{2}$, while the variances of $\hat{p}$ of the output modes are given by $\frac{\hbar}{2} \div 2 = \frac{1}{8}$. The other elements of the covariance matrix are equal to zero.

Table 9.10 Measurement results and theoretical values with $\eta = \frac{1}{2}$ (dB, error bars of measurement results are less than 0.05 dB)

Input	Output							
	$\hat{x}_1$(M)	$\hat{p}_1$(M)	$\hat{x}_3$(M)	$\hat{p}_3$(M)	$\hat{x}_1$(T)	$\hat{p}_1$(T)	$\hat{x}_3$(T)	$\hat{p}_3$(T)
Vacuum	2.62	0.74	2.91	0.71	3.01	0.93	3.01	0.93
Coherent x_α	17.46	5.54	2.91	6.02	17.02	6.29	3.01	6.29
Coherent p_α	3.70	11.65	2.91	1.21	3.01	11.24	3.01	0.93
Coherent x_β	2.71	9.50	20.40	9.72	3.01	8.71	20.04	8.71
Coherent p_β	2.66	1.02	4.18	13.74	3.01	0.93	3.01	14.15

Table 9.11 Measurement results and theoretical values with $\eta = \frac{1}{\sqrt{2}}$ (dB, error bars of measurement results are less than 0.05 dB)

Input	Output							
	$\hat{x}_1$(M)	$\hat{p}_1$(M)	$\hat{x}_3$(M)	$\hat{p}_3$(M)	$\hat{x}_1$(T)	$\hat{p}_1$(T)	$\hat{x}_3$(T)	$\hat{p}_3$(T)
Vacuum	2.59	1.64	2.81	1.89	3.01	2.10	3.01	2.10
Coherent x_α	17.37	8.05	2.87	8.94	17.02	8.84	3.01	8.84
Coherent p_α	4.12	11.35	2.83	2.63	3.01	11.37	3.01	2.10
Coherent x_β	2.74	11.54	20.35	12.81	3.01	11.46	20.04	11.46
Coherent p_β	2.68	2.34	3.96	14.16	3.01	2.10	3.01	14.21

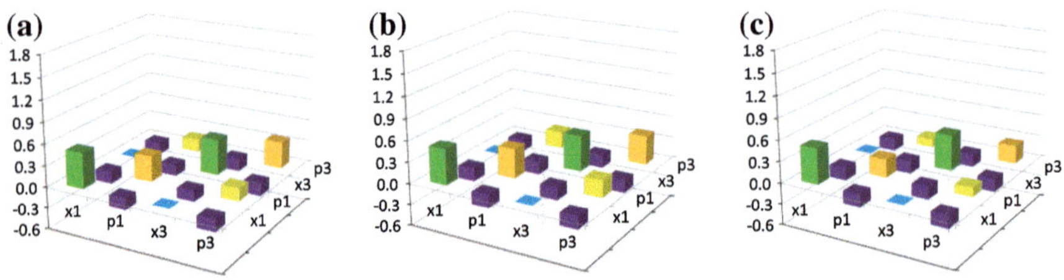

Fig. 9.28 Covariance matrices with $\eta = \frac{1}{\sqrt{2}}$. **a** Measurement result. **b** Theory (4.5 dB resource). **c** Theory (∞ dB resource)

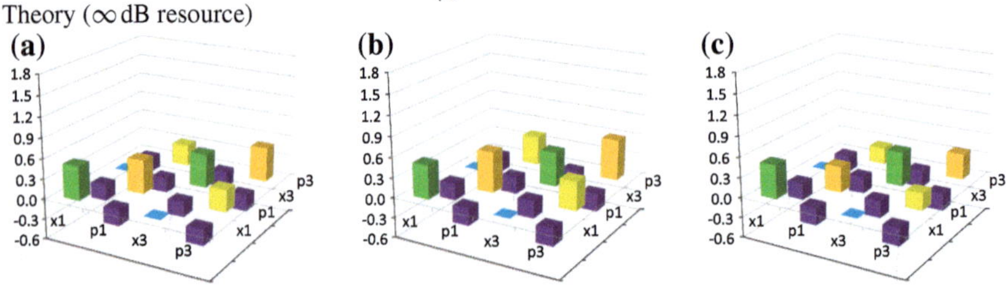

Fig. 9.29 Covariance matrices with $\eta = 1$. **a** Measurement result. **b** Theory (4.5 dB resource). **c** Theory (∞ dB resource)

Table 9.12 Measurement results and theoretical values with $\eta = 1$ (dB, error bars of measurement results are less than 0.05 dB)

	Output							
Input	$\hat{x}_1$(M)	$\hat{p}_1$(M)	$\hat{x}_3$(M)	$\hat{p}_3$(M)	$\hat{x}_1$(T)	$\hat{p}_1$(T)	$\hat{x}_3$(T)	$\hat{p}_3$(T)
Vacuum	2.60	3.09	2.71	3.64	3.01	3.78	3.01	3.78
Coherent x_α	17.14	10.39	2.74	11.75	17.02	11.60	3.01	11.60
Coherent p_α	4.09	12.02	2.71	4.02	3.01	11.60	3.01	3.78
Coherent x_β	2.71	14.20	19.99	15.69	3.01	14.34	20.04	14.34
Coherent p_β	2.85	3.81	3.71	14.39	3.01	3.78	3.01	14.34

Table 9.13 Measurement results and theoretical values with $\eta = \sqrt{2}$ (dB, error bars of measurement results are less than 0.05 dB)

	Output							
Input	$\hat{x}_1$(M)	$\hat{p}_1$(M)	$\hat{x}_3$(M)	$\hat{p}_3$(M)	$\hat{x}_1$(T)	$\hat{p}_1$(T)	$\hat{x}_3$(T)	$\hat{p}_3$(T)
Vacuum	2.70	4.88	2.63	5.71	3.01	5.93	3.01	5.93
Coherent x_α	17.12	13.07	2.91	14.56	17.02	14.48	3.01	14.48
Coherent p_α	4.39	11.61	2.95	6.54	3.01	12.04	3.01	5.93
Coherent x_β	2.85	17.48	19.81	18.70	3.01	17.28	20.04	17.28
Coherent p_β	2.61	6.15	3.53	14.65	3.01	5.93	3.01	14.57

Table 9.14 Measurement results and theoretical values with $\eta = 2$ (dB, error bars of measurement results are less than 0.06 dB)

	Output							
Input	$\hat{x}_1$(M)	$\hat{p}_1$(M)	$\hat{x}_3$(M)	$\hat{p}_3$(M)	$\hat{x}_1$(T)	$\hat{p}_1$(T)	$\hat{x}_3$(T)	$\hat{p}_3$(T)
Vacuum	2.45	7.69	2.05	8.69	3.01	8.44	3.01	8.44
Coherent x_α	16.67	16.19	3.50	17.72	17.02	17.43	3.01	17.43
Coherent p_α	4.10	12.07	2.08	9.20	3.01	12.80	3.01	8.44
Coherent x_β	2.78	19.65	18.85	21.43	3.01	20.25	20.04	20.25
Coherent p_β	2.33	8.21	3.35	15.01	3.01	8.44	3.01	15.01

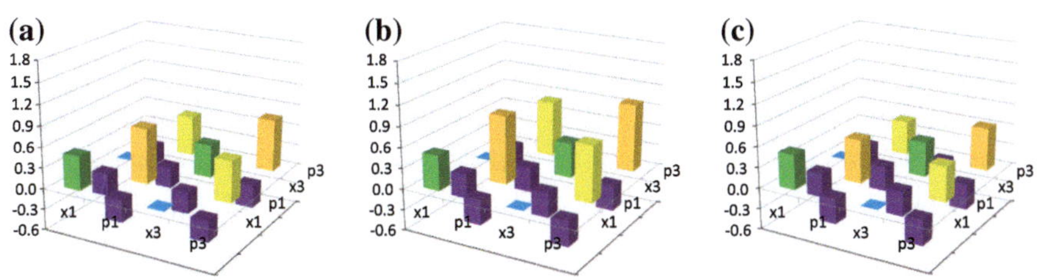

Fig. 9.30 Covariance matrices with $\eta = \sqrt{2}$. **a** Measurement result. **b** Theory (4.5 dB resource). **c** Theory (∞ dB resource)

In the realistic case with finitely squeezed state resources, V_{22}, V_{24}, V_{42}, and V_{44} have additional variances (Eq. 9.23). Measurement results agree with theoretical predictions with 4.5 dB squeezed state resources.

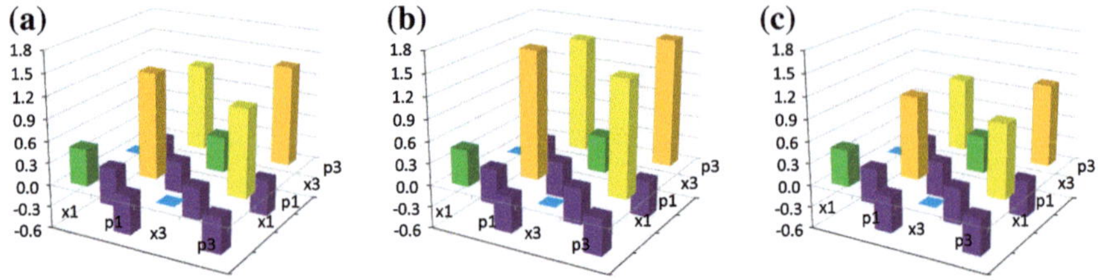

Fig. 9.31 Covariance matrices with $\eta = 2$. **a** Measurement result. **b** Theory (4.5 dB resource). **c** Theory (∞ dB resource)

We mention that the matrix elements V_{24} and V_{42}, which correspond to correlations between two output modes, have non-zero values. It derives from the fact that the resource cluster state is a Gaussian cluster state. Note that these elements are equal to zero if the resource is an ideal cluster state or a canonical cluster state. Although they have non-zero values when the Gaussian cluster state has a finite squeezing level, they vanish in the limit of infinite squeezing. This is because the nullifier $\hat{\delta}_j$ of the cluster state is given only by squeezing components $e^{-r_k}\hat{p}_k^{(0)}$, and because error of one-way quantum computation is formulated by $\hat{\delta}_j$ (δ representation, Sect. 5.6).

Case where operations are changed

We consider the case where operations are changed (in Figs. 9.25, 9.26, 9.27, 9.28, 9.29, 9.30 and 9.31). Behavior of measurement results can be categorized into the following five cases.

Variances of $\hat{x}_1$, $\hat{x}_3$ (shown in green in figures)

Because these values do not depend on η in the ideal case, and because errors derived from cluster imperfection are zero, the measurement results do not depend on η and they are equivalent to the ideal values.

Variances of $\hat{p}_1$, $\hat{p}_3$ (shown in orange in figures)

Because these values increase depending on the interaction gain η in the ideal case, and because errors derived from cluster imperfection also increase depending on η, the measurement results increase depending on η and they are larger than the ideal values.

Correlation between $(\hat{p}_1, \hat{p}_3)$ (shown in yellow in figures)

Because this value increases depending on the interaction gain η in the ideal case, and because errors derived from cluster imperfection also increase depending on η, the measurement results increase depending on η and they are larger than the ideal values.

Correlations between $(\hat{x}_1, \hat{p}_1)$, $(\hat{x}_1, \hat{p}_3)$, $(\hat{x}_3, \hat{p}_1)$, $(\hat{x}_3, \hat{p}_3)$ (shown in purple in figures)

Because these values are negative and their absolute values increase depending on the interaction gain η in the ideal case, and because errors derived from cluster imperfection are zero, the measurement results are negative and their absolute values increase depending on η, and they are equivalent to the ideal values.

Correlation between $(\hat{x}_1, \hat{x}_3)$ (shown in light blue in figures)

Because this value is zero in the ideal case, and because errors derived from cluster imperfection are zero as well, the measurement results are zero and they are equivalent to the ideal values.

In all cases, the experimental results agree well with their theoretical predictions.

9.7.2.3 Variables Acquired by Covariance Matrix

Table 9.15 shows several variables (Sects. 3.2.6 and 3.7.5) which are acquired by the covariance matrices. Here, det V_k represents the determinant of the kth order principal submatrix of a covariance matrix V, $\nu_\pm$ represent the symplectic eigenvalues, $\tilde{\nu}_\pm$ represent the PT symplectic eigenvalues, and E_N represents the logarithmic negativity.

Since all determinants of the kth order principal submatrices are positive: det $V_k > 0$ $(k = 1, \dots, 4)$, and the symplectic eigenvalue ν_- is $\nu_- \geq \frac{\hbar}{2}$, each symmetric matrix V certainly shows a covariance matrix of a physical state.

Figure 9.32 shows the PT symplectic eigenvalues $\tilde{\nu}_-$ with respect to η.

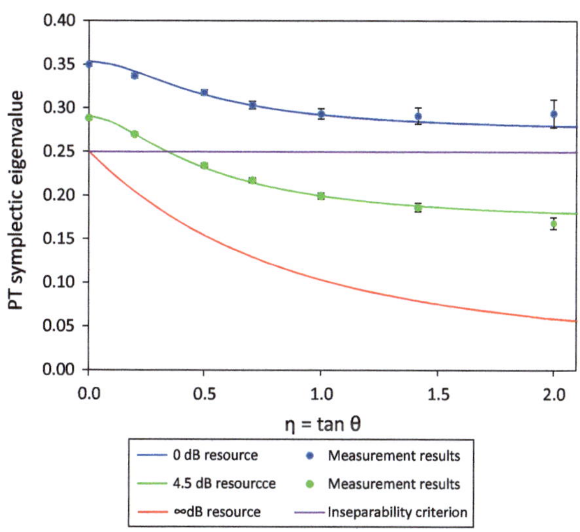

Fig. 9.32 PT symplectic eigenvalues $\tilde{\nu}_-$

When the PT symplectic eigenvalue $\tilde{\nu}_- < \frac{\hbar}{2} = \frac{1}{4}$, the output state is an entangled state. The closer to zero the PT symplectic eigenvalue $\tilde{\nu}_-$ is, the more the output state is entangled. Table 9.15 and Fig. 9.32 show that the output states with $\eta = 0$ and $\frac{1}{5}$ are not entangled since $\tilde{\nu}_- > \frac{1}{4}$, while those with $\eta = \frac{1}{2}, \frac{1}{\sqrt{2}}, 1, \sqrt{2}$, and 2 are entangled since $\tilde{\nu}_- < \frac{1}{4}$. We also find that the level of entanglement gets larger depending on the interaction gain η. Figure 9.32 shows that the measurement results agree well with theoretical predictions where 4.5 dB squeezed state resources are utilized.

Table 9.15 Variables acquired by covariance matrices

Variable	$\eta = 0$	$\eta = \frac{1}{5}$	$\eta = \frac{1}{2}$	$\eta = \frac{1}{\sqrt{2}}$	$\eta = 1$	$\eta = \sqrt{2}$	$\eta = 2$
det V_1	0.494 ± 0.005	0.489 ± 0.005	0.504 ± 0.005	0.497 ± 0.005	0.500 ± 0.005	0.484 ± 0.005	0.502 ± 0.005
det V_2	0.104 ± 0.002	0.108 ± 0.002	0.127 ± 0.002	0.148 ± 0.003	0.193 ± 0.004	0.284 ± 0.006	0.43 ± 0.02
det V_3	0.0514 ± 0.0008	0.0528 ± 0.0009	0.057 ± 0.002	0.061 ± 0.002	0.072 ± 0.003	0.088 ± 0.004	0.117 ± 0.007
det V_4	0.0104 ± 0.0002	0.0107 ± 0.0003	0.0122 ± 0.0003	0.0141 ± 0.0005	0.0177 ± 0.0007	0.024 ± 0.002	0.035 ± 0.004
ν_-	0.291 ± 0.002	0.289 ± 0.003	0.290 ± 0.003	0.299 ± 0.004	0.301 ± 0.005	0.307 ± 0.007	0.32 ± 0.02
ν_+	0.351 ± 0.002	0.358 ± 0.003	0.380 ± 0.004	0.397 ± 0.005	0.442 ± 0.007	0.507 ± 0.009	0.59 ± 0.02
$\tilde{\nu}_-$	0.288 ± 0.002	0.270 ± 0.003	0.234 ± 0.003	0.217 ± 0.003	0.199 ± 0.004	0.186 ± 0.005	0.169 ± 0.007
$\tilde{\nu}_+$	0.354 ± 0.002	0.384 ± 0.003	0.471 ± 0.003	0.546 ± 0.004	0.667 ± 0.005	0.835 ± 0.006	1.111 ± 0.009
E_N	0	0	0.06 ± 0.02	0.14 ± 0.02	0.23 ± 0.02	0.29 ± 0.03	0.39 ± 0.05

References

1. Ukai, R., Iwata, N., Shimokawa, Y., Armstrong, S.C., Politi, A., Yoshikawa, J., van Loock, P., Furusawa, A.: Demonstration of unconditional one-way quantum computations for continuous variables. Phys. Rev. Lett. **106**, 240504 (2011)
2. Ukai, R., Yoshikawa, J., Iwata, N., van Loock, P., Furusawa, A.: Universal linear Bogoliubov transformations through one-way quantum computation. Phys. Rev. A **81**, 032315 (2010)
3. Sparkes, B.M., Chrzanowski, H.M., Parrain, D.P., Buchler, B.C., Lam, P.K., Symul, T.: A scalable, self-analyzing digital locking system for use on quantum optics experiments. Rev. Sci. Instrum. **82**, 075113 (2011)

Chapter 10
Temporal-Mode Cluster States

10.1 Background

In the conventional demonstrations of one-way quantum computations using optical continuous-variable systems, each quantum mode of a cluster state is distinguished from the other modes by assigning it to a spatial location of its own [2–5]. Recently, one-way quantum computation using temporal modes is proposed [1, 6], where each mode is encoded temporally. In this section, we briefly refer to background of temporal-mode cluster states.

10.1.1 Temporal-Mode Cluster States

The discovery of beam-splitter based Gaussian cluster state scheme [3] made it easier to generate cluster states in laboratories (Sect. 1.2). Soon after the original proposal of Gaussian cluster states, experimental preparations of four-mode cluster states were reported [4, 5]. However, this traditional scheme was not suitable for large-scale one-way quantum computation experiments. In one-way quantum computation, we have to prepare an n-mode cluster state as a resource in order to achieve an n-step operation. For that purpose, we had to use n squeezed states generated by n optical parametric oscillators (OPOs). Thus, experimental setup becomes larger in proportion to the number of operations. Although generations of optical four-mode cluster states [4, 5] and eight-mode cluster states [7] for continuous variables are reported so far, we can hardly expect that more-mode cluster state would be generated with the traditional scheme.

The new scheme [1, 6] for generation of optical cluster states for continuous variables utilizes a new dimension "*time*", which was not utilized in the traditional scheme. As a reference, we first consider the traditional scheme from this point of view. In the traditional scheme, each resource squeezed state for a cluster state is generated *simultaneously* by using each OPO. These squeezed states are combined using a network of controlled-Z gates (canonical cluster states) or beam splitters (Gaussian cluster states) *simultaneously*. As a result, each mode of the cluster state exists *at the same time*.

R. Ukai, *Multi-Step Multi-Input One-Way Quantum Information Processing with Spatial and Temporal Modes of Light*, Springer Theses,
DOI 10.1007/978-4-431-55019-8_10

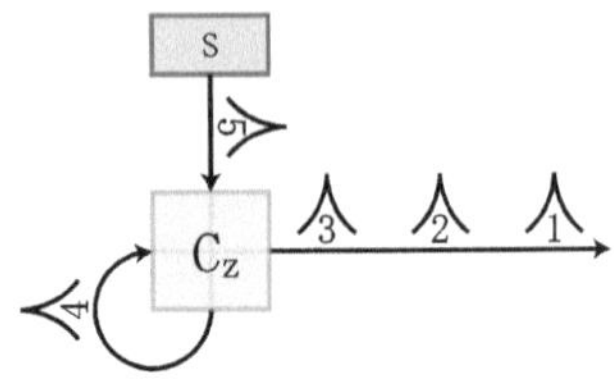

Fig. 10.1 Temporal-mode canonical cluster state using a controlled-Z gate

In the new scheme, on the other hand, squeezed beams are first generated by using a limited number of OPOs. Each squeezed beam is virtually split into a sequence of squeezed states with some time period T. By choosing T appropriately, two squeezed states at time t_1 and (t_1+T) can be almost orthogonal. A cluster state can be generated by combining squeezed states at different times. In this scheme, quantum modes of the cluster state do not exist simultaneously. Instead, although several modes of the cluster state might exist simultaneously, they are mainly assigned to different times. Since we can define multiple squeezed states from a squeezed beam, we do not have to prepare the same number of OPOs as the quantum modes. In principle, it is sufficient to use single OPO.[1]

The cluster state generated by assigning each mode to a different time is called the *temporal-mode* cluster state.

10.1.2 Temporal-Mode Canonical Cluster State

Based on the idea of the temporal-mode cluster state, the *temporal-mode canonical cluster state* is first proposed [6].

Figure 10.1 shows a schematic for generation of a multi-mode linear cluster state. In this figure, "S" shows an OPO which generates squeezed states periodically. Each double exponential function shows a quantum mode, which is defined within the duration T. Each mode is initially in a squeezed state. The n-th squeezed state generated by the OPO enters the controlled-Z gate. After passing through a delay line with the duration T, one of the two outputs of the controlled-Z gate goes into the other input port of the same controlled-Z gate. Since it has been delayed with the duration T, it is entangled with the $(n+1)$-th squeezed state. Temporal-mode linear canonical cluster state is acquired from the other output port of the controlled-Z gate. Each mode of the cluster state is outputted with a time interval of T.

It is reported that a controlled-Z gate can be experimentally implemented by using two squeezed states as resources [8, 9]. Therefore, a linear cluster state of an arbitrary mode length can be generated by using three OPOs.

[1] In reality, we can simplify experimental setup by using several OPOs.

10.1.3 Temporal-Mode Gaussian Cluster State

Similar to the case of spatial-mode cluster states, it is later proposed that the *temporal-mode Gaussian cluster states*, which are the theme in this chapter, can be generated by combining temporal-mode squeezed states on beams splitters [1]. In order to couple quantum modes at a different time, delay lines are also involved.

Figure 10.2 shows an experimental setup for generation of a temporal-mode Gaussian cluster state which can be utilized to implement one-mode quantum operations. The graph of the cluster state is shown in Fig. 10.3. In this scheme, the cluster state of an arbitrary mode length can be generated by using only two OPOs and two beam splitters. Details will be discussed in the succeeding sections.

10.1.4 Unsolved Problems and Contents of this Chapter

Although the original paper [1] proposed schematics for generation of temporal-mode Gaussian cluster states, there still exist several unsolved problems about temporal-mode Gaussian cluster states and their application to one-way quantum computation. They include:

1. why is the schematic of cluster state generation in this form?
2. a half or three fourths of the generated cluster states should be erased before they are used as resources for one-way quantum computations.

The former is related to theoretical consideration of the temporal-mode Gaussian cluster state scheme. The latter is an exploration of a theory for experimental demonstration of one-way quantum computation where a temporal-mode Gaussian cluster state is used as a resource. In the ideal case where an infinitely correlated entangled cluster state can be used, disposal of the half modes does not lead to errors in quantum computation. However, in the realistic case, we cannot utilize the ideal cluster state. The experimental cluster state is generated by combining squeezed states with finite squeezing levels. When it is used as a resource for one-way quantum computation, errors derived from its imperfection will accumulate gradually. We mention that error is indeed added during erasing of cluster modes. In order to minimize the effect of errors derived from the finite squeezing levels, quantum computation should be implemented without involving cluster mode erasings.

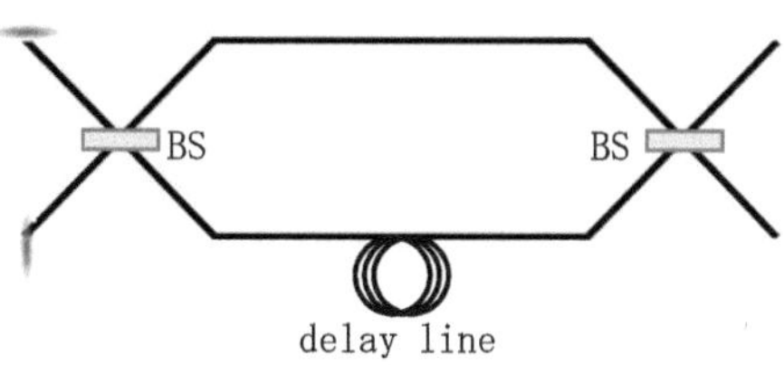

Fig. 10.2 Experimental setup for temporal-mode Gaussian cluster state

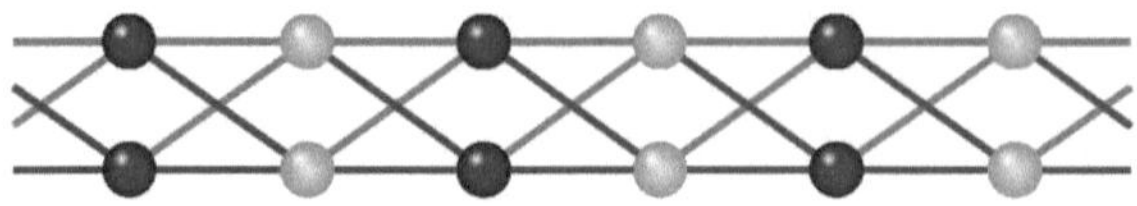

Fig. 10.3 Temporal-mode Gaussian cluster state

In this chapter, we discuss temporal-mode Gaussian cluster states and their application to one-way quantum computations, including solutions to the unsolved problems above. We show that quantum computation using a temporal-mode cluster state for one-mode operations is equivalent to a repetition of quantum teleportations (Sect. 10.3.3). Since we can implement one-mode Gaussian operations with two degrees of freedom by controlling the measurement bases in a quantum teleportation (Sect. 5.3.3), we can utilize all modes of the temporal-mode cluster state in one-way quantum computations without eliminating any of them. For example, when we implement an arbitrary one-mode Gaussian operation based on the original proposal, eight modes of the temporal-mode Gaussian cluster state should be used as a resource. On the contrary, by applying the scheme proposed in this chapter, it can be achieved by using four modes of the temporal-mode Gaussian cluster state. Therefore, the new scheme enables us to implement an arbitrary one-mode Gaussian operation with less errors.

In addition, we show that one-mode non-Gaussian operations and multi-mode Gaussian operations are also achieved without eliminating resource modes (Sects. 10.5 and 10.6) by considering concatenation of a one-way quantum computation circuit with the three-mode linear cluster state, and the quantum computation circuit of the controlled-Z gate experiment with the four-mode linear cluster state (Sect. 7). Multi-mode non-Gaussian operation can be achieved by combining these two components.

Before we move onto the detail, we mention selection of measurement bases. In the following discussion, we provide the explicit set of measurement bases for all one-mode Gaussian operations using a temporal-mode Gaussian cluster state. As for multi-mode Gaussian operations and one-mode non-Gaussian operations, it is not known how the explicit and optimum set of measurement bases can be chosen in order to implement a specific operation in the framework of one-way quantum computation. This thesis is not an exception. That is, in this chapter, we show (1) all components for universality can be implemented by using temporal-mode Gaussian cluster state (universality is achieved, it has already been proven by using a different temporal-mode cluster state in Ref. [1]), and (2) all the degrees of freedom derived from cluster-state modes in these components can be used for operations. Note that, since it is not known how a specific operation can be decomposed into a set of known operations, it is not proven that universal quantum computation could be achieved by using all degrees of freedom of cluster-state modes. That is, one might have to remove several modes of a temporal-mode cluster state in order to follow the decomposition of quantum computation, although the modes to be removed have abilities to implement some operations.

10.2 Nullifiers

We here acquire nullifiers of the two-mode cluster state. The following discussion is derived from Sects. 3.2.4 and 3.3.4 [10].

10.2.1 EPR State and Two-Mode Cluster State

We start our discussion with a quantum state $|p_1 = 0\rangle|p_2 = 0\rangle$. The set of its nullifiers is given by

$$\{\hat{p}_1, \hat{p}_2\}. \tag{10.1}$$

By applying a Fourier operator $\hat{F}_2$ on mode 2, the set of nullifiers for the second quantum state becomes

$$\{\hat{p}_1, -\hat{x}_2\}. \tag{10.2}$$

We then apply a beam splitter operator $\hat{B}_{12}^{(2)}(\frac{1}{\sqrt{2}})$. By using the nullifier transformation rules

$$\hat{p}_1 \to \frac{1}{\sqrt{2}}\hat{p}_1 + \frac{1}{\sqrt{2}}\hat{p}_2, \quad \hat{x}_2 \to -\frac{1}{\sqrt{2}}\hat{x}_1 + \frac{1}{\sqrt{2}}\hat{x}_2, \tag{10.3}$$

for $\hat{B}_{12}^{(2)}(\frac{1}{\sqrt{2}})$, we get the following set of nullifiers for the third quantum state:

$$\left\{\frac{1}{\sqrt{2}}\hat{p}_1 + \frac{1}{\sqrt{2}}\hat{p}_2, \frac{1}{\sqrt{2}}\hat{x}_1 - \frac{1}{\sqrt{2}}\hat{x}_2\right\}. \tag{10.4}$$

Since an operator which is acquired by multiplying a nullifier by a constant is also a nullifier, we get the revised set of nullifiers for the *third* state:

$$\{\hat{p}_1 + \hat{p}_2, \hat{x}_1 - \hat{x}_2\}, \tag{10.5}$$

which represents the Einstein-Podolsky-Rosen state (EPR state) [11, 12]. We apply another Fourier operator $\hat{F}_2$ on mode 2. By using the nullifier transformation rules

$$\hat{x}_2 \to \hat{p}_2, \quad \hat{p}_2 \to -\hat{x}_2 \tag{10.6}$$

for $\hat{F}_2$, we get the set of nullifiers for the fourth quantum state:

$$\{\hat{p}_1 - \hat{x}_2, \hat{x}_1 - \hat{p}_2\}. \tag{10.7}$$

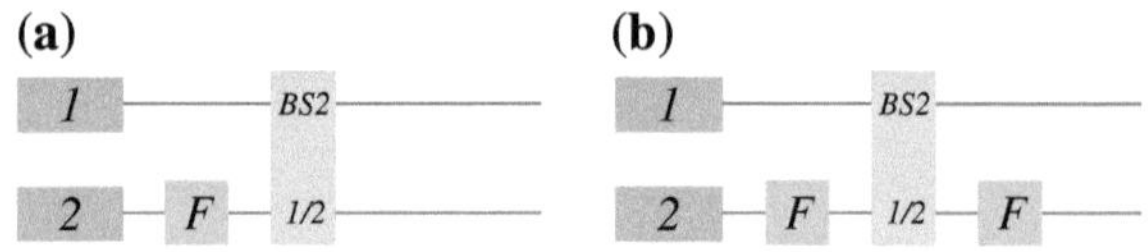

Fig. 10.4 EPR state and two-mode cluster state. **a** EPR state, **b** Two-mode cluster state

Since an operator which is acquired by multiplying a nullifier by a constant is also a nullifier, we get the revised set of nullifiers for the *fourth* state:

$$\{\hat{p}_1 - \hat{x}_2, \hat{p}_2 - \hat{x}_1\}, \tag{10.8}$$

which represents the two-mode cluster state [2, 3].

10.3 One-Mode Gaussian Operations Using Temporal-Mode Gaussian Cluster States

In this section, we describe how we can implement one-mode Gaussian operations using temporal-mode Gaussian cluster states. Although we use ideal quantum resource states $|p = 0\rangle$ in all discussion for simplicity, it can easily be extended to realistic cases where finitely squeezed states are utilized as resources.

10.3.1 Teleportation-Based Input-Coupling Scheme

Implementation methodology of one-mode Gaussian operations using a temporal-mode Gaussian cluster state will be given by considering the concatenation of teleportation-based input-coupling circuits (Sect. 5.3.3) [13]. The schematic of the teleportation-based input-coupling scheme is shown in Fig. 10.5. Although we have assumed that modes 1 and 2 are in a two-mode Gaussian cluster state given by Fig. 10.4b, we can apply discussion in Sect. 5.3.3 to Fig. 10.5 because it becomes identical to the ideal two-mode cluster state in the limit of infinite squeezing: $r \to \infty$ (Sect. 5.6).

In this figure, mode *in* shows the input mode, while we assume that modes 1 and 2 are initially in $|p = 0\rangle_1 |p = 0\rangle_2$. From Sect. 5.3.3, the input-output relation of this quantum circuit is given by

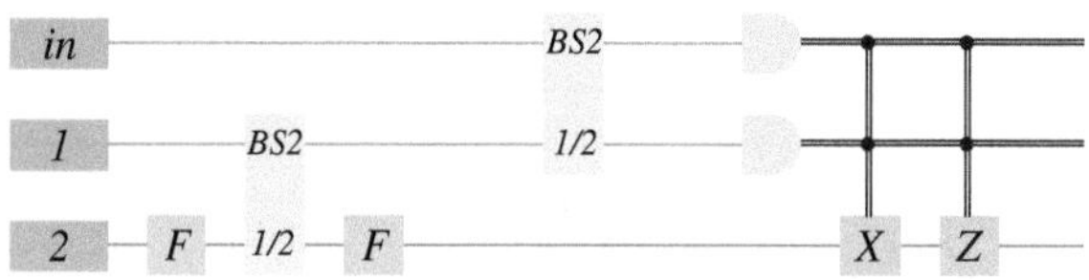

Fig. 10.5 Teleportation-based input-coupling scheme

$$\begin{pmatrix} \hat{x}_{out} \\ \hat{p}_{out} \end{pmatrix} = M_{tele}(\theta_+, \theta_-) \begin{pmatrix} \hat{x}_{in} \\ \hat{p}_{in} \end{pmatrix}, \tag{10.9}$$

where $\theta_\pm = \theta_{in} \pm \theta_1$, while θ_{in} and θ_1 represent the measurement bases (the relative phases between the signal beams and the local oscillator beams for homodyne measurements in modes *in* and 1). Here, the matrix $M_{tele}(\theta_+, \theta_-)$ is given by

$$M_{tele}(\theta_+, \theta_-) = \frac{1}{\sin\theta_-} \begin{pmatrix} \cos\theta_+ + \cos\theta_- & \sin\theta_+ \\ -\sin\theta_+ & \cos\theta_+ - \cos\theta_- \end{pmatrix}. \tag{10.10}$$

It can be decomposed into

$$M_{tele}(\theta_+, \theta_-) = R\Big(-\frac{1}{2}\theta_+\Big) S\Big(\ln\tan\frac{1}{2}\theta_-\Big) R\Big(-\frac{1}{2}\theta_+\Big), \tag{10.11}$$

where

$$R(\theta) = \begin{pmatrix} \cos\theta & -\sin\theta \\ \sin\theta & \cos\theta \end{pmatrix}, \quad S(r) = \begin{pmatrix} e^{-r} & 0 \\ 0 & e^{r} \end{pmatrix}. \tag{10.12}$$

10.3.2 Concatenation of Teleportation-Based Input-Coupling Circuits and Its Operation

10.3.2.1 Concatenation of Teleportation-Based Input-Coupling Circuits

As a next step, we consider concatenation of the teleportation-based input-coupling circuits. Figure 10.6 shows the circuit consisting of four elementary circuits.

Since the input-output relationship of the teleportation-based input-coupling circuit is given by Eq. (10.9), we can easily get the input-output relationship of the four-step circuit:

$$\begin{pmatrix} \hat{x}_{out} \\ \hat{p}_{out} \end{pmatrix} = M_{tele}(\theta_{+4}, \theta_{-4}) M_{tele}(\theta_{+3}, \theta_{-3}) M_{tele}(\theta_{+2}, \theta_{-2}) M_{tele}(\theta_{+1}, \theta_{-1}) \begin{pmatrix} \hat{x}_{in} \\ \hat{p}_{in} \end{pmatrix}. \tag{10.13}$$

Note that we can determine all $\{\theta_{+k}, \theta_{-k}\}$ independently of each other.

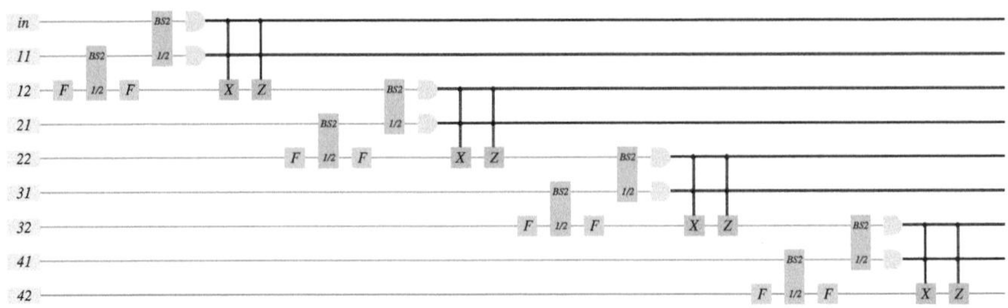

Fig. 10.6 Concatenation of teleportation-based input-coupling circuits

In general, the operation achieved by the n-step teleportation-based input-coupling circuit can be formulated as a product of $M_{tele}(\theta_{+k}, \theta_{-k})$, where we can determine all $\{\theta_{+k}, \theta_{-k}\}$ independently of each other.

10.3.2.2 Universal One-Mode Gaussian Operations

We consider how we can implement an arbitrary one-mode Gaussian operation by using the concatenation of teleportation-based input-coupling circuits.[2]

The basic case is the *one-step* teleportation-based input-coupling circuit. By considering that the degree of freedom of one-mode Gaussian operation is three, while that of the one-step teleportation-based input-coupling circuit is two, we can easily find that universality cannot be achieved since the one-step circuit does not have enough degrees of freedom.

The second basic case is the *two-step* teleportation-based input-coupling circuit. Since its degree of freedom is four, there is possibility that universality is achieved. By using Eq. (10.13), we get the input-output relationship of the two-step circuit:

$$\begin{pmatrix}\hat{x}_{out}\\ \hat{p}_{out}\end{pmatrix} = M_{tele}(\theta_{+2}, \theta_{-2}) M_{tele}(\theta_{+1}, \theta_{-1}) \begin{pmatrix}\hat{x}_{in}\\ \hat{p}_{in}\end{pmatrix} \tag{10.14}$$

$$= R\Big(-\frac{1}{2}\theta_{+2}\Big) S\Big(\ln\tan\frac{1}{2}\theta_{-2}\Big) R\Big(-\frac{1}{2}\theta_{+2}\Big) R\Big(-\frac{1}{2}\theta_{+1}\Big) \times S\Big(\ln\tan\frac{1}{2}\theta_{-1}\Big) R\Big(-\frac{1}{2}\theta_{+1}\Big) \begin{pmatrix}\hat{x}_{in}\\ \hat{p}_{in}\end{pmatrix}. \tag{10.15}$$

By taking $\theta_{-1} = \frac{\pi}{2}$, we get $S\Big(\ln\tan\frac{1}{2}\theta_{-1}\Big) = I$, and thus

$$\begin{pmatrix}\hat{x}_{out}\\ \hat{p}_{out}\end{pmatrix} = R\Big(-\frac{1}{2}\theta_{+2}\Big) S\Big(\ln\tan\frac{1}{2}\theta_{-2}\Big) R\Big(-\frac{1}{2}\theta_{+2} - \theta_{+1}\Big) \begin{pmatrix}\hat{x}_{in}\\ \hat{p}_{in}\end{pmatrix}. \tag{10.16}$$

[2] To be precise, the operations we here consider are members of the Symplectic group. Although displacements in phase space are excluded from the Symplectic group, it is known that they can be implemented by using one-step one-mode teleportation circuit (Sect. 5.5).

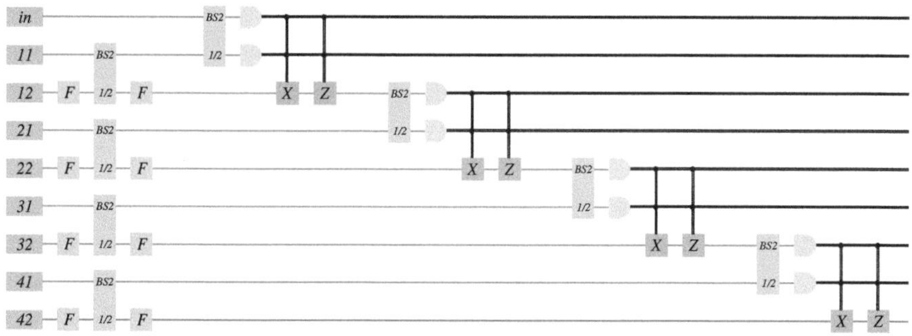

Fig. 10.7 Gathering all Fourier transformations and beam splitters which are used to generate two-mode cluster states

First, the angle of the last (left) rotation operator can be determined arbitrarily by choosing θ_{+2} appropriately. Next, the squeezing parameter of the squeezing operator can be determined arbitrarily by choosing θ_{-2} appropriately. Finally, the angle of the first (right) rotation operator can also be determined arbitrarily by choosing θ_{+1} appropriately, after θ_{+2} has been set. By using the Bloch-Messiah reduction [14] that an arbitrary one-mode Gaussian operation can be decomposed into

$$R(\theta_2)S(r)R(\theta_1), \tag{10.17}$$

we find that an arbitrary one-mode Gaussian operation can be implemented by the two-step teleportation-based input-coupling circuit.

10.3.3 Transformation of Circuit

In this subsection, we consider transformation of the circuit shown in Fig. 10.6 without changing the input-output relation given by Eq. (10.13).

First, all Fourier transformations and beam splitters which are used to generate two-mode cluster states can be gathered at the beginning of the quantum circuit (Fig. 10.7).

Next, we change the order of the first feed-forward operations and the beam splitter for input coupling at the second-step computation. In general, a feed-forward displacement operator $\hat{D}_a$ and a beam splitter operator $\hat{B}_{ab}$ do not commute, where subscripts represent modes on which the operators work. However, since $\hat{D}_a$ is a member of the Pauli group and $\hat{B}_{ab}$ is a member of the Clifford group [15, 16], there exists a Pauli operator $\hat{D}'_{ab}$ which satisfies

$$\hat{B}_{ab}\hat{D}_a\hat{B}^{\dagger}_{ab} = \hat{D}'_{ab}, \tag{10.18}$$

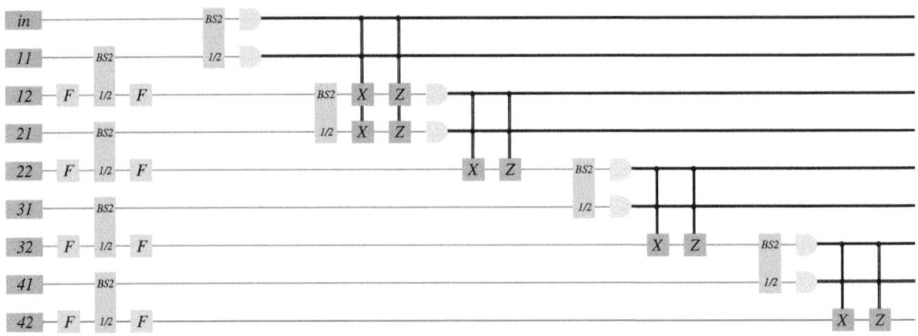

Fig. 10.8 Changing the order of the first feed-forward operations and the beam splitter for input coupling at the second-step computation

leading to

$$\hat{B}_{ab}\hat{D}_a = \hat{D}'_{ab}\hat{B}_{ab}. \tag{10.19}$$

Therefore, we can change the order of the feed-forward operations and the beam splitter by adding another feed-forward operation which works on mode 21, without changing the beam-splitter operation (Fig. 10.8).

Then, we change the order of the new feed-forward operations and the measurements at the second-step computation. By considering that the feed-forward operations are members of the Pauli group which show displacements in phase space, and that the measurement variables in the homodyne measurements are linear combinations of quadrature operators, we find that the set of the feed-forwards and the measurements is equivalent to the same homodyne measurements followed by addition of the previous measurement results to the current measurement results with appropriate gains. By considering that the current measurement results are used for the next feed-forwards, it is also equivalent to add another feed-forward from the previous measurements to mode 22 (Fig. 10.9).

By repeating the procedure above, the quantum circuit can be transformed into Figs. 10.10 and 10.11.

It can be considered as a one-way quantum computation where the resource is a quantum state (shown in Fig. 10.12) which is acquired by entangling two-mode cluster states sequentially by using beam splitters.

Although we have so far discussed by using four-step quantum computation, we can easily extend its discussion to the general n-step case.

10.3.4 Temporal-Mode Gaussian Cluster State Generation

10.3.4.1 Mode Length

In the temporal-mode Gaussian cluster-state scheme, a squeezed beam is virtually split into a sequence of squeezed states with some time period. We define t_c, c, and n to be the time period, the speed of light, and the refractive index. Within this time period, the light propagates

$$l = \frac{ct_c}{n}, \tag{10.20}$$

which we name the *mode length*.

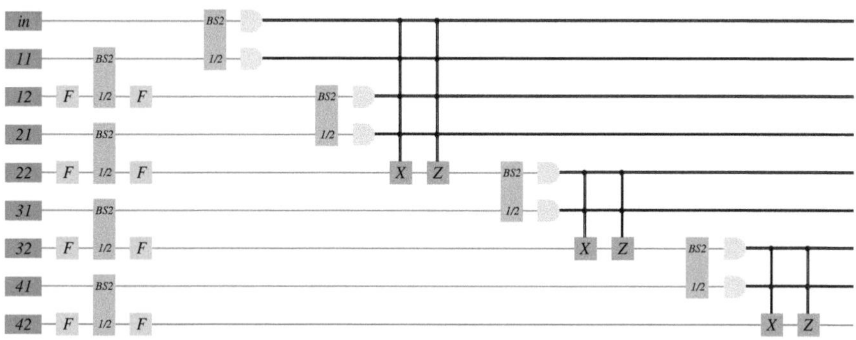

Fig. 10.9 Change the order of the new feed-forward operations and the measurements at the second-step computation

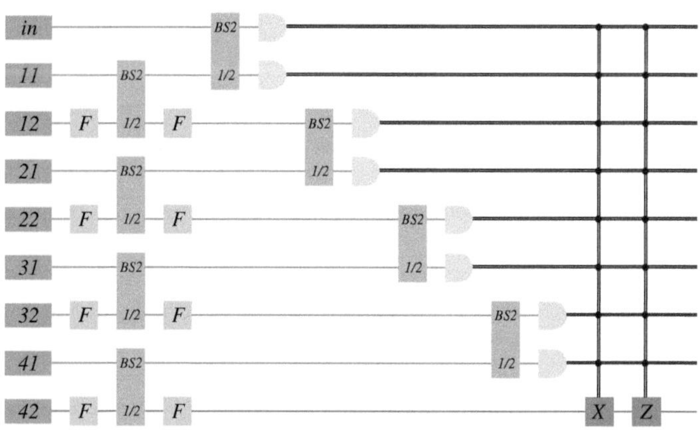

Fig. 10.10 Gathering all feed-forwards

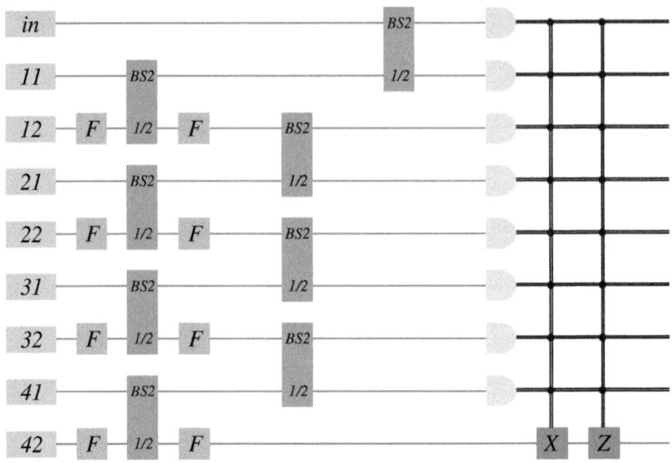

Fig. 10.11 Sorting the beam splitters

Fig. 10.12 Cluster state

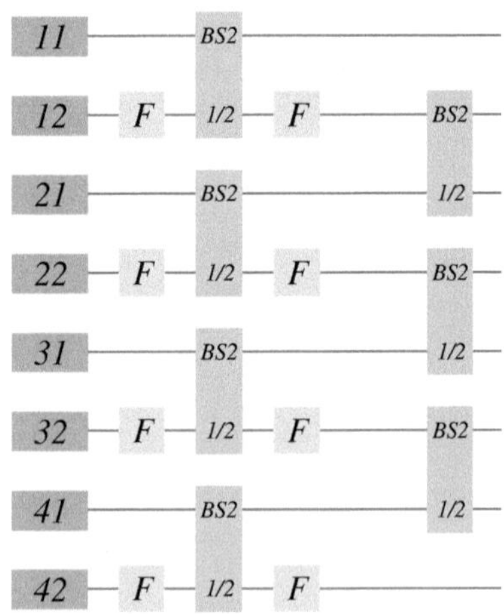

10.3.4.2 Temporal-Mode Gaussian Cluster State Generation

We consider generation of temporal-mode Gaussian cluster states of infinite length based on Fig. 10.12.

The parts of Fig. 10.4b in Fig. 10.12 show generation of two-mode cluster states consisting of modes $\{n1, n2\}$. As is clearly understood from the quantum circuits, a two-mode cluster state can be generated by combining $|p = 0\rangle^{\otimes 2}$ on a half beam splitter. In experiment, $|p = 0\rangle^{\otimes 2}$ are approximated by two p-squeezed states.

We assume that the two-mode cluster state $\{(n+1)1, (n+1)2\}$ is generated after the two-mode cluster state $\{n1, n2\}$. On the latter array of beam splitters in Fig. 10.12, we find that the mode 2 at time $(n-1)$, or equivalently mode $(n-1)2$, and the mode 1 at time n, or equivalently mode $n1$, are combined. Since we cannot combine two modes which do not exist simultaneously, we have to adjust the timing so that these two modes exist simultaneously by using a delay line for mode $(n-1)2$. These two modes can be then combined on the beam splitter after this time delay. Figure 10.13 shows the procedure above.

Figure 10.14 shows where each mode exists at each time. In this figure, we have assumed that both the lengths of the diagonal lines which correspond to the beam splitters, and the length of the upper horizontal line are equal to the *mode length*. Therefore, the length of the lower path including the delay line is twice the length of the mode length.

By using this setup, we can generate the temporal-mode Gaussian cluster state of an arbitrary length.

Although we have assumed in Fig. 10.14 that the lengths of the each diagonal line for the beam splitters and the upper horizontal line are equivalent to the mode length, we do not have to follow this assumption in experiment. For example, we can utilize an experimental setup such as Fig. 10.15, where each line is shortened compared to the mode length.

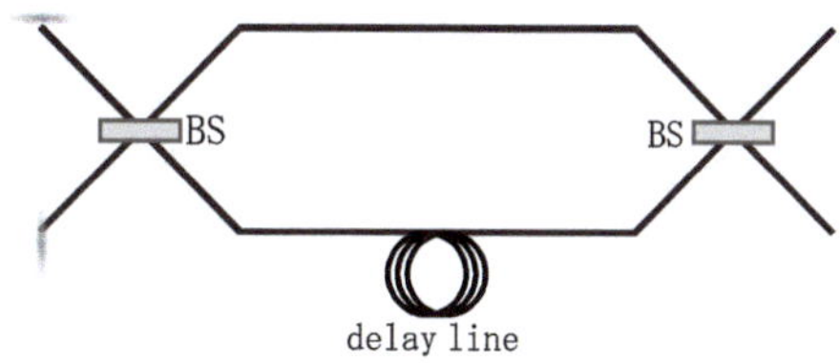

Fig. 10.13 Temporal-mode Gaussian cluster state generation

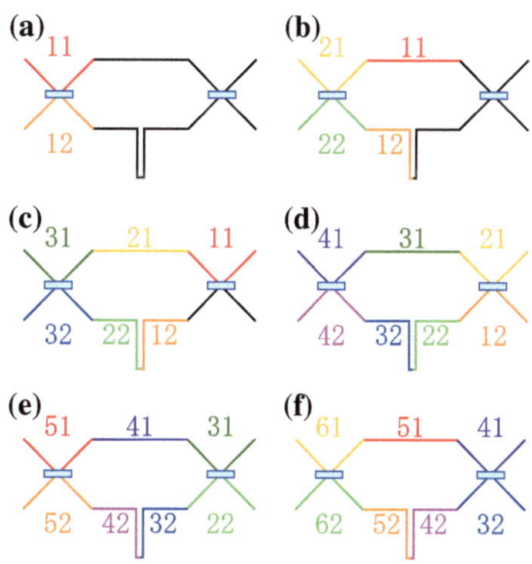

Fig. 10.14 Mode location at each time. **a** Time 1, **b** Time 2, **c** Time 3, **d** Time 4, **e** Time 5, **f** Time 6

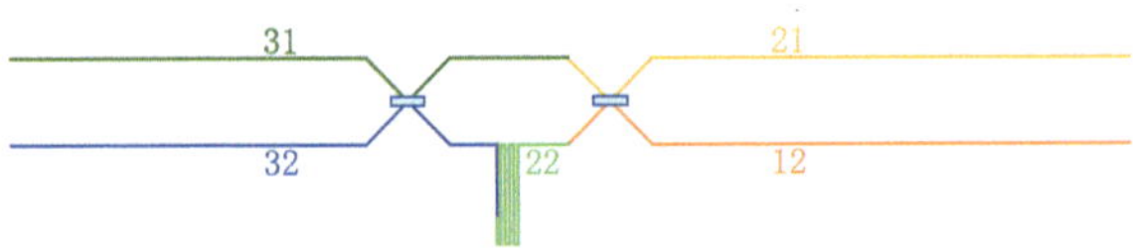

Fig. 10.15 Mode location at each time

10.3.4.3 Equivalence Between the Original Paper and Concatenation of Teleportation-Based Input-Coupling Circuits

The schematic of Fig. 10.13 is nothing but the generation scheme of the temporal-mode Gaussian cluster state proposed in Ref. [1] which can be used to implement one-mode operations. This equivalence between the original paper and the concatenation of the teleportation-based input-coupling circuit provides efficient implementation protocols of one-mode Gaussian operations using the temporal-mode Gaussian cluster state. That is, since Fig. 10.13 shows generation of the resource state which is acquired by transformation of the concatenated teleportation-based input-coupling circuits (Fig. 10.6), the quantum computation using the cluster state shown in Fig. 10.13 is equivalent to that shown in Fig. 10.6. We have already mentioned that the operation achieved by Fig. 10.6 is given by Eq. (10.13), where all the degrees of freedom of cluster-state modes can be utilized for operations. Therefore, we do not have to remove a half of the cluster-state modes as proposed in cite [1], but we can implement one-mode Gaussian operations using all the degrees of freedom of homodyne measurements for one-way quantum computation.

10.3.5 Nullifiers of the Cluster State

In this subsection, we get the nullifiers of the temporal-mode Gaussian cluster state. For simplicity, we will ignore the effect of the both ends of the cluster state.

The quantum state acquired by Fig. 10.4b is the two-mode cluster state, whose nullifiers are given by

$$\{\hat{p}_1 - \hat{x}_2, \hat{p}_2 - \hat{x}_1\}. \tag{10.21}$$

In this subsection, we write the step number explicitly:

$$\{\hat{p}_{n1} - \hat{x}_{n2}, \hat{p}_{n2} - \hat{x}_{n1}\}, \tag{10.22}$$

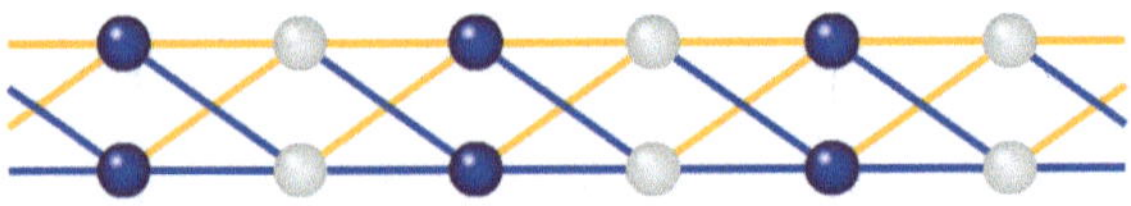

Fig. 10.16 Graph of the temporal-mode Gaussian cluster state

where n represents that it is the set of nullifiers of the two-mode cluster state which is utilized in the n-th teleportation-based input-coupling circuit as a resource. We then apply beam splitter operators $\hat{B}^{(2)}_{(k-1)2,k1}(\frac{1}{\sqrt{2}})$ sequentially on the series of the two-mode cluster states. The effect of the beam splitter operator $\hat{B}^{(2)}_{(n-1)2,n1}(\frac{1}{\sqrt{2}})$ on the set of nullifiers is given by,

$$\hat{x}_{n1} \to -\frac{1}{\sqrt{2}}\hat{x}_{(n-1)2} + \frac{1}{\sqrt{2}}\hat{x}_{n1}, \quad \hat{p}_{n1} \to -\frac{1}{\sqrt{2}}\hat{p}_{(n-1)2} + \frac{1}{\sqrt{2}}\hat{p}_{n1}, \tag{10.23}$$

while that of $\hat{B}^{(2)}_{n2,(n+1)1}(\frac{1}{\sqrt{2}})$ is given by

$$\hat{x}_{n2} \to \frac{1}{\sqrt{2}}\hat{x}_{n2} + \frac{1}{\sqrt{2}}\hat{x}_{(n+1)1}, \quad \hat{p}_{n2} \to \frac{1}{\sqrt{2}}\hat{p}_{n2} + \frac{1}{\sqrt{2}}\hat{p}_{(n+1)1}. \tag{10.24}$$

As a result, we get the new set of nullifiers:

$$\{-\hat{p}_{(n-1)2} + \hat{p}_{n1} - \hat{x}_{n2} - \hat{x}_{(n+1)1}, \hat{p}_{n2} + \hat{p}_{(n+1)1} + \hat{x}_{(n-1)2} - \hat{x}_{n1}\}. \tag{10.25}$$

By replacing n in the second nullifier by $n-1$, we get

$$\{-\hat{p}_{(n-1)2} + \hat{p}_{n1} - \hat{x}_{n2} - \hat{x}_{(n+1)1}, \hat{p}_{(n-1)2} + \hat{p}_{n1} + \hat{x}_{(n-2)2} - \hat{x}_{(n-1)1}\}. \tag{10.26}$$

Since linear combinations of the nullifiers are also nullifiers, we get the final form of the nullifiers by taking the sum and the difference of the nullifiers above:

$$\{2\hat{p}_{n1} - \hat{x}_{n2} - \hat{x}_{(n+1)1} + \hat{x}_{(n-2)2} - \hat{x}_{(n-1)1}, 2\hat{p}_{(n-1)2} + \hat{x}_{(n-2)2} - \hat{x}_{(n-1)1} + \hat{x}_{n2} + \hat{x}_{(n+1)1}\}. \tag{10.27}$$

The graph of the temporal-mode Gaussian cluster state is given by Fig. 10.16, where yellow lines show positive-weight C_Z interactions (the sign is minus in the nullifier), while blue lines show negative-weight C_Z interactions (the sign is positive in the nullifier).

We have so far defined the suffixes so that the n represents timing of the initial two-mode cluster state before the time delay. However, we can also define the suffixes so that it shows timing of the modes of the temporal-mode Gaussian cluster state. Since mode $n1$ and mode $(n-1)2$ exist simultaneously, we redefine the suffixes as $n1 = T1$, $(n-1)2 = T2$, leading to

$$\begin{aligned}&\{2\hat{p}_{T1} - \hat{x}_{(T+1)2} - \hat{x}_{(T+1)1} + \hat{x}_{(T-1)2} - \hat{x}_{(T-1)1},\\ &2\hat{p}_{T2} + \hat{x}_{(T-1)2} - \hat{x}_{(T-1)1} + \hat{x}_{(T+1)2} + \hat{x}_{(T+1)1}\}.\end{aligned} \tag{10.28}$$

In the following discussion, we will use the original n-based suffixes.

10.3.6 Implementation of Quantum Computation with One-by-One Feed-Forward

In this subsection, we propose an experimental setup to implement quantum computation using the temporal-mode scheme. We assume that each feed-forward operation is carried out in each step of the teleportation-based input-coupling circuit.

10.3.6.1 Beginning of Input Coupling

First, two-mode cluster states are generated by the left beam splitter. Since the input mode *in* should be combined with mode 11, the timing of mode *in* is adjusted so that it arrives at the right beam splitter simultaneously with mode 11 (Fig. 10.17).

The square box at the head of mode *in* is a switch which selectively outputs one of two inputs: the input mode or the cluster mode. At this step, it is adjusted so that the input mode is perfectly transmitted to the output port. The displacement system is set so that it has no effect although it is not shown explicitly in Fig. 10.17.

10.3.6.2 Switching From Input to Cluster Path

Figure 10.18 shows the arrangement of modes when the end of the input mode has just passed through the switch for input coupling. At this timing, we change the switch so that the cluster mode is perfectly transmitted to the output port.

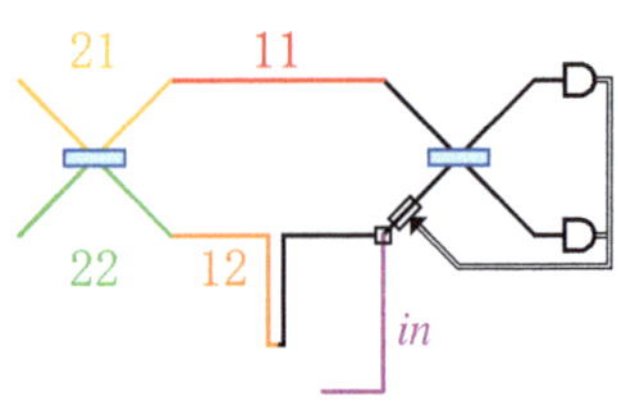

Fig. 10.17 Quantum computation with one-by-one feed-forward 1

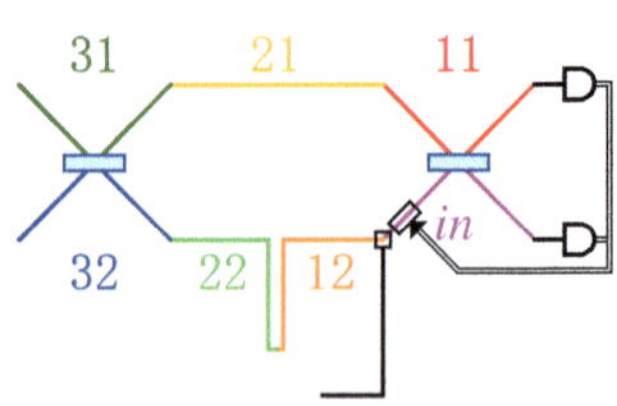

Fig. 10.18 Quantum computation with one-by-one feed-forward 2

10.3.6.3 Beginning of the First Operation

Figure 10.19 shows the beginning of the first operation when the heads of mode *in* and mode 11 have just arrived at the detectors. In the following figures, we assume that the measurement results are instantaneously transmitted to the displacement system without any time delay. Therefore, the head of mode 12, or equivalently the end of mode *in*, is located at the displacement system.

At this timing, we enable the feed-forward system. The measurement results of the heads of mode *in* and mode 11 are feed-forwarded to the head of mode 12.

10.3.6.4 End of the First Operation, Beginning of the Second Operation

Figure 10.20 shows the end of the first operation, or equivalently the beginning of the second operation. Therefore, the ends of mode *in* and mode 11, or equivalently the heads of mode 12 and mode 21, have just arrived at the detectors. In addition, the end of mode 12 has just passed through the displacement system.

The succeeding operations can be achieved by repeating from Figs. 10.19 to 10.20.

10.3.6.5 Beginning of the Last Operation

In the rest of this subsection, we will discuss the end of the whole quantum computation. In order to show its procedure, we use an example of a four-step computation.

The end of computation can be discussed by considering the last-step operation. In our case, we consider the fourth step.

Figure 10.21 shows the beginning of the fourth-step operation. When we implement the four-step computation, we will not use mode 51, mode 52, and later modes. Therefore these modes are described by black letters.

10.3.6.6 Switching to Output the Computation Result

Figure 10.22 shows the arrangement of modes when the head of mode 42, which is the output of the computation, has just arrived at the right beam splitter. At this

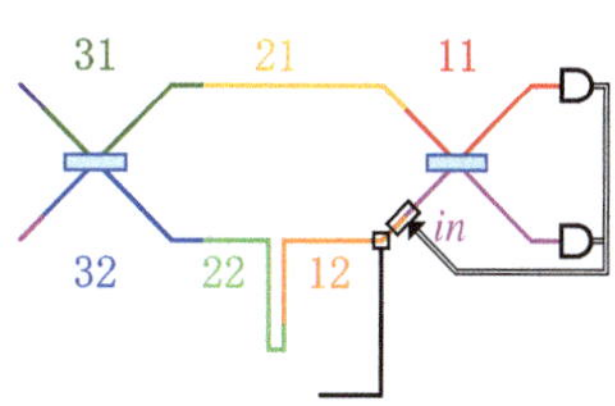

Fig. 10.19 Quantum Computation with one-by-one feed-forward 3

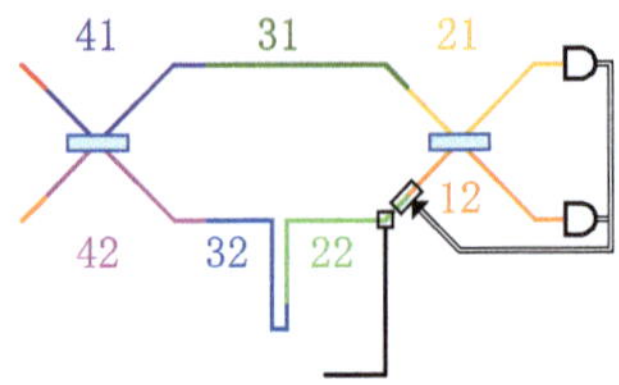

Fig. 10.20 Quantum computation with one-by-one feed-forward 4

timing, we replace the right half beam splitter with a total-reflection mirror. As a result, mode 42 will not be entangled with mode 51.

10.3.6.7 End of Computation

Figure 10.23 shows the end of the whole computation. At this timing, mode 42 has just arrived at the detector. If we want to verify the output state, we can use the existing homodyne measurement system. If we want to use the output quantum state in the succeeding computation, it can be taken out from the circuit by preparing a switch in front of the homodyne detection system.

In this way, we can implement quantum computation with one-by-one feed-forwards.

10.3.7 Implementation of Quantum Computation with Gathered Feed-Forward

In Sect. 10.3.6, we have proposed an experimental setup to implement quantum computation using the temporal-mode scheme, where each feed-forward operation is

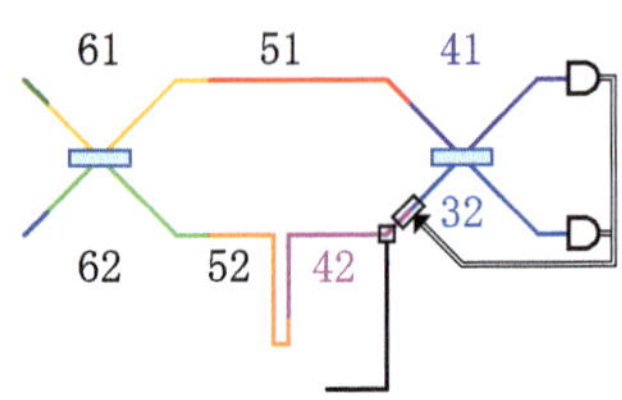

Fig. 10.21 Quantum computation with one-by-one feed-forward 5

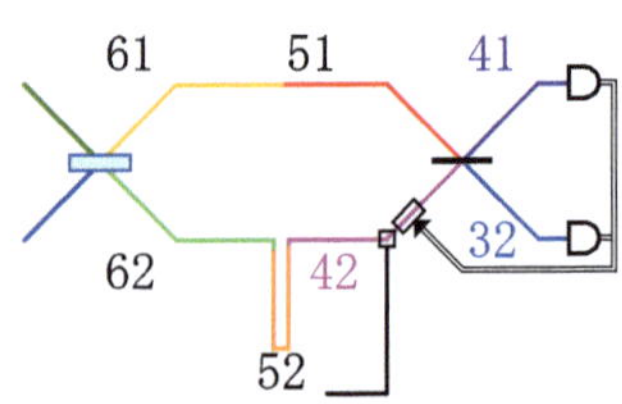

Fig. 10.22 Quantum computation with one-by-one feed-forward 6

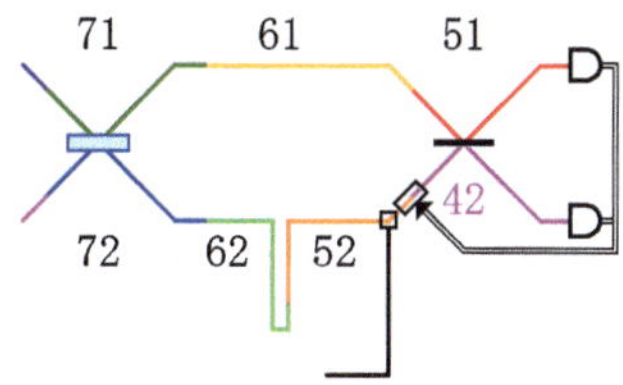

Fig. 10.23 Quantum computation with one-by-one feed-forward 7

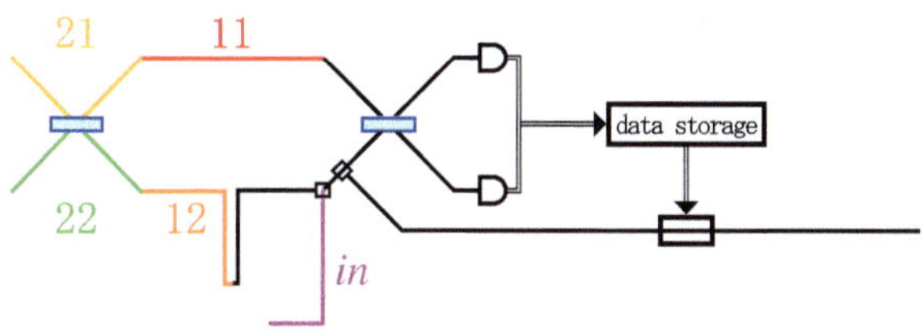

Fig. 10.24 Quantum computation with gathered feed-forward 1

carried out in each step of the teleportation-based input-coupling circuit. In this subsection, we propose another experimental setup where all feed-forwards are gathered at the end of quantum computation. All measurement results are first recorded in a memory. Feed-forward operations are carried out at a stroke by using the all results. This procedure is based on Fig. 10.11.

10.3.7.1 Beginning of Input Coupling

Basic structure of an experimental setup with gathered feed-forward (Fig. 10.24) is similar to that with one-by-one feed-forward. However, there exist several differences. One is that the feed-forward system working in each step of computation is removed from the setup. Instead, a switch is inserted which enables us to pick out the final quantum state from the circuit. During the computation, it is adjusted so that the input of the switch is transmitted to the right beam splitter. At the end of the computation, it is changed so that the input of the switch is transmitted to the feed-forward system specially prepared on the different path.

The input coupling for quantum computation with gathered feed-forward can be achieved in the same way as that with one-by-one feed-forward.

10.3.7.2 Beginning of the First Operation

Figure 10.25 shows the beginning of the first operation.

10.3.7.3 End of the First Operation, Beginning of the Second Operation

Figure 10.26 shows the end of the first operation, or equivalently the beginning of the second operation. The measurement results are stored in a data storage.

All results of the succeeding measurements will also be stored in the data storage.

10.3.7.4 Beginning of the Last Operation

In the rest of this subsection, we will discuss the end of the whole quantum computation. Figure 10.27 shows the beginning of the last operation, where we have assumed that we implement a four-step computation.

Similar to the one-by-one feed-forward case, mode 51, mode 52, and later are not utilized for computation, thus they are described by black letters.

Before the head of mode 42 arrives at the right beam splitter, we have to pick it out and send it to the feed-forward system specially prepared on the different path. Although its switching can be done at an arbitrary timing, we choose that it is carried out at this moment for simplicity.

10.3.7.5 End of the Last Measurement

Figure 10.28 shows the arrangement of modes when the last measurement for the last operation has just been finished.

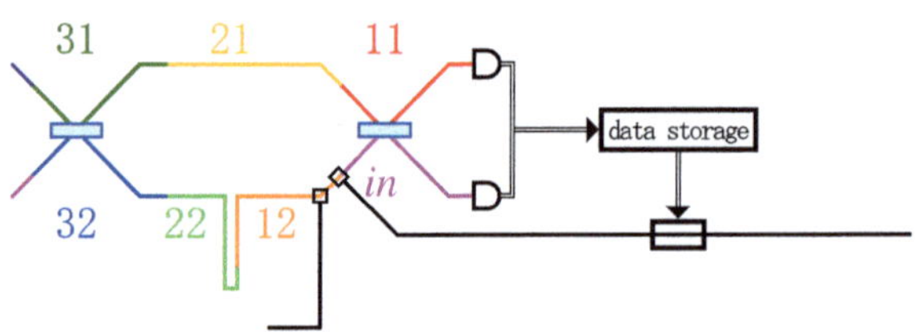

Fig. 10.25 Quantum computation with gathered feed-forward 2

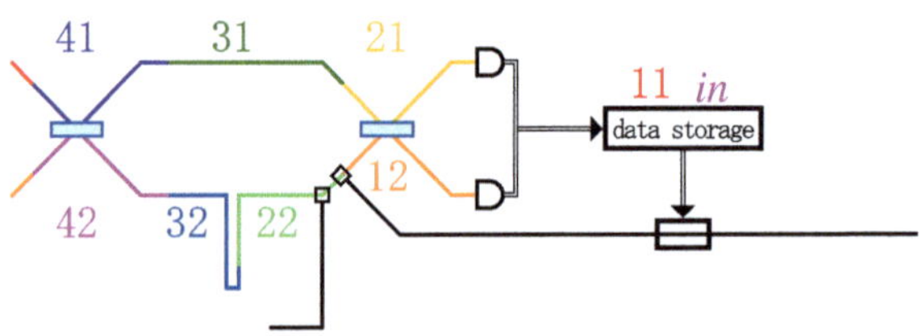

Fig. 10.26 Quantum computation with gathered feed-forward 3

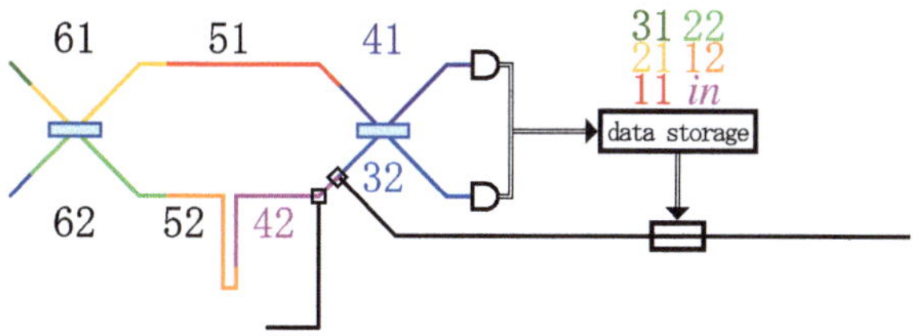

Fig. 10.27 Quantum computation with gathered feed-forward 4

The ends of mode 41 and 32 have arrived at the detectors. All measurement results are now stored in the data storage. The end of mode 42 has just passed through the switch.

10.3.7.6 Feed-Forward

Feed-forward is finally performed on mode 42 (Figs. 10.29, 10.30). It can be achieved at a stroke by adding all measurement results with appropriate gains in advance.

In this way, we can implement quantum computation with gathered feed-forward.

10.4 One-Mode Gaussian Operations Using Temporal-Mode Extended EPR States

In Sect. 10.3, we have described how one-mode Gaussian operations can be achieved by using temporal-mode Gaussian cluster states. The essential resource state was a two-mode linear cluster state. In this section, we discuss implementation of one-mode Gaussian operations using temporal-mode *extended EPR states*, whose essential resource state is an EPR state. Although almost all theoretical discussion in this section is the same to that in Sect. 10.3, it is easier to generate the extended EPR states experimentally than the temporal-mode Gaussian cluster states given in Sect. 10.3.

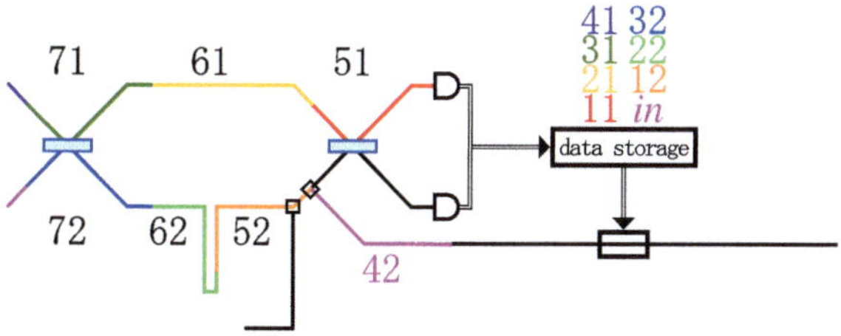

Fig. 10.28 Quantum computation with gathered feed-forward 5

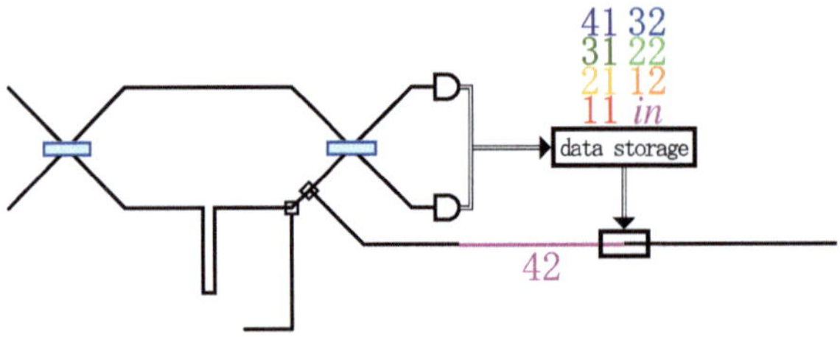

Fig. 10.29 Quantum computation with gathered feed-forward 6

10.4.1 Elementary Circuit

The elementary circuit for this EPR-based scheme is shown in Fig. 10.31.

It is similar to Fig. 10.5, but the latter Fourier transformation is removed. Although it is not shown in Fig. 10.31 explicitly, feed-forward operations are changed as well.

We assume that mode *in* represents the input mode, and mode 1 and 2 are initially in $|p=0\rangle_1|p=0\rangle_2$. The input-output relationship of this quantum circuit is given by

$$\begin{pmatrix}\hat{x}_{out}\\ \hat{p}_{out}\end{pmatrix} = M_{teleE}(\theta_+,\theta_-)\begin{pmatrix}\hat{x}_{in}\\ \hat{p}_{in}\end{pmatrix}, \tag{10.29}$$

where $\theta_\pm = \theta_{in} \pm \theta_1$, while θ_{in} and θ_1 represent the measurement bases for modes *in* and 1. Here, $M_{teleE}(\theta_+,\theta_-)$ is given by

$$M_{teleE}(\theta_+,\theta_-) = \frac{1}{\sin\theta_-}\begin{pmatrix}-\sin\theta_+ & \cos\theta_+ - \cos\theta_-\\ -\cos\theta_+ - \cos\theta_- & -\sin\theta_+\end{pmatrix}. \tag{10.30}$$

Since we have

$$M_{teleE}(\theta_+,\theta_-) = \begin{pmatrix}0 & 1\\ -1 & 0\end{pmatrix} M_{tele}(\theta_+,\theta_-) = R\Big(-\frac{\pi}{2}\Big)M_{tele}(\theta_+,\theta_-), \tag{10.31}$$

thus it can be decomposed into

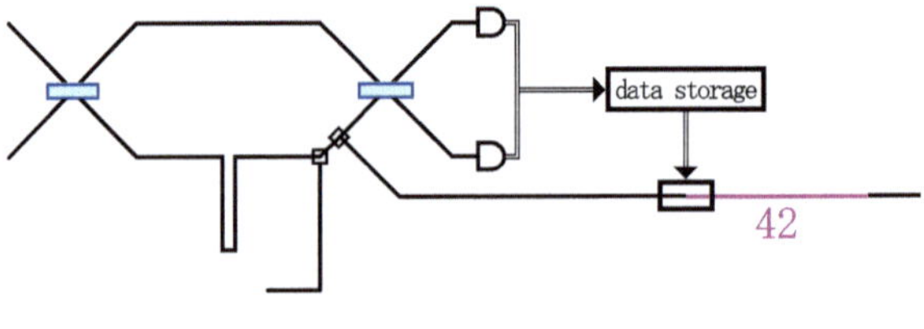

Fig. 10.30 Quantum computation with gathered feed-forward 7

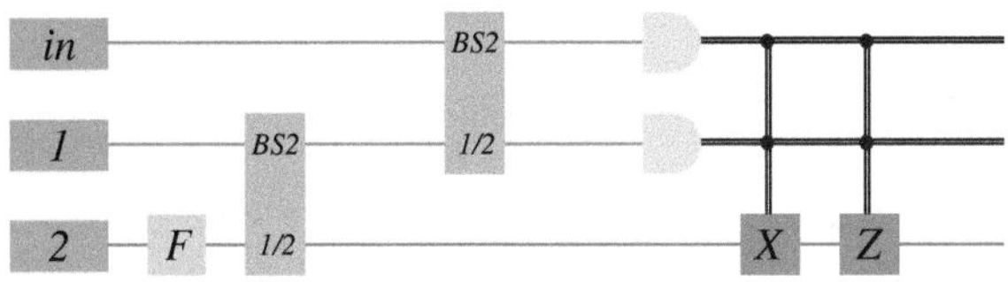

Fig. 10.31 Elementary circuit

$$M_{teleE}(\theta_+, \theta_-) = R\Big(-\frac{\pi}{2} - \frac{1}{2}\theta_+\Big) S\Big(\ln \tan \frac{1}{2}\theta_-\Big) R\Big(-\frac{1}{2}\theta_+\Big). \tag{10.32}$$

10.4.2 Transformation of Concatenated Circuit

10.4.2.1 Concatenation of Elementary Circuits

As a next step, we consider concatenation of the elementary circuits. We here take an example where four steps of the elementary circuits are concatenated. The total input-output relationship is given by

$$\begin{pmatrix} \hat{x}_{out} \\ \hat{p}_{out} \end{pmatrix} = M_{teleE}(\theta_{+4}, \theta_{-4}) M_{teleE}(\theta_{+3}, \theta_{-3}) M_{teleE}(\theta_{+2}, \theta_{-2}) M_{teleE}(\theta_{+1}, \theta_{-1}) \begin{pmatrix} \hat{x}_{in} \\ \hat{p}_{in} \end{pmatrix}, \tag{10.33}$$

where we can determine all $\{\theta_{+k}, \theta_{-k}\}$ independently of each other.

In general, the operation achieved by the n-step elementary circuits can be formulated as a product of $M_{teleE}(\theta_{+k}, \theta_{-k})$, where we can determine all $\{\theta_{+k}, \theta_{-k}\}$ independently of each other.

10.4.2.2 Universal One-Mode Gaussian Operations

We then consider how we can implement an arbitrary one-mode Gaussian operation. Similar to the case in Sect. 10.3, a one-step circuit does not have enough degrees of freedom. The operation by a two-step circuit is

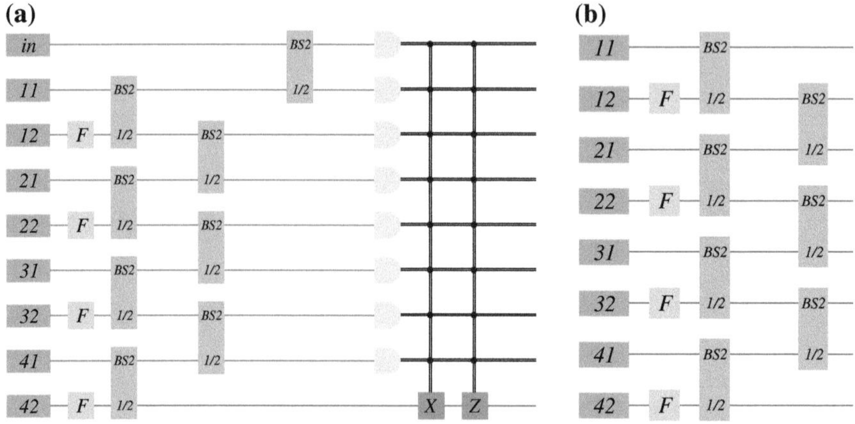

Fig. 10.32 Transformation of EPR-based elementary circuit. **a** Transformation of circuit, **b** Resource state

$$\begin{pmatrix}\hat{x}_{out}\\ \hat{p}_{out}\end{pmatrix} = M_{teleE}(\theta_{+2},\theta_{-2})M_{teleE}(\theta_{+1},\theta_{-1})\begin{pmatrix}\hat{x}_{in}\\ \hat{p}_{in}\end{pmatrix} \tag{10.34}$$

$$= R\Big(-\frac{\pi}{2}-\frac{1}{2}\theta_{+2}\Big)S\Big(\ln\tan\frac{1}{2}\theta_{-2}\Big)R\Big(-\frac{1}{2}\theta_{+2}\Big)R\Big(-\frac{\pi}{2}-\frac{1}{2}\theta_{+1}\Big) \tag{10.35}$$

$$\times\, S\Big(\ln\tan\frac{1}{2}\theta_{-1}\Big)R\Big(-\frac{1}{2}\theta_{+1}\Big)\begin{pmatrix}\hat{x}_{in}\\ \hat{p}_{in}\end{pmatrix}.$$

By taking $\theta_{-1}=\frac{\pi}{2}$, we get $S\Big(\ln\tan\frac{1}{2}\theta_{-1}\Big)=I$, and thus

$$\begin{pmatrix}\hat{x}_{out}\\ \hat{p}_{out}\end{pmatrix} = R\Big(-\frac{\pi}{2}-\frac{1}{2}\theta_{+2}\Big)S\Big(\ln\tan\frac{1}{2}\theta_{-2}\Big)R\Big(-\frac{\pi}{2}-\frac{1}{2}\theta_{+2}-\theta_{+1}\Big)\begin{pmatrix}\hat{x}_{in}\\ \hat{p}_{in}\end{pmatrix}. \tag{10.36}$$

Since it is equivalent to Eq. (10.16) except for fixed rotations, we can easily find that an arbitrary one-mode Gaussian operation can be implemented by the two-step elementary circuits.

10.4.2.3 Transformation

By following the similar discussion in Sect. 10.3.2, the four-step circuit can be transformed into Fig. 10.32a.

Figure 10.32a can be considered as a measurement-based quantum computation where a quantum state shown in Fig. 10.32b is utilized as a resource state. The new resource state is not a member of cluster states since it does not have nullifiers given by $\hat{p}_a - \sum_{a' \in N(a)} g_{a'} \hat{x}_{a'}$. However, it is known that it is equivalent to a cluster state except for local Fourier transformations [17].

10.4.3 Nullifiers of the Extended EPR State

In this subsection, we get the nullifiers of the temporal-mode extended EPR state. For simplicity, we will ignore the effect of the both ends of the cluster state.

The quantum state acquired by Fig. 10.4a is the EPR state, whose nullifiers are given by

$$\{\hat{x}_1 - \hat{x}_2, \hat{p}_1 + \hat{p}_2\}. \tag{10.37}$$

In this subsection, we write the step number explicitly:

$$\{\hat{x}_{n1} - \hat{x}_{n2}, \hat{p}_{n1} + \hat{p}_{n2}\}, \tag{10.38}$$

where n shows that it is the set of nullifiers of the EPR state which is utilized in the n-th circuit as a resource. We then apply beam splitter operators $\hat{B}^{(2)}_{(k-1)2,k1}(\frac{1}{\sqrt{2}})$ sequentially on the series of the EPR states. As a result, we get

$$\{\hat{x}_{(n-1)2} - \hat{x}_{n1} + \hat{x}_{n2} + \hat{x}_{(n+1)1}, -\hat{p}_{(n-1)2} + \hat{p}_{n1} + \hat{p}_{n2} + \hat{p}_{(n+1)1}\}. \tag{10.39}$$

This is the nullifiers of the extended EPR state.

We have so far defined the suffixes so that the n represents timing of the initial EPR state before the time delay. However, we can also define the suffixes so that it shows timing of the modes of the temporal-mode extended EPR state. Since mode $n1$ and mode $(n-1)2$ exist simultaneously, we redefine the suffixes as $n1 = T1$, $(n-1)2 = T2$, leading to

$$\{\hat{x}_{T2} - \hat{x}_{T1} + \hat{x}_{(T+1)2} + \hat{x}_{(T+1)1}, -\hat{p}_{T2} + \hat{p}_{T1} + \hat{p}_{(T+1)2} + \hat{p}_{(T+1)1}\}. \tag{10.40}$$

10.5 One-Mode Non-Gaussian Operation Using Temporal-Mode Gaussian Cluster State

We have shown that we can implement an arbitrary one-mode Gaussian operation by means of the temporal-mode cluster scheme, by using all the degrees of freedom of cluster modes for quantum computation. In this section, we mention one-mode non-Gaussian operations with temporal-mode Gaussian cluster states.

In Ref. [1], it has already been proposed that one can implement an arbitrary one-mode operations, including non-Gaussian operations, by using the same temporal-mode Gaussian cluster state for Gaussian operations. A half modes of the cluster state is removed via cluster-mode erasing in advance, leading to a universal resource for one-mode unitary operations: a linear cluster state.

In this section, on the other hand, we propose a protocol to implement one-mode non-Gaussian operations, where all the degrees of freedom of cluster modes can be utilized for quantum computation.[3]

10.5.1 Elementary Circuit

10.5.1.1 Elementary Circuit

The elementary circuit for one-mode non-Gaussian operations is shown in Fig. 10.33, where mode *in* represents the mode of an input state, while the initial states in mode 1, 2, and 3 are zero eigenstates of the momentum operators: $|p = 0\rangle$.

Figure 10.33 consists of two parts: a teleportation-based input-coupling circuit followed by an elementary one-mode one-way gate (one-mode teleportation circuit, half teleportation circuit). The operator $\hat{D}$ is given by $\hat{D} = e^{\frac{i}{\hbar} f(\hat{x})}$, where the order of a polynomial function $f(x)$ can be more than or equal to three.

The operation by the teleportation-based input-coupling circuit is given by

$$\begin{pmatrix} \hat{x}' \\ \hat{p}' \end{pmatrix} = M_{tele}(\theta_+, \theta_-) \begin{pmatrix} \hat{x}_{in} \\ \hat{p}_{in} \end{pmatrix}, \tag{10.41}$$

$$M_{tele}(\theta_+, \theta_-) = \frac{1}{\sin\theta_-} \begin{pmatrix} \cos\theta_+ + \cos\theta_- & \sin\theta_+ \\ -\sin\theta_+ & \cos\theta_+ - \cos\theta_- \end{pmatrix}, \tag{10.42}$$

where $\theta_\pm = \theta_{in} \pm \theta_1$ (5.3.3). Here, θ_{in} and θ_1 are measurement bases (relative phases between signal beams and local oscillator beams) in homodyne detections [13].

The input-output relationship of the succeeding elementary one-mode teleportation circuit is given by

$$\hat{x}_{out} + i\,\hat{p}_{out} = -\hat{p}' - f'(\hat{x}) + i\hat{x}', \tag{10.43}$$

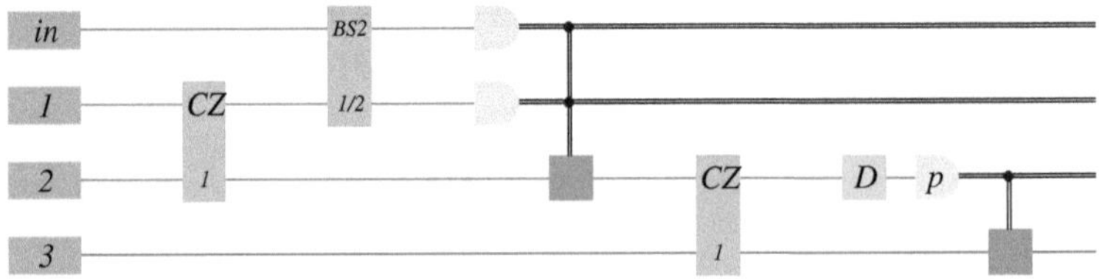

Fig. 10.33 Elementary circuit for one-mode non-Gaussian operation 1

[3] Note again that one might have to remove several modes of a temporal-mode cluster state although the modes to be removed have abilities to implement some operations. See Sect. 10.1.4.

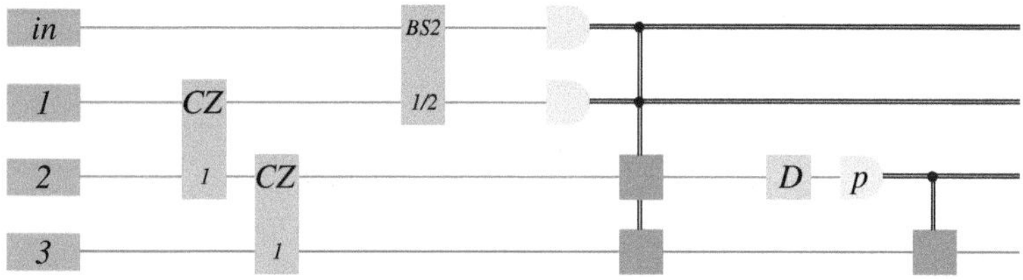

Fig. 10.34 Elementary circuit for one-mode non-Gaussian operation 2

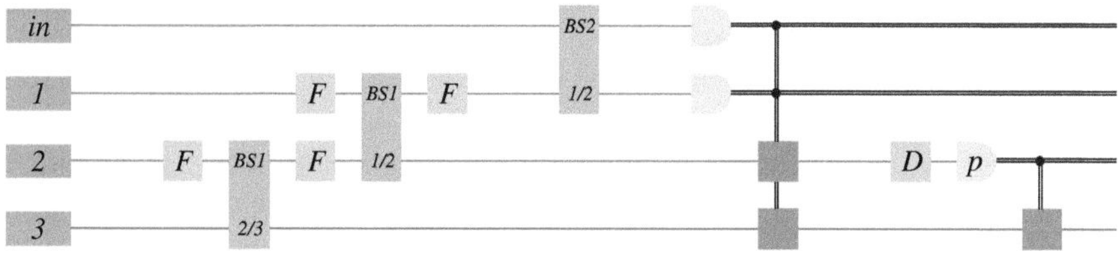

Fig. 10.35 Elementary circuit for one-mode non-Gaussian operation 3

where $f'(x) = \dfrac{\mathrm{d}f(x)}{\mathrm{d}x}$ is the first derivative of the function $f(x)$ with respect to x (Eq. 4.58). If the order of $f(x)$ is more than or equal to three, the order of $f'(x)$ is at least two. Therefore, non-Gaussian operations can be implemented [16].

10.5.1.2 Transformation of Elementary Circuit

We consider transformation of the elementary circuit shown in Fig. 10.33 without changing the input-output relation. The revised circuit will be the elementary circuit for one-mode non-Gaussian operations using the temporal-mode scheme.

We first consider a change of the order of the first feed-forward and the succeeding controlled-Z (C_Z) gate. Since $\hat{D}_a$ is a member of the Pauli group and $\hat{C}_{Zab}$ is a member of the Clifford group, there exists a Pauli group operator $\hat{D}'_{ab}$ which satisfies

$$\hat{C}_{Zab}\hat{D}_a\hat{C}^{\dagger}_{Zab} = \hat{D}'_{ab}, \tag{10.44}$$

leading to

$$\hat{C}_{Zab}\hat{D}_a = \hat{D}'_{ab}\hat{C}_{Zab}. \tag{10.45}$$

Thus, we can change the order of the feed-forward and the C_Z gate by adding another feed-forward operation which works on mode 3 (Fig. 10.34).

We then note that the quantum state acquired by entangling $|p=0\rangle^{\otimes 3}$ in mode 1 , 2, and 3 by using two C_Z gates is nothing but the three-mode linear cluster state. Since the same state can be generated by using an appropriate network of beam splitters (Gaussian cluster state, Sect. 5.2.2, Eq. 6.43), we can replace the two C_Z gates with Fourier operators and beam splitters (Fig. 10.35).

We use the revised circuit as the elementary circuit for one-mode non-Gaussian operations using the temporal-mode scheme.

10.5.2 Concatenation of Revised Elementary Circuits

We consider concatenation of the revised elementary circuits. Although we consider three steps of the elementary circuits as an example in the following discussion (Fig. 10.36), it can easily be extended to the n-step case in general. In this figure, the network of beam splitters including Fourier transformations for generation of a three-mode linear cluster state is abbreviated to "3 mode linear cluster state".

We transform the concatenated circuit without changing the input-output relation. First, all beam splitter networks can be gathered at the beginning of the circuit because they don't have interaction with the other modes (Fig. 10.37).

Next, we consider a change of the order of the beam splitter for the second input coupling on mode 13 and 21, and the feed-forward operation next to it (feed-forward from mode 12 to mode 13). Similar to the previous discussion, since the feed-forward operator is a member of the Pauli group, while the beam splitter operator is a member of the Clifford group, we can change the circuit into Fig. 10.38 by adding another displacement feed-forward on mode 21.

Then, we change the order of this beam splitter and the first feed-forward operators on mode 12 and 13. It can be achieved by adding another displacement feed-forward on mode 21 (Fig. 10.39).

Therefore, we can move the beam splitter for the second input coupling to the front of the first-step measurements and feed-forwards.

In a similar manner, we can move the beam splitter for the third input coupling (Figs. 10.40 and 10.41).

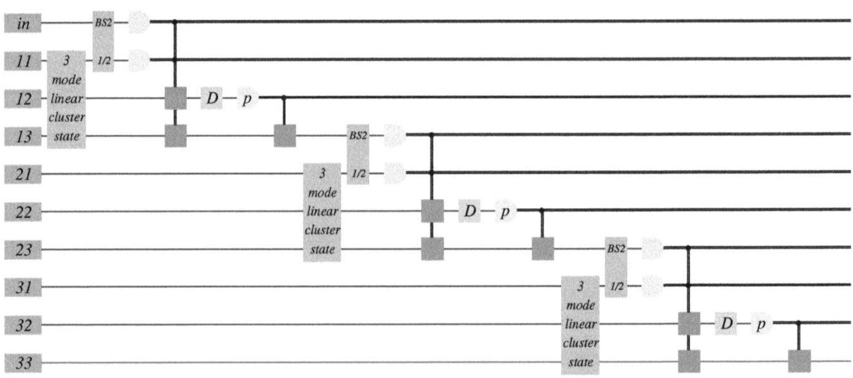

Fig. 10.36 Concatenation of revised elementary circuits 1

In Fig. 10.41, all beam splitters are gathered at the beginning of the circuit. It can be regarded as a one-way quantum computation where a Gaussian cluster state shown in Fig. 10.42 is utilized as a resource, and an input quantum state is coupled with the cluster state via quantum teleportation. A Non-Gaussian operation can be achieved via the operator $\hat{D}$ at the third measurement in each step. In addition, all the degrees of freedom of all cluster modes can be used for operations. Furthermore, an arbitrary one-mode non-Gaussian operation can be implemented by repeating this procedure.

In an n-step case, the resource is the state acquired by sequentially entangling modes 1 and modes 3 of n three-mode linear cluster states using $n-1$ beam splitters.

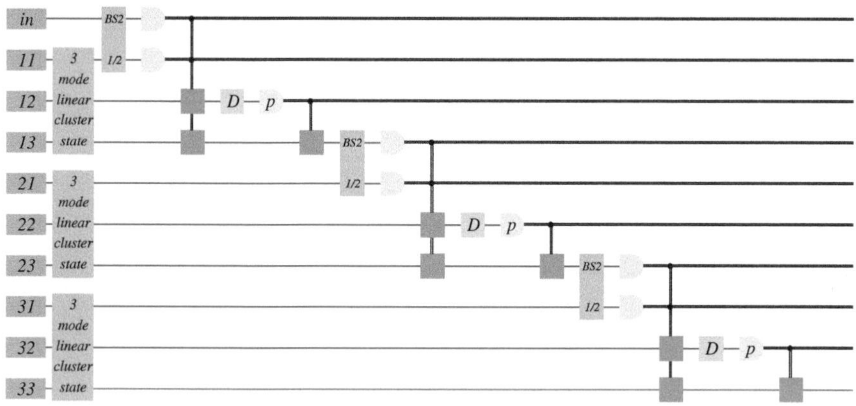

Fig. 10.37 Concatenation of revised elementary circuits 2

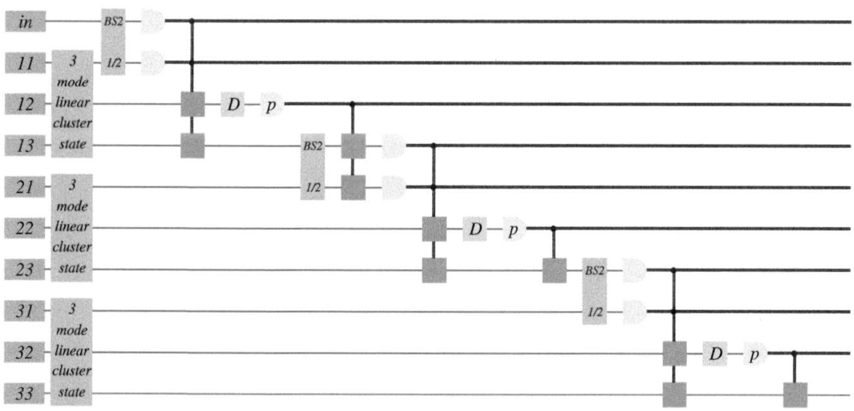

Fig. 10.38 Concatenation of revised elementary circuits 3

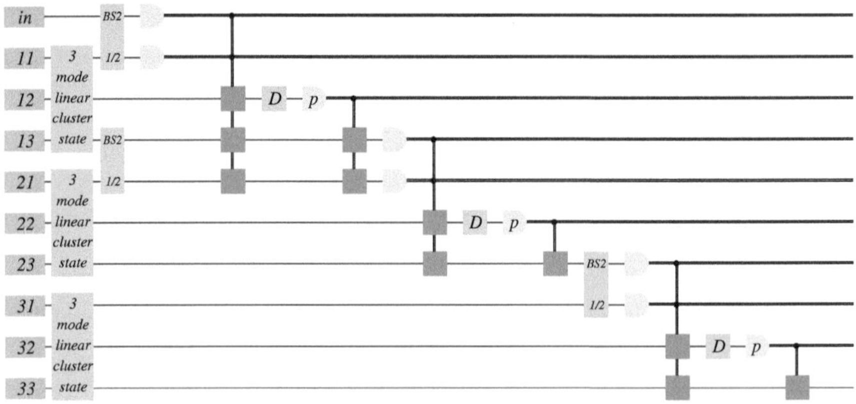

Fig. 10.39 Concatenation of revised elementary circuits 4

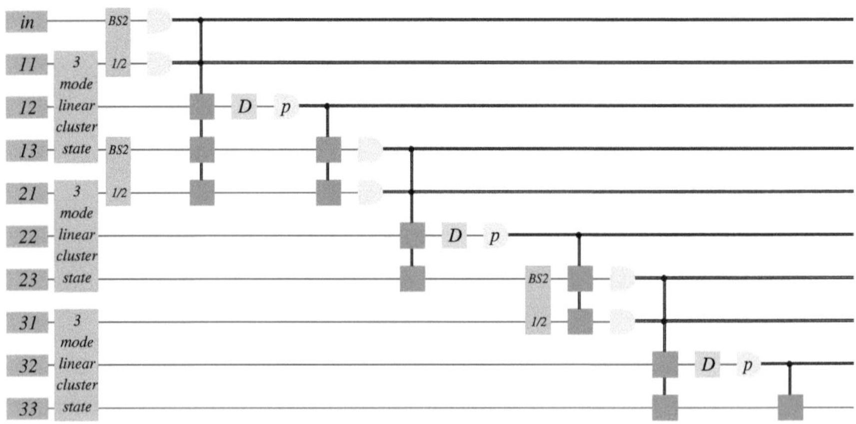

Fig. 10.40 Concatenation of revised elementary circuits 5

10.5.3 Generation of Temporal-Mode Gaussian Cluster State

Figure 10.43 shows an experimental setup for generation of the temporal-mode Gaussian cluster state which can be utilized for implementation of one-mode non-Gaussian operations. In this figure, we have explicitly shown the arrangement of cluster modes at each time. Note that mode 21 and mode 22 have no interaction in Fig. 10.43b although they cross each other.

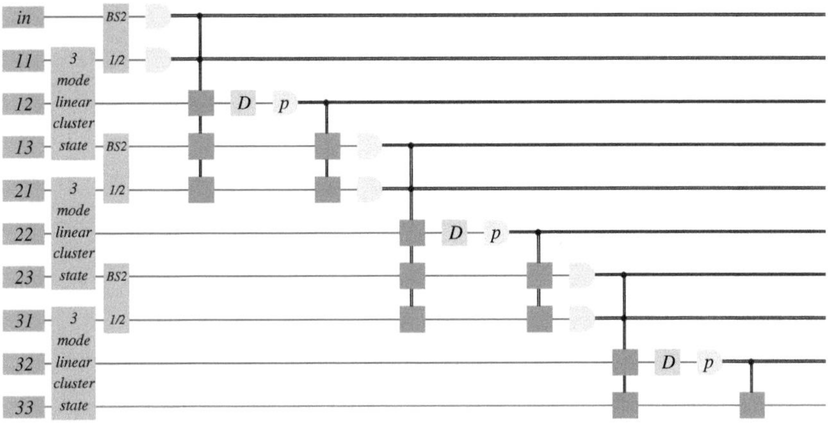

Fig. 10.41 Concatenation of revised elementary circuits 6

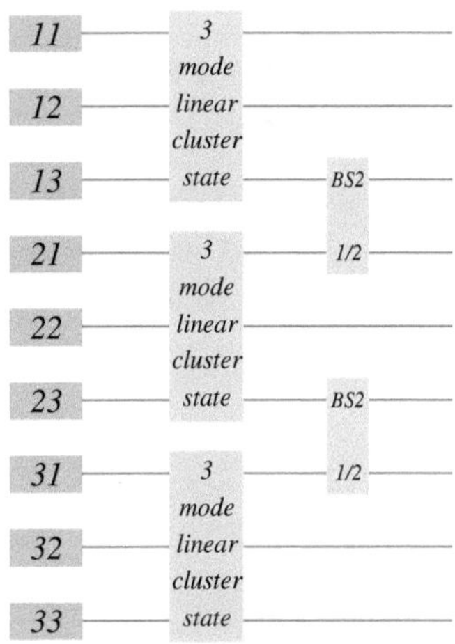

Fig. 10.42 Concatenation of revised elementary circuits 7

10.5.4 Nullifiers of the Cluster State

In this subsection, we get the nullifiers of the temporal-mode Gaussian cluster state. We will ignore the effect of the both ends of the cluster state.

First, the nullifiers of a three-mode linear cluster state are given by

$$\{\hat{p}_1 - \hat{x}_2, \hat{p}_2 - \hat{x}_1 - \hat{x}_3, \hat{p}_3 - \hat{x}_2\}. \tag{10.46}$$

In this subsection, we write the step number explicitly:

$$\{\hat{p}_{n1} - \hat{x}_{n2}, \hat{p}_{n2} - \hat{x}_{n1} - \hat{x}_{n3}, \hat{p}_{n3} - \hat{x}_{n2}\}. \tag{10.47}$$

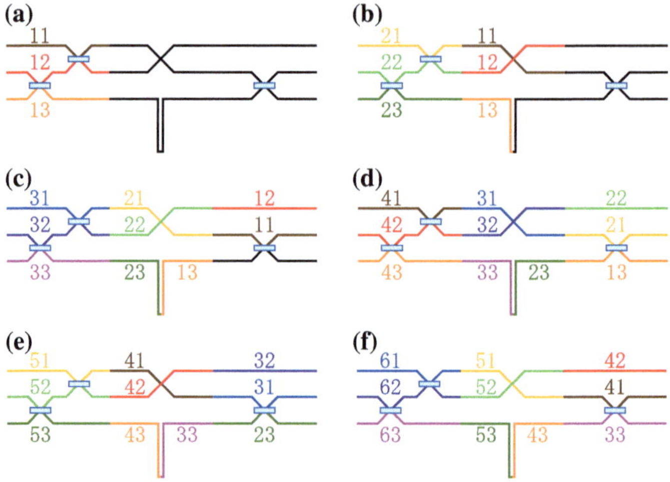

Fig. 10.43 Generation of temporal-mode Gaussian cluster state. **a** Time 1, **b**, Time 2, **c** Time 3, **d** Time 4, **e** Time 5, **f** Time 6

where n shows that it is the set of nullifiers of the three-mode linear cluster state which is utilized in the n-th step as a resource. We then apply beam splitter operators $\hat{B}^{(2)}_{(k-1)3,k1}(\frac{1}{\sqrt{2}})$ sequentially on the series of the three-mode linear cluster states. The effect of the beam splitter operator $\hat{B}^{(2)}_{(k-1)3,k1}(\frac{1}{\sqrt{2}})$ on the set of nullifiers is given by,

$$\hat{x}_{n1} \rightarrow -\frac{1}{\sqrt{2}}\hat{x}_{(n-1)3} + \frac{1}{\sqrt{2}}\hat{x}_{n1}, \quad \hat{p}_{n1} \rightarrow -\frac{1}{\sqrt{2}}\hat{p}_{(n-1)3} + \frac{1}{\sqrt{2}}\hat{p}_{n1}, \tag{10.48}$$

while that of $\hat{B}^{(2)}_{n3,(n+1)1}(\frac{1}{\sqrt{2}})$ is given by

$$\hat{x}_{n3} \rightarrow \frac{1}{\sqrt{2}}\hat{x}_{n3} + \frac{1}{\sqrt{2}}\hat{x}_{(n+1)1}, \quad \hat{p}_{n3} \rightarrow \frac{1}{\sqrt{2}}\hat{p}_{n3} + \frac{1}{\sqrt{2}}\hat{p}_{(n+1)1}. \tag{10.49}$$

As a result, we get the new set of nullifiers:

$$\{-\hat{p}_{(n-1)3} + \hat{p}_{n1} - \sqrt{2}\hat{x}_{n2}, \sqrt{2}\hat{p}_{n2} + \hat{x}_{(n-1)3} - \hat{x}_{n1} - \hat{x}_{n3} - \hat{x}_{(n+1)1}, \hat{p}_{n3} + \hat{p}_{(n+1)1} - \sqrt{2}\hat{x}_{n2}\}. \tag{10.50}$$

By replacing n in the third nullifier by $n-1$, we get

$$\{-\hat{p}_{(n-1)3} + \hat{p}_{n1} - \sqrt{2}\hat{x}_{n2}, \sqrt{2}\hat{p}_{n2} + \hat{x}_{(n-1)3} - \hat{x}_{n1} - \hat{x}_{n3} - \hat{x}_{(n+1)1}, \hat{p}_{(n-1)3} + \hat{p}_{n1} - \sqrt{2}\hat{x}_{(n-1)2}\}. \tag{10.51}$$

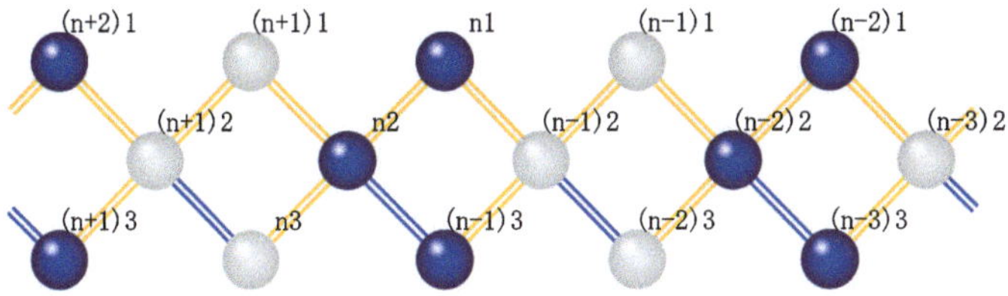

Fig. 10.44 Graph of cluster state

By taking linear combinations of nullifiers, we get

$$\{\sqrt{2}\hat{p}_{n1} - \hat{x}_{n2} - \hat{x}_{(n-1)2}, \sqrt{2}\hat{p}_{n2} + \hat{x}_{(n-1)3} - \hat{x}_{n1} - \hat{x}_{n3} - \hat{x}_{(n+1)1}, \sqrt{2}\hat{p}_{(n-1)3} + \hat{x}_{n2} - \hat{x}_{(n-1)2}\}. \tag{10.52}$$

By replacing $n-1$ in the third nullifier by n, we get

$$\{\sqrt{2}\hat{p}_{n1} - \hat{x}_{n2} - \hat{x}_{(n-1)2}, \sqrt{2}\hat{p}_{n2} + \hat{x}_{(n-1)3} - \hat{x}_{n1} - \hat{x}_{n3} - \hat{x}_{(n+1)1}, \sqrt{2}\hat{p}_{n3} + \hat{x}_{(n+1)2} - \hat{x}_{n2}\}, \tag{10.53}$$

which is the final form of the set of nullifiers. The graph of this state is shown in Fig. 10.44. Three modes $\{k1, k2, (k-1)3\}$ shown in the same color are generated simultaneously.

10.5.5 Using Multi-Mode Linear Cluster State

10.5.5.1 Using Multi-Mode Linear Cluster State

We have so far utilized two-mode and three-mode linear cluster states as initial cluster resource states for temporal-mode schemes. By increasing the number of modes from two to three, we can add a one-mode teleportation circuit. As a result, teleportation-based input-coupling operations with two degrees of freedom as well as a one-mode non-Gaussian operation can be implemented.

We can easily extend this discussion to the cases where larger multi-mode linear cluster states are utilized as initial cluster resources for temporal-mode schemes. By using an N-mode linear cluster state, we can implement teleportation-based input-coupling operations with two degrees of freedom, as well as one-mode non-Gaussian operations with $n-2$ degrees of freedom. Therefore, non-Gaussian operations can be implemented more efficiently.

A compensation for this is that we have to prepare N squeezed states in order to generate an N-mode linear cluster state. It might lead to a larger-scale experimental setup. However, it is known how an N-mode linear Gaussian cluster state can be generated, that is, it can be generated by combining N squeezed states on $N-1$

beam splitters sequentially. By using a similar concept of the temporal-mode scheme, we can also prepare N squeezed states from a squeezed beam by splitting it with some time period T.

10.5.5.2 Nullifiers of the Cluster State

The nullifiers of an N-mode linear cluster state with $N \geq 4$ are given by

$$\{\hat{p}_{n1} - \hat{x}_{n2}, \hat{p}_{n2} - \hat{x}_{n1} - \hat{x}_{n3}, \hat{p}_{ni} - \hat{x}_{n(i-1)} - \hat{x}_{n(i+1)}, \\ \hat{p}_{n(N-1)} - \hat{x}_{n(N-2)} - \hat{x}_{nN}, \hat{p}_{nN} - \hat{x}_{n(N-1)}\}, \quad i = 3, \ldots, N-2. \tag{10.54}$$

where n shows that it is the set of nullifiers of the N-mode linear cluster state which is utilized in the n-th operations as a resource. Note that there exists no term with i when $N = 4$. We then apply beam splitter operators $\hat{B}^{(2)}_{(k-1)N,k1}(\frac{1}{\sqrt{2}})$ sequentially on the series of the N-mode cluster state. The effect of the beam splitter operator $\hat{B}^{(2)}_{(n-1)N,n1}(\frac{1}{\sqrt{2}})$ on the set of nullifiers is given by,

$$\hat{x}_{n1} \to -\frac{1}{\sqrt{2}}\hat{x}_{(n-1)N} + \frac{1}{\sqrt{2}}\hat{x}_{n1}, \quad \hat{p}_{n1} \to -\frac{1}{\sqrt{2}}\hat{p}_{(n-1)N} + \frac{1}{\sqrt{2}}\hat{p}_{n1}, \tag{10.55}$$

while that of $\hat{B}^{(2)}_{nN,(n+1)1}(\frac{1}{\sqrt{2}})$ is given by

$$\hat{x}_{nN} \to \frac{1}{\sqrt{2}}\hat{x}_{nN} + \frac{1}{\sqrt{2}}\hat{x}_{(n+1)1}, \quad \hat{p}_{nN} \to \frac{1}{\sqrt{2}}\hat{p}_{nN} + \frac{1}{\sqrt{2}}\hat{p}_{(n+1)1}. \tag{10.56}$$

As a result, we get the new set of nullifiers:

$$\Bigg\{-\frac{1}{\sqrt{2}}\hat{p}_{(n-1)N} + \frac{1}{\sqrt{2}}\hat{p}_{n1} - \hat{x}_{n2}, \hat{p}_{n2} + \frac{1}{\sqrt{2}}\hat{x}_{(n-1)N} - \frac{1}{\sqrt{2}}\hat{x}_{n1} - \hat{x}_{n3}, \\ \hat{p}_{ni} - \hat{x}_{n(i-1)} - \hat{x}_{n(i+1)}, \hat{p}_{n(N-1)} - \hat{x}_{n(N-2)} - \frac{1}{\sqrt{2}}\hat{x}_{nN} - \frac{1}{\sqrt{2}}\hat{x}_{(n+1)1}, \\ \frac{1}{\sqrt{2}}\hat{p}_{nN} + \frac{1}{\sqrt{2}}\hat{p}_{(n+1)1} - \hat{x}_{n(N-1)}\Bigg\}, \quad i = 3, \ldots, N-2, \tag{10.57}$$

leading to,

$$\left\{\hat{p}_{n1} - \frac{1}{\sqrt{2}}\hat{x}_{n2} - \frac{1}{\sqrt{2}}\hat{x}_{(n-1)(N-1)}, \hat{p}_{n2} + \frac{1}{\sqrt{2}}\hat{x}_{(n-1)N} - \frac{1}{\sqrt{2}}\hat{x}_{n1} - \hat{x}_{n3}, \right.$$
$$\hat{p}_{ni} - \hat{x}_{n(i-1)} - \hat{x}_{n(i+1)}, \hat{p}_{n(N-1)} - \hat{x}_{n(N-2)} - \frac{1}{\sqrt{2}}\hat{x}_{nN} - \frac{1}{\sqrt{2}}\hat{x}_{(n+1)1},$$
$$\left. \hat{p}_{nN} - \frac{1}{\sqrt{2}}\hat{x}_{n(N-1)} + \frac{1}{\sqrt{2}}\hat{x}_{(n+1)2}\right\}, \quad i = 3, \ldots, N-2. \tag{10.58}$$

It shows that modes $1, 2, N-1$, and N of the N-mode linear cluster state have correlations with the previous or the next cluster state.

The graphs of $N = 4$ and $N = 5$ are shown in Figs. 10.45 and 10.46, respectively. Four modes $\{k1, k2, k3, (k-1)4\}$ and five modes $\{k1, k2, k3, k4, (k-1)5\}$ shown in the same color are generated simultaneously.

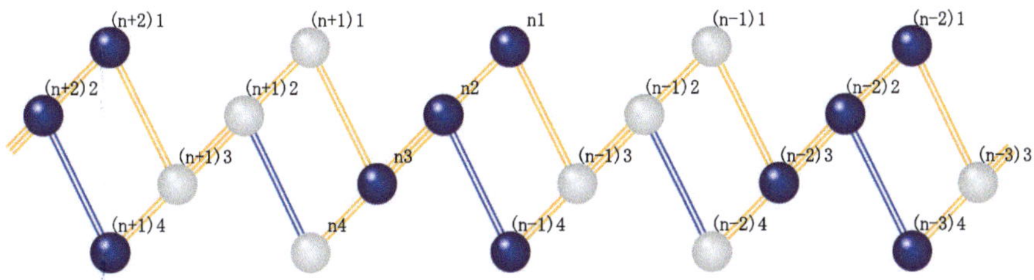

Fig. 10.45 Cluster state generated from four-mode linear cluster state

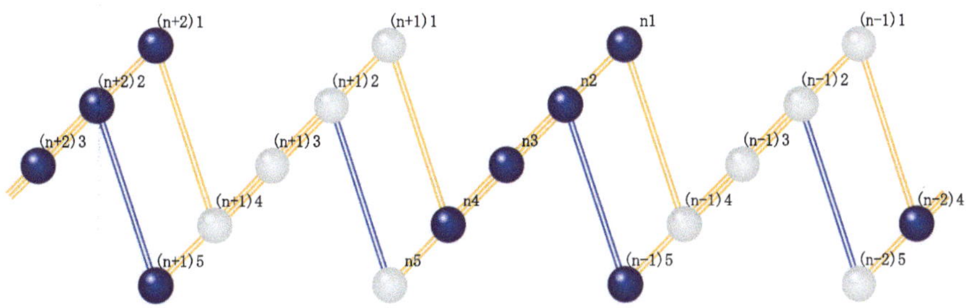

Fig. 10.46 Cluster state generated from five-mode linear cluster state

10.6 Multi-Mode Gaussian Operation Using Temporal-Mode Gaussian Cluster State

We have shown that we can implement an arbitrary one-mode operation by means of the temporal-mode cluster scheme by using all the degrees of freedom of cluster modes for quantum computation. The only missing element for universal quantum computation is a two-mode Gaussian operation [20].

In this section, we propose a protocol to implement two-mode Gaussian operations with temporal-mode Gaussian cluster states, where all the degrees of freedom of cluster modes can be utilized for quantum computation.[4] We choose the C_Z gate as a two-mode Gaussian operation. Although the cluster state we propose in this section is different from that in Ref. [1], the essence is the same.

10.6.1 Elementary Circuit

The elementary circuit for multi-mode Gaussian operations is shown in Fig. 10.47, where mode *in*1 and *in*2 show the modes of a two-mode input state, while the initial states in mode 1, 2, 3, and 4 are zero eigenstates of momentum operators: $|p = 0\rangle$.

Figure 10.47 is nothing but the quantum circuit of the C_Z gate where a four-mode linear cluster state is utilized as a resource for one-way quantum computation, which we have experimentally demonstrated and reported in Sect. 7 and Ref. [21].

In this circuit, two teleportation-based input-coupling protocols are utilized for the two-mode input state labeled by *in*1 and *in*2. The input-output relationship in the Heisenberg picture is given by

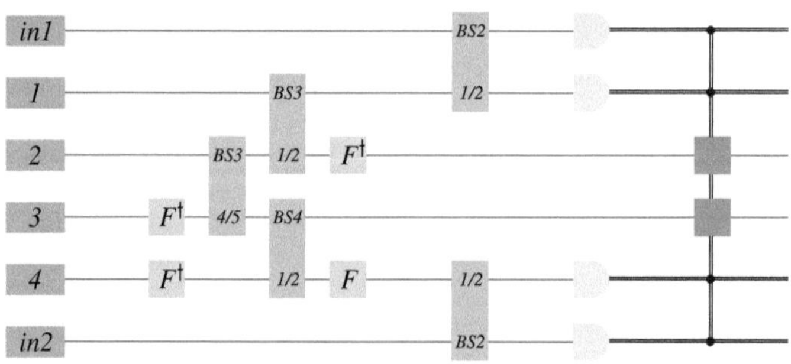

Fig. 10.47 Elementary circuit for multi-mode Gaussian operation

[4] Note again that one might have to remove several modes of a temporal-mode cluster state although the modes to be removed have abilities to implement some operations. See Sect. 10.1.4.

$$\begin{pmatrix} \hat{x}_{out1} \\ \hat{p}_{out1} \\ \hat{x}_{out2} \\ \hat{p}_{out2} \end{pmatrix} = C_Z \begin{pmatrix} M_{tele}(\theta_{+1}, \theta_{-1}) & O_2 \\ O_2 & M_{tele}(\theta_{+2}, \theta_{-2}) \end{pmatrix} \begin{pmatrix} \hat{x}_{in1} \\ \hat{p}_{in1} \\ \hat{x}_{in2} \\ \hat{p}_{in2} \end{pmatrix}. \tag{10.59}$$

We write it as

$$C_Z \begin{pmatrix} M_{tele}(\theta_{+1}, \theta_{-1}) & O_2 \\ O_2 & M_{tele}(\theta_{+2}, \theta_{-2}) \end{pmatrix} = C_{Z12} M_{tele12}, \tag{10.60}$$

for simplicity, where the subscripts 12 show that the gate works on mode 1 and mode 2. Here, $\theta_{\pm 1} = \theta_{in1} \pm \theta_1, \theta_{\pm 2} = \theta_{in2} \pm \theta_4$. The matrix C_Z:

$$C_Z = \begin{pmatrix} 1 & 0 & 0 & 0 \\ 0 & 1 & 1 & 0 \\ 0 & 0 & 1 & 0 \\ 1 & 0 & 0 & 1 \end{pmatrix} \tag{10.61}$$

represents a C_Z gate, while

$$O_2 = \begin{pmatrix} 0 & 0 \\ 0 & 0 \end{pmatrix} \tag{10.62}$$

is the 2×2 zero matrix. Therefore, the operation consists of the following two parts: two teleportation-based input-coupling circuits; followed by entangling of their outcomes via the C_Z gate. Since the resource for the operation is a four-mode linear cluster state, and the degrees of freedom of the operation is four, we find that all the degrees of freedom of cluster modes can be utilized for cluster-based quantum computation.

10.6.2 Concatenation of Elementary Circuits

We consider concatenation of the elementary circuits.

Although we consider five steps of the elementary circuits as an example in the following discussion (Fig. 10.36), it can easily be extended to the n-step case in general.

Figure 10.48 shows an example where we use a three-mode input state. In this figure, the network of beam splitters including Fourier transformations for generation of a four-mode linear cluster state is abbreviated to "4 mode linear cluster state". Each blue dotted rectangle shows a C_Z operation with one-mode Gaussian operations. Two input modes of the gate are described by red circles, while the output modes by blue circles. First, the set of operations is applied to mode 1 and mode 2. Second, another set is applied to mode 2 and mode 3. Third, mode 3 and mode 1. in the fourth step, the target modes are mode 1 and mode 2 again. Thereafter operations are applied cyclically. It is obvious that the operation achieved by Fig. 10.48 is

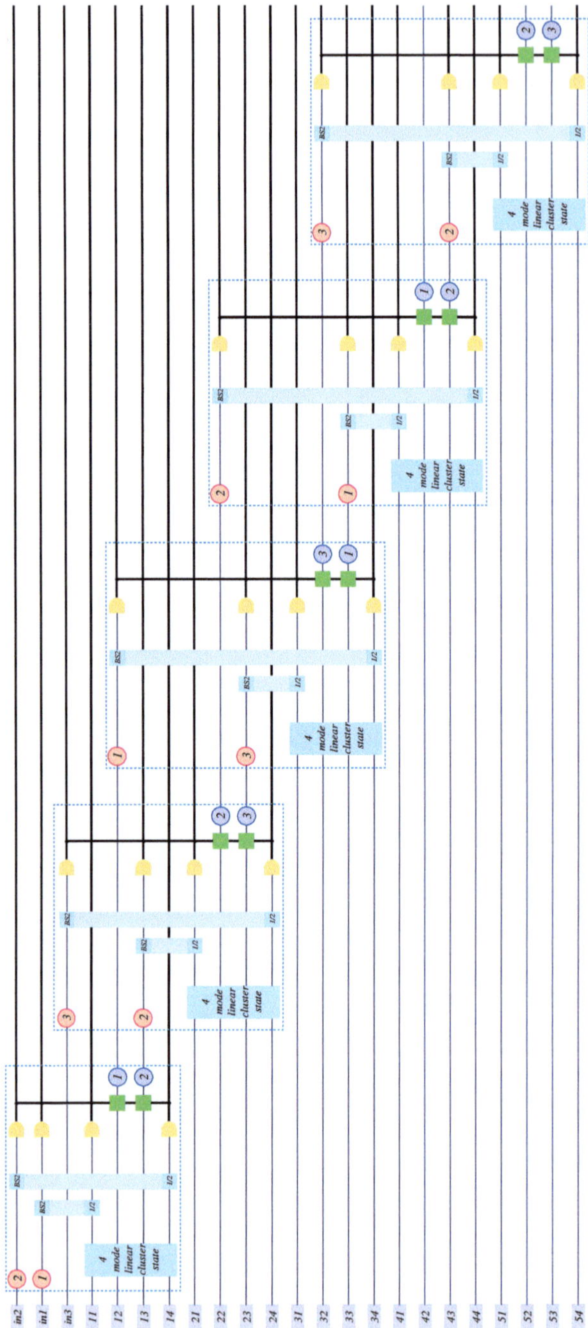

Fig. 10.48 Concatenation of elementary circuits

$$\begin{pmatrix} \hat{x}_{out1} \\ \hat{p}_{out1} \\ \hat{x}_{out2} \\ \hat{p}_{out2} \\ \hat{x}_{out3} \\ \hat{p}_{out3} \end{pmatrix} = C_{Z23} M_{tele23} C_{Z12} M_{tele12} C_{Z31} M_{tele31} C_{Z23} M_{tele23} C_{Z12} M_{tele12} \begin{pmatrix} \hat{x}_{in1} \\ \hat{p}_{in1} \\ \hat{x}_{in2} \\ \hat{p}_{in2} \\ \hat{x}_{in3} \\ \hat{p}_{in3} \end{pmatrix}. \tag{10.63}$$

10.6.3 Transformation of the Concatenated Circuit

By following the same strategy in Sects. 10.3.3 and 10.5.2, we can transform Fig. 10.48 into 10.49. Although we can gather all the feed-forwards at the end of the circuit since we only consider Gaussian operations, we omitted this process for simplicity.

Figure 10.49 shows a one-way quantum computation where a cluster state shown in Fig. 10.50 is utilized as a resource.

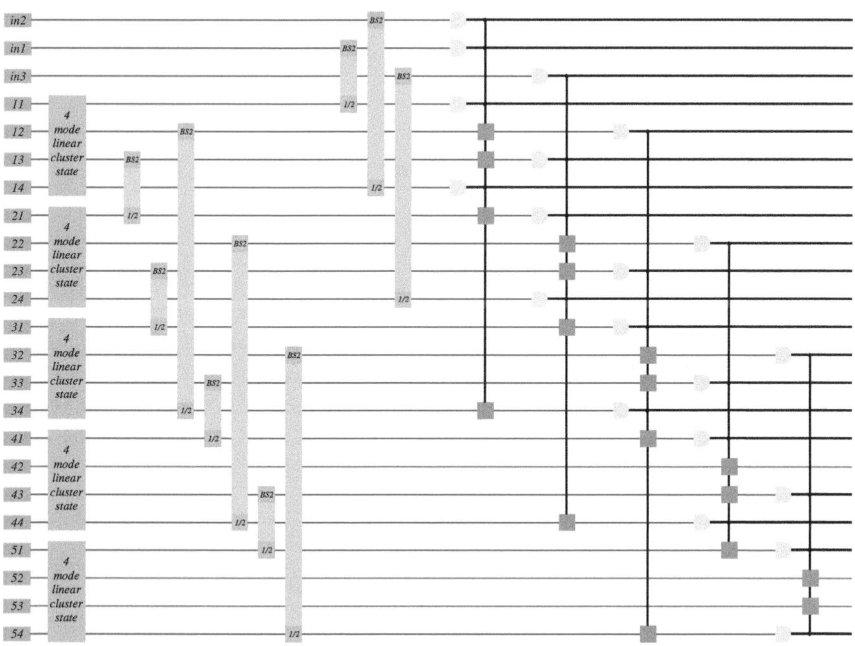

Fig. 10.49 Transformation of the concatenated circuit

Fig. 10.50 Cluster state

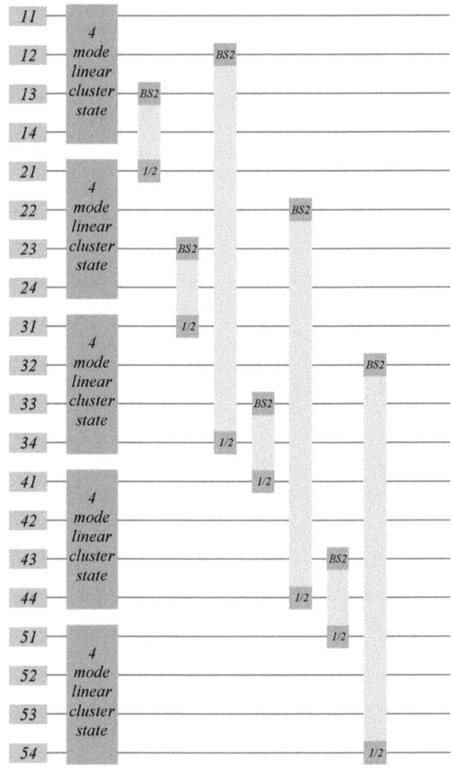

10.6.4 Generation of Temporal-Mode Gaussian Cluster State

Figure 10.51 shows an experimental setup for generation of the temporal-mode Gaussian cluster state. Since we have shown generation of the temporal-mode cluster state which can be used for three-mode Gaussian operations, the length of the delay line for mode $n2$ is twice as long as the mode length. In this figure, the arrangement of modes at each time is also shown. In order to generate a cluster state for N-mode Gaussian operations, the length of the delay line for mode $n2$ should be changed to $(N-1)$ times as long as the mode length. On the other hand, the length of the delay line for mode $n3$ remains the same to the mode length.

10.6.5 Nullifiers and Graph of the Cluster State

Before we conclude this section, we get the nullifiers and the graph of the temporal-mode Gaussian cluster state. We will ignore the effect of the both ends of the cluster state. Although we have considered the cluster state for three-mode Gaussian

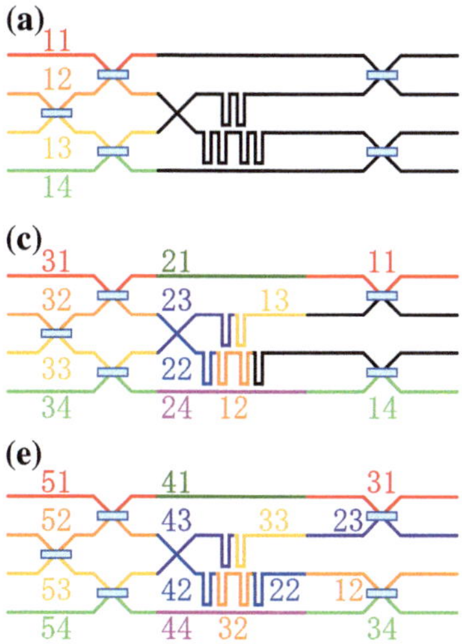

Fig. 10.51 Generation of temporal-mode Gaussian cluster state which can be utilized for three-mode Gaussian operations. **a** Time 1, **b** Time 2, **c** Time 3, **d** Time 4, **e** Time 5, **f** Time 6

operations in Fig. 10.50, we will get the general form of the nullifiers of the cluster state which can be used for N-mode Gaussian operations.

First, the nullifiers of a four-mode linear cluster state is given by

$$\{\hat{p}_{n1} - \hat{x}_{n2}, \hat{p}_{n2} - \hat{x}_{n1} - \hat{x}_{n3}, \hat{p}_{n3} - \hat{x}_{n2} - \hat{x}_{n4}, \hat{p}_{n4} - \hat{x}_{n3}\}. \tag{10.64}$$

We then apply beam splitter operators $\hat{B}^{(2)}_{(k-1)3,k1}(\frac{1}{\sqrt{2}})$ and $\hat{B}^{(2)}_{(k-N+1)2,k4}(\frac{1}{\sqrt{2}})$ sequentially. Since modes on which these operators work are different from each other, the order of beam splitters do not affect the final state. As a result, we get the set of nullifiers:

$$\begin{aligned}
\Big\{&\Big(-\frac{1}{\sqrt{2}}\hat{p}_{(n-1)3} + \frac{1}{\sqrt{2}}\hat{p}_{n1}\Big) - \Big(\frac{1}{\sqrt{2}}\hat{x}_{n2} + \frac{1}{\sqrt{2}}\hat{x}_{(n+N-1)4}\Big),\\
&\Big(\frac{1}{\sqrt{2}}\hat{p}_{n2} + \frac{1}{\sqrt{2}}\hat{p}_{(n+N-1)4}\Big) - \Big(-\frac{1}{\sqrt{2}}\hat{x}_{(n-1)3} + \frac{1}{\sqrt{2}}\hat{x}_{n1}\Big) - \Big(\frac{1}{\sqrt{2}}\hat{x}_{n3} + \frac{1}{\sqrt{2}}\hat{x}_{(n+1)1}\Big),\\
&\Big(\frac{1}{\sqrt{2}}\hat{p}_{n3} + \frac{1}{\sqrt{2}}\hat{p}_{(n+1)1}\Big) - \Big(\frac{1}{\sqrt{2}}\hat{x}_{n2} + \frac{1}{\sqrt{2}}\hat{x}_{(n+N-1)4}\Big) - \Big(-\frac{1}{\sqrt{2}}\hat{x}_{(n-N+1)2} + \frac{1}{\sqrt{2}}\hat{x}_{n4}\Big),\\
&\qquad\Big(-\frac{1}{\sqrt{2}}\hat{p}_{(n-N+1)2} + \frac{1}{\sqrt{2}}\hat{p}_{n4}\Big) - \Big(\frac{1}{\sqrt{2}}\hat{x}_{n3} + \frac{1}{\sqrt{2}}\hat{x}_{(n+1)1}\Big)\Big\}.
\end{aligned} \tag{10.65}$$

By taking linear combinations of these nullifiers, we get the final form of nullifiers:

$$\begin{aligned}\Big\{&2\hat{p}_{n1} - \hat{x}_{n2} - \hat{x}_{(n+N-1)4} - \hat{x}_{(n-1)2} - \hat{x}_{(n+N-2)4} + \hat{x}_{(n-N)2} - \hat{x}_{(n-1)4},\\ &2\hat{p}_{n2} + \hat{x}_{(n-1)3} - \hat{x}_{n1} - \hat{x}_{n3} - \hat{x}_{(n+1)1} + \hat{x}_{(n+N-1)3} + \hat{x}_{(n+N)1},\\ &2\hat{p}_{n3} - \hat{x}_{n2} - \hat{x}_{(n+N-1)4} + \hat{x}_{(n-N+1)2} - \hat{x}_{n4} + \hat{x}_{(n+1)2} + \hat{x}_{(n+N)4},\\ &2\hat{p}_{n4} - \hat{x}_{n3} - \hat{x}_{(n+1)1} + \hat{x}_{(n-N)3} - \hat{x}_{(n-N+1)1} - \hat{x}_{(n-N+1)3} - \hat{x}_{(n-N+2)1}\Big\}.\end{aligned} \tag{10.66}$$

Although we can draw the graph of the cluster state by directly using the set of nullifiers above, it might not be the best plan since each mode is entangled with others in a complex form. Instead, we draw a graph of the cluster state by gathering several modes which are used for quantum computation as a set. We name this form of graph as a *macro-node* graph, while we name the original form as a *micro-node* graph.

First, we consider macro nodes by using an example of Fig. 10.48, where operations on a three-mode input state are implemented. The procedure is summarized in the following:

- First, operations on mode *in*1 are carried out by using mode 11 and 12, while those on mode *in*2 by using mode 13 and 14.
- Second, operations on mode *in*2 are carried out by using mode 21 and 22, while those on mode *in*3 by using mode 23 and 24.
- Third, operations on mode *in*3 are carried out by using mode 31 and 32, while those on mode *in*1 by using mode 33 and 34.
- Fourth, operations on mode *in*1 are carried out by using mode 41 and 42, while those on mode *in*2 by using mode 43 and 44.
- Finally, operations on mode *in*2 are carried out by using mode 51 and 52, while those on mode *in*3 by using mode 53 and 54.

In these cyclical operations, four modes $\{n3, n4, (n+1)1, (n+1)2\}$ are utilized in a set for operations on one of the input modes. The same goes for the case of an N-mode input computation. In the following, we define a macro node $M(n)$ as

$$M(n) = \{n3, n4, (n+1)1, (n+1)2\}. \tag{10.67}$$

Next, we consider macro-node connections from macro-node $M(n)$. The nullifier $2\hat{p}_{n3} - \hat{x}_{n2} - \hat{x}_{(n+N-1)4} + \hat{x}_{(n-N+1)2} - \hat{x}_{n4} + \hat{x}_{(n+1)2} + \hat{x}_{(n+N)4}$ gives us connections from micro-node $n3$. The other micro nodes $(\hat{x}_{n2}, \ldots, \hat{x}_{(n+N)4})$ in this nullifier belong to

$$M(n-1), M(n+N-1), M(n-N), M(n), M(n), M(n+N). \tag{10.68}$$

In a similar manner, the nullifier $2\hat{p}_{n4} - \cdots$ shows that the micro nodes to which the micro node $n4$ is connected belong to

$$M(n), M(n), M(n-N), M(n-N), M(n-N+1), M(n-N+1). \tag{10.69}$$

The nullifier $2\hat{p}_{(n+1)1} - \cdots$ shows that the micro nodes to which the micro node $(n+1)1$ is connected belong to

$$M(n), M(n+N), M(n-1), M(n+N-1), M(n-N), M(n). \tag{10.70}$$

The nullifier $2\hat{p}_{(n+1)2} - \cdots$ shows that the micro nodes to which the micro node $(n+1)2$ is connected belong to

$$M(n), M(n), M(n+1), M(n+1), M(n+N), M(n+N). \tag{10.71}$$

Since micro nodes $n3$, $n4$, $(n+1)1$, and $(n+1)2$ belong to macro node $M(n)$, we find $M(n)$ is connected to

$$M(n-N), M(n-(N-1)), M(n-1), M(n), M(n+1), M(n+(N-1)), M(n+N). \tag{10.72}$$

It shows that a macro node is connected to macro nodes which are located at distances of ± 1, $\pm(N-1)$ and $\pm N$.

Figure 10.52 shows the graph of the cluster state in the macro-node representation which can be used for six-mode Gaussian operations.

In this figure, each green circle corresponds to a macro node. Blue lines show connections from macro node $M(n)$ to macro nodes $M(n-1)$ and $M(n+1)$. Since Fig. 10.52 shows a cluster state for six-mode operations, it has a spiral structure consisting of six macro nodes. This structure is derived from the original strategy that C_Z gate are applied cyclically. Horizontal purple thick lines show connections from macro node $M(n)$ to macro nodes $M(n-N)$ and $M(n+N)$, while purple thin lines show connections from macro node $M(n)$ to macro nodes $M(n-(N-1))$ and $M(n+(N-1))$. Each mode of an input state is transmitted horizontally as operations are carried out.

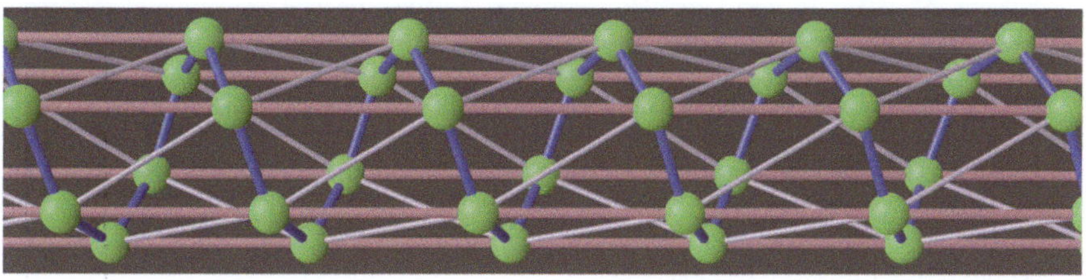

Fig. 10.52 Graph of cluster state which can be used for six-mode Gaussian operations (in macro-node representation)

10.7 Multi-Mode Non-Gaussian Operation

We have so far shown the following three protocols for one-way quantum computations using temporal-mode Gaussian cluster states:

1. One-mode Gaussian operations where the two-mode cluster state is the elementary resource.
2. One-mode non-Gaussian operations where the three-mode linear cluster state is the elementary resource.
3. Multi-mode Gaussian operations where the four-mode linear cluster state is the elementary resource.

By combining these procedures, we can implement multi-mode non-Gaussian operations which correspond to universal quantum computation. Since 1 is a part of 2 or 3, it is sufficient to choose 2 and 3 as elementary components. Furthermore, we can easily merge these two components into one component. That is, by only using a six-mode linear cluster state as the elementary resource, we can implement multi-mode non-Gaussian operations.

References

1. Nicolas, C.: Menicucci, temporal-mode continuous-variable cluster states using linear optics. Phys. Rev. A **83**, 062314 (2011)
2. Zhang, J., Braunstein, S.L.: Continuous-variable Gaussian analog of cluster states. Phys. Rev. A **73**, 032318 (2006)
3. van loock, P., Weedbrook, C., Gu, M.: Building Gaussian cluster states by linear optics. Phys. Rev. A **76**, 032321 (2007)
4. Su, X., Tan, A., Jia, X., Zhang, J., Xie, C., Peng, K.: Experimental preparation of quadripartite cluster and Greenberger-Horne-Zeilinger Entangled states for continuous variables. Phys. Rev. Lett. **98**, 070502 (2007)
5. Yukawa, M., Ukai, R., van Loock, P., Furusawa, A.: Experimental generation of four-mode continuous-variable cluster states. Phys. Rev. A **78**, 012301 (2008)
6. Menicucci, N.C., Ma, X., Ralph, T.C.: Arbitrarily large continuous-variable cluster states from a single quantum nondemolition gate. Phys. Rev. Lett. **104**, 250503 (2010)
7. Su X., Zhao Y., Hao S., Jia X., Xie C., Peng K.: Experimental preparation of eight-partite linear and two-diamond shape cluster states for photonic qumodes, arXiv:1205.0590 [quant-ph] (2012)
8. Filip, R., Marek, P., Andersen, U.L.: Measurement-induced continuous-variable quantum interactions. Phys. Rev. A **71**, 042308 (2005)
9. Yoshikawa, J., Miwa, Y., Huck, A., Andersen, U.L., van Loock, P., Furusawa, A.: Demonstration of a quantum nondemolition sum gate. Phys. Rev. Lett. **101**, 250501 (2008)
10. Gu, M., Weedbrook, C., Menicucci, N.C., Ralph, T.C., van Loock, P.: Quantum computing with continuous-variable clusters. Phys. Rev. A **79**, 062318 (2009)
11. Einstein, A., Podolsky, B., Rosen, N.: Can quantum-mechanical description of physical reality be considered complete? Phys. Rev. **47**, 777 (1935)
12. Furusawa A., van Loock P.: Quantum teleportation and entanglement, WILEY-VCH, Weinheim (2011)
13. Ukai, R., Yoshikawa, J., Iwata, N., van Loock, P., Furusawa, A.: Universal linear Bogoliubov transformations through one-way quantum computation. Phys. Rev. A **81**, 032315 (2010)

14. Braunstein, S.L.: Squeezing as an irreducible resource. Phys. Rev. A **71**, 055801 (2005)
15. Bartlett, S.D., Sanders, B.C., Braunstein, S.L., Nemoto, K.: Efficient classical simulation of continuous variable quantum information processes. Phys. Rev. Lett. **88**, 097904 (2002)
16. van Loock, P.: Examples of Gaussian cluster computation. J. Opt. Soc. Am. B **24**, 340 (2007)
17. Yokoyama, S., Ukai, R., Armstrong, S.C., Sornphiphatphong, C., Kaji, T., Suzuki, S., Yoshikawa, J., Yonezawa, H., Menicucci, N.C., Furusawa, A.: Ultra-large-scale continuous-variable cluster states multiplexed in the time domain. Nat. Photonics **7**, 982 (2013)
18. Takeda, S., Mizuta, T., Fuwa, M., Yoshikawa, J., Yonezawa, H., Furusawa, A.: Generation and eight-port homodyne characterization of time-bin qubits for continuous-variable quantum information processing. Phys. Rev. A **87**, 043803 (2013)
19. Takeda, S., Mizuta, T., Fuwa, M., van Loock, P., Furusawa, A.: Deterministic quantum teleportation of photonic quantum bits by a hybrid technique. Nature **500**, 315 (2013)
20. Lloyd, S., Braunstein, S.L.: Quantum computation over continuous variables. Phys. Rev. Lett. **82**, 1784 (1999)
21. Ukai, R., Yokoyama, S., Yoshikawa, J., van Loock, P., Furusawa, A.: Demonstration of a controlled-phase gate for continuous-variable one-way quantum computation. Phys. Rev. Lett. **107**, 250501 (2011)

Chapter 11
Summary

11.1 Conclusions

In this thesis, we have demonstrated three one-way quantum computation protocols over two-mode input states.

In Chap. 7, we have experimentally demonstrated a unity-gain controlled-Z gate for optical continuous-variable one-way quantum computation. We have utilized a four-mode linear Gaussian cluster state as a resource for the experiment. It can also be considered as an experimental demonstration of gate teleportations [2, 3] on a two-mode system.

In Chap. 8, we have experimentally demonstrated an optimum nonlocal controlled-Z gate for optical continuous variables. It has been achieved by using a bipartite entangled state called a two-mode Gaussian cluster state, shared in advance, and one classical channel in each direction. They are the minimum requirements for a nonlocal controlled-Z gate.

In Chap. 9, We have experimentally demonstrated a gain-tunable entangling gate for optical continuous-variable one-way quantum computation, where a three-mode linear Gaussian cluster state is utilized as a resource state. In contrast to the other two experiments where the operations were fixed, this gate had the ability of on-off switching of two-mode interactions, as well as interaction-gain tuning. Control of the operation was precisely achieved by adjusting the relative phase in a homodyne detection, which represents the feature of one-way quantum computation: the selection of measurement bases determines the quantum operations.

In all the experimental demonstrations presented in this thesis, the general input-output relations have been verified by appropriate transmission of input coherent-state signals. The genuine quantum character of these gates became manifest and was verified through the existence of entanglement at the output modes. We have utilized the generalized van Loock-Furusawa entanglement criteria (Sect. 3.7.4), as well as the logarithmic negativities of the output states.

In this thesis, one-way quantum computation using temporal-mode cluster states was also studied [4, 5]. We showed that quantum computation using a temporal-mode cluster state for one-mode operations is equivalent to a repetition of quantum teleportations (Sect. 10.3.3). Since we can implement one-mode Gaussian operations with two DOF by controlling the measurement basis in a quantum teleportation

R. Ukai, *Multi-Step Multi-Input One-Way Quantum Information Processing with Spatial and Temporal Modes of Light*, Springer Theses,
DOI 10.1007/978-4-431-55019-8_11

(Sect. 5.3.3), we can utilize all modes of the temporal-mode cluster state in one-way quantum computations without eliminating any of them. In addition, we showed that non-Gaussian operations and multi-mode Gaussian operations are also achieved without eliminating resource modes (Sects. 10.5 and 10.6) by considering repetitions of a one-way quantum computation circuit with the three-mode linear cluster state, and the quantum computation circuit of the controlled-Z gate experiment (Sect. 7).

11.2 Future Prospects

We have previously reported one-mode Gaussian operation experiments using a four-mode linear Gaussian cluster state as a resource in Ref. [1]. By combining the two-mode gate experiments in this thesis with the one-mode gate experiments in Ref. [1], we now have all the tools required to implement an arbitrary multi-mode Gaussian operation in a framework of one-way quantum computation. Our gates can be directly incorporated into a large-scale multi-mode one-way quantum computation.

The only missing element for an arbitrary multi-mode unitary operation is a non-Gaussian operation. In our laboratory, considerable efforts are now being devoted to the implementation of non-Gaussian operations [6–8]. Experimental generations of non-Gaussian states and experimental quantum teleportations of non-Gaussian states will lead to non-Gaussian operations in the near future.

We next consider the number of operations we have implemented. In our experiments in this thesis, up to four-mode entangled states were utilized as resources for one-way quantum computations. It shows that a four-step operation has been experimentally demonstrated. As a promising scheme for many-step operations, temporal-mode cluster states have been proposed. Recently, we have experimentally generated ultra-large-scale entangled states where more than 10,000 modes are entangled (extended EPR state, Sect. 10.4) [9].

Although we can generate ultra-large-scale entangled states in continuous variables, we lack an appropriate algorithm that can efficiently utilize such *ultra-large-scale* but *finitely correlated* entangled states. In Chap. 10, we have proposed how to duplicate the number of operations when we implement quantum computation using temporal-mode cluster states. However, our proposal is not yet sufficient to fully utilize the ultra-large-scale entangled states we have created. The vast unfulfilled potential of the ultra-large-scale entangled states will no doubt be explored by more sophisticated quantum computation algorithms.

References

1. Ukai, R., Iwata, N., Shimokawa, Y., Armstrong, S.C., Politi, A., Yoshikawa, J., van Loock, P., Furusawa, A.: Demonstration of unconditional one-way quantum computations for continuous variables. Phys. Rev. Lett. **106**, 240504 (2011)
2. Gottesman, D., Chuang, I.L.: Demonstrating the viability of universal quantum computation using teleportation and single-qubit operations. Nature **402**, 390 (1999)

3. Bartlett, S.D., Munro, W.J.: Quantum teleportation of optical quantum gates. Phys. Rev. Lett. **90**, 117901 (2003)
4. Menicucci, N.C., Ma, X., Ralph, T.C.: Arbitrarily large continuous-variable cluster states from a single quantum nondemolition gate. Phys. Rev. Lett. **104**, 250503 (2010)
5. Nicolas, C.: Menicucci, temporal-mode continuous-variable cluster states using linear optics. Phys. Rev. A **83**, 062314 (2011)
6. Yukawa, M., Miyata, K., Mizuta, T., Yonezawa, H., Marek, P., Filip, R., Furusawa, A.: Generating superposition of up-to three photons for continuous variable quantum information processing. Opt. Express **21**, 5529 (2013)
7. Yoshikawa, J., Makino, K., Kurata, S., van Loock, P., Furusawa, A.: Creation, storage, and on-demand release of optical quantum states with a negative Wigner function. Phys. Rev. X **3**, 041028 (2013)
8. Lee, N., Benichi, H., Takeno, Y., Takeda, S., Webb, J., Huntington, E., Furusawa, A.: Teleportation of non-classical wave-packets of light. Science **332**, 330 (2011)
9. Yokoyama, S., Ukai, R., Armstrong, S.C., Sornphiphatphong, C., Kaji, T., Suzuki, S., Yoshikawa, J., Yonezawa, H., Menicucci, N.C., Furusawa, A.: Ultra-Large-Scale continuous-variable cluster states multiplexed in the time domain. Nat. Photonics **7**, 982 (2013)

Appendix A
Photos And Schematic

A.1 Photos

Figure A.1 shows photos of our optical table and classical channels

(a) (b)

Fig. A.1 Photos; **a** Optical table. **b** Classical channels

A.2 Schematic of Optical Table

Figure A.2 shows the schematic of our optical table for the controlled-Z gate experiment

R. Ukai, *Multi-Step Multi-Input One-Way Quantum Information Processing with Spatial and Temporal Modes of Light*, Springer Theses,
DOI 10.1007/978-4-431-55019-8

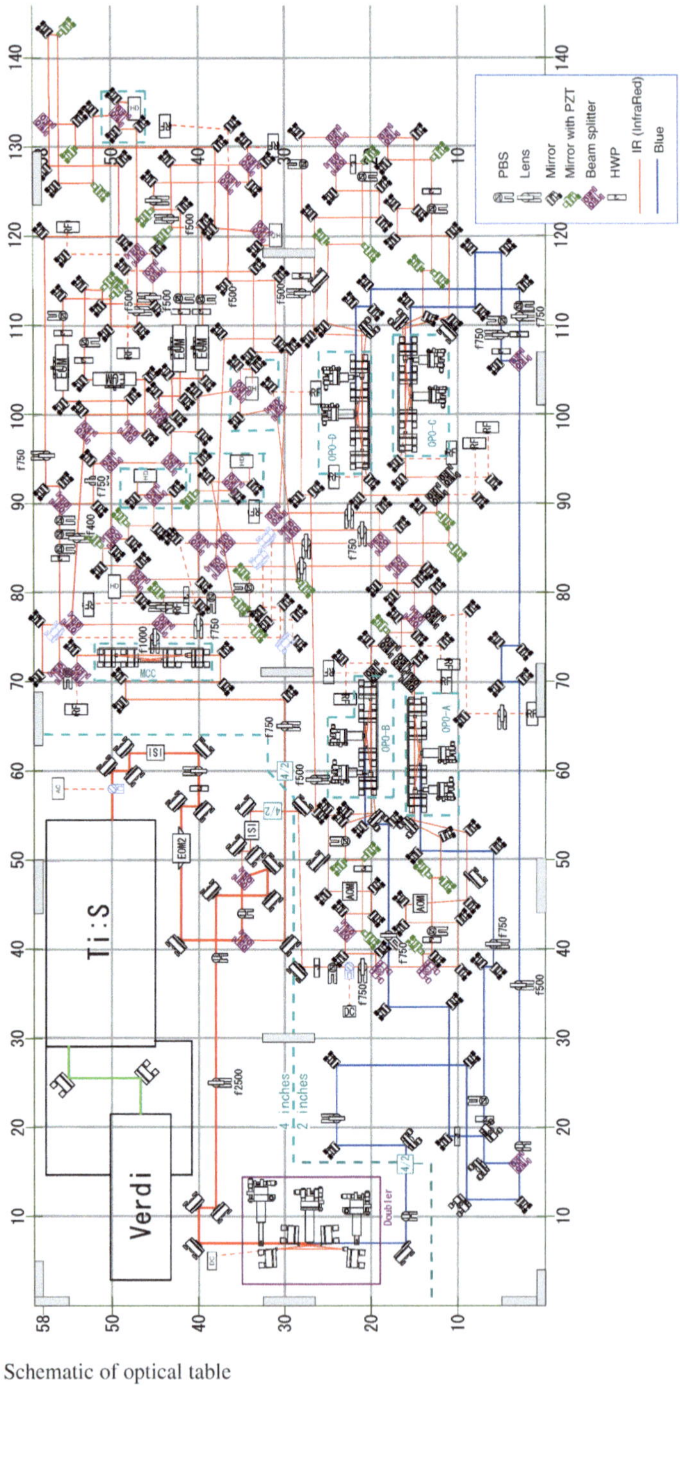

Fig. A.2 Schematic of optical table

Appendix B
Programs

B.1 Ray Tracer

This is a program with a graphical user interface for calculations of waist positions and waist sizes of 00-mode Gaussian beams based on ABCD matrices.

This program shows beam sizes of multiple 00-mode Gaussian beams at each position when we place multiple optical objects such as lenses on their propagation paths.

The main features of this program are listed below.

- Calculate the visibility and mode match of two beams.
- Place a lens automatically at an appropriate position so that the beam has the desired waist size.
- Place two lenses automatically at appropriate positions so that the beam has the desired waist size and the desired waist position.
- Design a cavity by imposing the condition that the initial waist size is equivalent to the last waist size.

B.2 Quantum Computation Builder

This is a program with a graphical user interface for calculations of multi-mode Gaussian operations in the Heisenberg picture.

It is an example of classical simulations of Gaussian operations described in Sect. 3.6.2. This program generates a Mathematica (Wolfram Research) code, which is processed by a Mathematica macro programed by S. Yokoyama.

The main features of this program are listed below.

R. Ukai, *Multi-Step Multi-Input One-Way Quantum Information Processing with Spatial and Temporal Modes of Light*, Springer Theses,
DOI 10.1007/978-4-431-55019-8

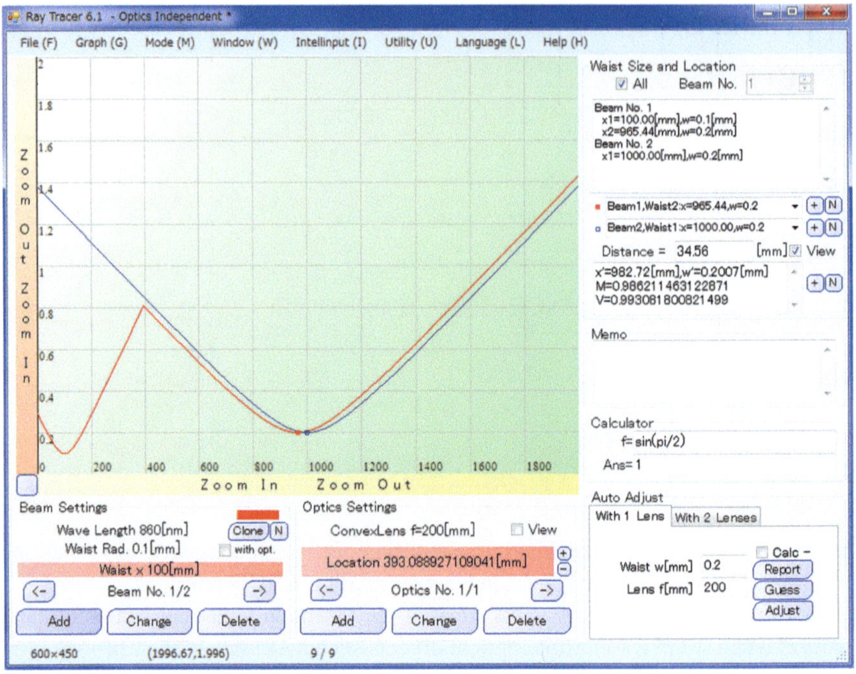

Fig. B.1 Program "Ray Tracer"

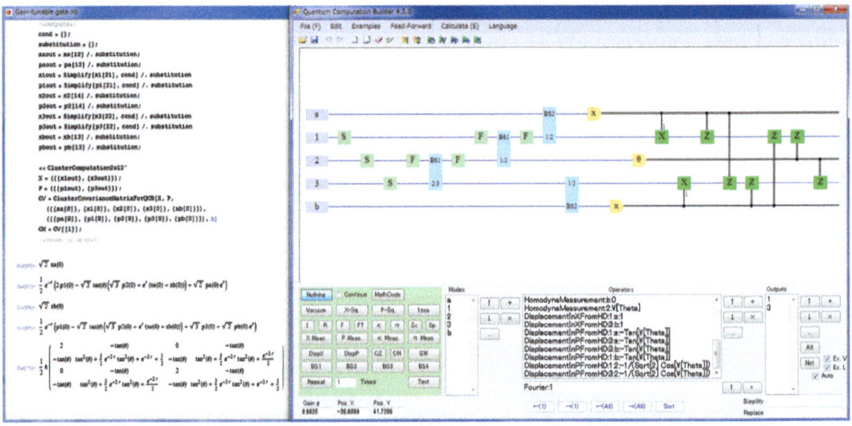

Fig. B.2 Program "Quantum Computation Builder"

- Simulate major Gaussian operations (such as squeezing operations and controlled-Z gates) as well as homodyne measurements and displacement operations based on measurement results.

- Calculate feed-forward gains for Gaussian operations so that the output-mode quadratures do not have anti-squeezing components of resource modes.
- Calculate input-output relationships in the Heisenberg picture.
- Evaluate entanglement criteria for the two-mode output states.

Curriculum Vitae

Ryuji Ukai
Department of Applied Physics,
School of Engineering,
The University of Tokyo,
7-3-1 Hongo, Bunkyo-ku,
Tokyo 113-8656, Japan

Education

- Doctor of Philosophy (Ph.D.) in Applied Physics, The University of Tokyo (2010–2013)
- Master of Science (M.Sc.) in Applied Physics, The University of Tokyo (2008–2010)
- Bachelor of Science (B.Sc.) in Applied Physics, The University of Tokyo (2004–2008)

Research Interests

My research aims to explore how we can handle optical quantum states experimentally. Quantum physics proposes a lot of mysterious phenomena which are contradictory to our feeling. As an undergraduate, I started my exploration into quantum physics. I was excited at generation and measurement of quantum entangled states. Since then, I become captivated by quantum physics. My interest covers quantum state generation, quantum state manipulation, and quantum information processing. Quantum physics would open the door to full understanding of our actual world.

R. Ukai, *Multi-Step Multi-Input One-Way Quantum Information Processing with Spatial and Temporal Modes of Light*, Springer Theses,
DOI 10.1007/978-4-431-55019-8

MIX
Papier aus verantwortungsvollen Quellen
Paper from responsible sources
FSC® C105338

If you have any concerns about our products,
you can contact us on
ProductSafety@springernature.com

In case Publisher is established outside the EU,
the EU authorized representative is:
Springer Nature Customer Service Center GmbH
Europaplatz 3, 69115 Heidelberg, Germany

Printed by Libri Plureos GmbH
in Hamburg, Germany